A Logical Theory of Causality

A Logical Theory of Causality

Alexander Bochman

The MIT Press
Cambridge, Massachusetts
London, England

The MIT Press would like to thank the anonymous peer reviewers who provided comments on drafts of this book. The generous work of academic experts is essential for establishing the authority and quality of our publications. We acknowledge with gratitude the contributions of these otherwise uncredited readers.

This book was set in Times Roman by Alexander Bochman. Printed and bound in the United States of America.

Library of Congress Cataloging-in-Publication Data

Names: Bochman, Alexander, 1955 author.
Title: A logical theory of causality / Alexander Bochman.
Description: Cambridge, Massachusetts : The MIT Press, 2021. | Includes bibliographical references and index.
Identifiers: LCCN 2020037281 | ISBN 9780262045322 (paperback)
Subjects: LCSH: Nonmonotonic reasoning. | Causation. | Logistics.
Classification: LCC Q339.2 .B628 2021 | DDC 122–dc23
LC record available at https://lccn.loc.gov/2020037281

10 9 8 7 6 5 4 3 2 1

Contents

Preface

The subject of causation is not a virgin land. Rather, today it is more like a jungle that has grown on a rich soil after centuries of philosophical wars, occupation, and even exile. A significant part of our task in this book will involve archaeological excavations and restorations using modern logical tools.

Still, this is not a philosophical or historical book. Our main objective will be to build a formal theory of causal reasoning on this territory and explore its applications in areas ranging from artificial intelligence to legal theory and dynamic reasoning. In this respect, our study will follow a broadly axiomatic, "Euclidean" approach to causation. We will leave outside the scope of the book whether our theory is compatible with the wealth of other philosophical views of causation, though we will make an effort to bring historical evidence that support, or agree with, our approach.

Causation is a fundamental concept of our reasoning, but even general studies of this notion have always been highly susceptible of falling into the parable of the blind men and the elephant; too often they have chosen only one, however important, aspect of the notion as a key to the whole concept. In this respect, our aim in this book has also consisted in showing how many diverse and even undiscovered parts this "elephant" has as well as a broad range of topics that ought to be pursued in its study, such as laws, counterfactuals, actual and indeterminate causation, abduction, structural equations and logic programs, and even general dynamic reasoning.

In recent years, we have been witnessing a revival of interest in the concept of causation, accompanied by new, more formal, but also more practical insights about its role in our reasoning. Most prominent in this respect is undoubtedly Pearl's theory of causal reasoning in the mathematical framework of structural equations and its applications in statistics, economics, and cognitive and social sciences (see Pearl 2000).

A more "logic-oriented" approach to causality has also emerged in artificial intelligence (AI). Basically, it considers causal reasoning as a particular kind of nonmonotonic reasoning, a kind especially suitable for explanation, diagnosis, abduction, and reasoning about actions and change. By its very nature, it brings the promise of an explainable AI.

This book could be naturally affiliated to this latter approach, though we see its scope as a general theory of causal reasoning that is not restricted to problems of AI alone. A large portion of examples and notions that will be studied in what follows will be driven by problems of causal reasoning in the legal theory, general dynamic formalisms, and even linguistic semantics.

The theory that will be developed in this book will also aim to transform causality "from a concept shrouded in mystery" into a formal object "with well-defined semantics and well-founded logic" (see the preface to Pearl 2000). However, in contrast to the mathematization of causality that has been successfully implemented by Pearl, the present book suggests a truly *logical* theory of causality as a study of causal *models, reasoning, and inference*, just as it was envisaged in the title of Pearl's book.

A Road Map

To a large extent, each of the four parts of this book, and even most of the chapters, could be read independently of one another. To facilitate such readings for those who are only interested in particular areas or problems of causal reasoning, we will provide below the main shortcut paths through the book.

To begin with, the main purpose of the unusually large prolegomena to the book (part I) consists in reshaping the standard picture of reasoning that has been solidified in the past century, a picture that has been dominated by deductive logic and its possible world semantics. In particular, these prolegomena lay down two logical prerequisites, or foundations, on which we build our theory of causal reasoning. The first is a local, situation-based semantics for (classical) logic that provides the required "logical space" for our causal formalism; it is described in chapter 2. It also establishes a subtle boundary between what can be viewed as a purely logical reasoning and what falls beyond it. The second prerequisite of our formalism is a general theory of nonmonotonic reasoning that is described in chapter 3. As we show in due course, our main formalism for causal reasoning turns out to be a special instance of this general theory. Still, chapters 2 and 3 are not formally required for reading subsequent chapters of the book, so they could be skipped (at least on a first reading) by readers who are ready to take the existence of such foundations for granted.

The main formalism of this book, the causal calculus, is described in chapter 4. This chapter also provides a formal background for subsequent chapters in parts II and III. Still, there are no further dependencies among most of these chapters, except for chapters 8 and 9, which form, in effect, two successive stages of a single theory of actual (or token) causality. However, for readers that are especially (or exclusively) interested in this latter notion, most of the content of these two chapters revolves around well-known examples from the literature, and we could even recommend here a "reverse" reading order that starts with the examples and proceeds to the formal justifications we give for our solutions.[1] Actually, this way of acquiring knowledge could be viewed as a truly Aristotelian scientific method (see section 1.1 in chapter 1).

Finally, part IV of the book presents a generalization of causal reasoning to dynamic domains. It is described, however, as a self-contained theory that does not require any previous knowledge of causal reasoning, though it presupposes some knowledge of propositional dynamic logic (PDL) and related logical formalisms. But the last chapter, chapter 12, presupposes again some knowledge of the causal calculus from chapter 4.

A Detailed Plan

A more detailed plan of our study is as follows.

In chapter 1, we discuss the general structure of our approach and the very possibility of a logical analysis of causation. We will argue that causal reasoning is not a competitor of logical reasoning but its complement for situations where we do not have logically sufficient data. We suggest a "two-tier" construction of such a reasoning in which the causal component is built on top of an underlying logical formalism. We will show that this two-layer scheme of causal reasoning can be found already in Aristotle's *Analytics*. We will argue also that causal claims should be viewed as default assumptions we make about the world, making causal reasoning a particular form of assumption-based, nonmonotonic reasoning that has been developed in AI.

As a logical preparation, chapter 2 describes an alternative, situation-based semantics for classical logic. We will show that this semantics creates a powerful representation framework that allows us to express some basic forms of dependence and independence among propositions without changing the logic. One of the important implications of these extended expressive capabilities will be that this semantic framework encompasses David Lewis's possible world semantics of comparative similarity. As a somewhat surprising consequence of this embedding, we will show that counterfactuals on Lewis's interpretation of the latter, and thereby his theory of causation, are confined to this logical level.

Chapter 3 provides a summary of those parts of general theory of nonmonotonic reasoning that are relevant for our suggested theory of causal reasoning. We first describe Reiter's default logic and show that it is reducible to a relatively simple formalism that contains only logical inference rules and default assumptions. This reduction will display default logic as a principal instantiation of general assumption-based framework for nonmonotonic reasoning. Then we consider modern formal argumentation theory that is based on Dung's abstract argumentation frameworks. These frameworks will be extended to a full-fledged logical formalism of propositional argumentation in which assumptions play the role of (reified) arguments. Later we will show that this formalism subsumes the main

---

1. One of the reviewers of the book has successfully explored this route.

formalism of this book, the causal calculus. Finally, we consider the problem of defeasible entailment and its solutions that have been suggested in AI.

Chapter 4 describes the main causal formalism of the book, the causal calculus. This calculus can be seen as a natural generalization of classical logic that allows for causal reasoning. From a purely logical point of view, this generalization amounts to dropping the reflexivity postulate of classical inference. The associated inference systems of production and causal inference will be assigned both a standard logical semantics (that will give a semantic interpretation to causal rules) and a natural nonmonotonic semantics, which will provide a representation framework for causal reasoning. It will be shown also that the causal calculus constitutes a special kind of propositional argumentation (described in the previous chapter) and subsumes, in turn, logic programming as a special case.

Chapter 5 provides a logical representation for Pearl's structural equation models in the causal calculus. It will be shown that, under this representation, the nonmonotonic semantics of the causal calculus will describe precisely the solutions of the structural equations. We also introduce a formal definition of counterfactuals in the causal calculus that will correspond to their original definition in the structural approach and will study their properties. It will be shown, in particular, that structural equations and counterfactuals are essentially equivalent formalisms. We will demonstrate, however, that the causal calculus is a more expressive formalism than the latter; this difference in expressivity will turn out to be crucial in representing actual causality in chapter 8. Finally, we will show that Pearl's causal models are intimately related to abstract dialectical frameworks (ADFs) in the formal argumentation theory. The material in this chapter is based on Bochman and Lifschitz (2015) and Bochman (2016a, 2018b).

Chapter 6, which is based on Bochman (2016b), extends the causal calculus to indeterminate causation by using disjunctive (multiple conclusion) causal rules. First, we provide a logical formalization of such rules in the form of a disjunctive inference relation and describe its logical semantics. Then, we consider a nonmonotonic semantics for such rules that has been suggested in H. Turner (1999). It will be shown, however, that under this semantics, these rules are reducible to ordinary, singular causal rules. This semantics also tends to give an exclusive interpretation of disjunctive causal effects. To overcome these shortcomings, we will introduce an alternative nonmonotonic semantics for disjunctive causal rules, called a covering semantics, that permits an inclusive interpretation of indeterminate causal information. Still, it will be shown that even in this case there exists a systematic procedure that we will call a normalization that allows us to capture the covering semantics using only singular causal rules. This normalization procedure generalizes established ways of representing indeterminate effects in current theories of action in AI.

Chapter 7, which is based on Bochman (2007a), describes an important application area of causal reasoning in AI: namely, abduction and diagnosis. We describe in this chapter a causal approach to abductive reasoning. The resulting formalism will subsume "classical" abductive reasoning as a special case, and we will give precise conditions when it is appropriate. The formalism will provide us, however, with additional representation capabilities that will encompass important alternative forms of abduction, such as abductive reasoning suitable for theories of actions and change and for abductive logic programming. As a result, we obtain a generalized theory of abduction that covers in a single formal framework practically all kinds of abductive reasoning studied in the AI literature.

In chapter 8, we will describe a (causal) regularity approach to actual causation that is based on the necessary element of a sufficient set (NESS) test of Richard Wright and contrast it with currently dominant counterfactual approaches to this notion. After establishing some theoretical features of our definition, we will apply this approach to wide range of examples and counterexamples suggested in the literature, including overdetermination, switches, and preemption. Then we will turn to more difficult cases of negative causality and show the necessity of a more articulated causal representation that uses names of underlying mechanisms. This "deep" representation will allow us to resolve, in effect, the problem of structural equivalents that has plagued the structural accounts of causality. The examples discussed will also show the need for a further enhancement of our approach, which will be described in the next chapter on relative causality.

Chapter 9 introduces the concept of relative causation that takes into account a distinction between causes and conditions in causal descriptions as well as the role of normality and defaults. A relativized notion of a *proximate* cause will provide a formalization of this idea. The relative notion of causation will also give us an opportunity to reconsider some general properties of the causal relation, such as transitivity of causation, as well as the relations among causality, interventions, and counterfactuals. As before, the new definition of causality will be tested on a range of examples, including examples that have already been discussed in the preceding chapter as well as on a number of further well-known and controversial examples from the legal literature.

In part IV, we present a dynamic generalization of the concept of causation and describe its applications in particular to reasoning about action and change in AI.

Chapter 10 provides a *causal* description of dynamic formalisms used in logic and AI. We introduce first a structural calculus of dynamic causal inference that is defined on sequences of propositions (called processes) and satisfies certain "sequential" variants of familiar inference rules such as monotonicity and cut (though not reflexivity!). This calculus will be first assigned an abstract monoid-based dynamic semantics, and the completeness results will be proved. Then we will show that the latter semantics can be transformed into more familiar relational, or transition, models. Moreover, it will be shown that our dynamic inference constitutes in this respect a generalization of the inference relation of dynamic predicate logic (DPL). After that, we will introduce sequential dynamic logic (SDL) that subsumes the propositional part of dynamic predicate logic and will be shown to be expressively equivalent to propositional dynamic logic (PDL). Completeness of SDL with respect to the intended relational semantics will be established. The material of this chapter originates in (Bochman and Gabbay 2012b, 2012a), though it provides a somewhat different, causal interpretation of the corresponding constructions and results.

Chapter 11, based on Bochman (2013), contains a formal description of the dynamic Markov principle that plays an important role in reasoning about action and change in AI. This principle will be shown to constitute a nonmonotonic, abductive assumption that justifies the actual restrictions on action descriptions in these AI theories, as well as constraints on allowable queries. It will be shown, in particular, that the well-known regression principle is a consequence of the dynamic Markov principle, and it is valid also for nondeterministic domains. We will provide both a semantic characterization of the dynamic Markov principle in terms of transparent transition models and its syntactic description as a kind of interpolation rule.

Finally, in chapter 12 we describe a dynamic causal calculus and its relation to the action description languages in AI. On a logical side, this formalism will incorporate and combine in an abstract form the basic insights of the preceding two chapters, and it will be based on dynamic causal rules of the form $B . C \Rightarrow E$ on classical propositions having an informal meaning *"After B, C causes E."* Dynamic causal theories containing such rules will be assigned both a nonmonotonic transition semantics and a standard logical, possible worlds semantics with ternary accessibility relations. It will be shown that the resulting formalism is sufficiently expressive to capture the main high-level action description languages in AI that have been described in the literature.

Acknowledgments

I would like to express my gratitude to a number of people for their help, both direct and indirect, during the writing of this book. First, I want to thank Vladimir Lifschitz for numerous discussions on the subject of causality and multiple improvements he has suggested as well as his important contribution to what has become chapter 5 on structural equation models. Past discussions with Hudson Turner have also been most instructive in formulating the final version of chapter 6 on indeterminate causation. I am also grateful to Dov Gabbay, who has encouraged me to go beyond existing approaches to dynamic modeling; our collaboration has led to the formal systems of dynamic causal reasoning described in chapter 10. Finally, my deep thanks go to Marc Denecker for his interest in my work and for the opportunity to discuss fundamental questions about causation despite (and maybe even due to) conceptual differences and especially for creating wonderful working conditions for writing this book during my sabbatical stay in Groot Begijnhof.

Alexander Bochman
September 2020

I LOGICAL PROLEGOMENA

1 A Two-Tier System of Causal Reasoning

All reasonings concerning matter of fact seem to be founded on the relation of Cause *and* Effect.
—David Hume, *An Enquiry Concerning Human Understanding* (1748)

Causation plays an essential role in our view of the world, from commonsense reasoning to natural and social sciences and up to legal theory, linguistic semantics, and artificial intelligence. It has always been one of the central and perhaps most discussed concepts in the history of human thought. It is also intimately related to practically every notion that is important for both naive commonsense and mature scientific reasoning, such as laws, counterfactuals, and explanation.

Causation and related notions have shown themselves, however, to be notoriously elusive and problematic concepts. Efforts of many philosophers and logicians in the past have been focused on precise, formal explication of these notions, but the task has turned out to be surprisingly difficult. These difficulties have obviously contributed to emerged discrepancies and tensions between commonsense causal reasoning and the mainstream understanding of reasoning informed by formal deductive logic.

In this book, causation will be primarily viewed as a particular *system of reasoning* we use in describing the world. This view implies that, just as logic, causation has neither fixed form nor fixed material.[1] Instead, its main subject should consist in establishing domain-specific rules of causal reasoning, describing its models, and studying inferences and the conclusions we can derive from them. Whether causation is a feature of the world, it is definitely a feature of how we view, represent, and reason about the world.

We will argue, however, that a large part of the difficulties in understanding and formal interpretation of the concept of causation are due to the fact that causal reasoning is fundamentally distinct from (narrowly) logical, or deductive, reasoning, and consequently it cannot be fully captured using the tools and constructions of deductive logic.

---

1. It is what Glymour (2016) has called a "shapeshifter."

The understanding of causation as a reasoning tool that aims to describe the world immediately brings to the fore an apparently inevitable clash between causation and logic since they seem to share the same aim. Moreover, it seems that it is this alleged clash between two competing forms of reasoning (sometimes explicit, but mostly implicit) that has played a crucial role in the emergence of an influential line of thought at the beginning of the twentieth century that causation should be expelled not only from the language of logic, but also from the language of science (see, e.g., Russell 1912). In a nutshell, by this (imperialistic) line of thought, there is no "logical space" for causation in our reasoning, at least as a non-reductive concept, because "Logic pervades the world: the limits of the world are also its limits" (Wittgenstein [1921]1961, 5.61).

A hundred years later, however, we are going to suggest in this book a reconciliation between these two kinds of reasoning. The reconciliation will be based on the claim that causal reasoning is not a rival, or competitor, of logical reasoning but its *complement* for ubiquitous reasoning situations where we do not have complete data or logically sufficient knowledge. Further, on this view, causal reasoning *presupposes* logical reasoning and should even be constructed on top of an underlying logical formalism. As we will see in the next section, the origins of such a two-layer system of reasoning can actually be found in Aristotle's *Analytics*, so it is almost as old as logic itself.

The suggested reconciliation will impose, however, some nontrivial requirements on both sides of the agreement. To begin with, it will require that the logical part should be formulated in such a way as to make it receptive to the use of causal assertions. We will find the standard possible worlds semantics for classical logic not suitable for this role, so in chapter 2 we will describe an alternative, situation-based semantics for classical logic and explore its representation capabilities. This will also provide us with a better understanding of what can be captured by purely logical means before introducing causal concepts. We will discover, in particular, that David Lewis's semantics of counterfactuals, and thereby his theory of causation, belong entirely to this logical level.

On the other side of the agreement, causal claims that form the basis of causal reasoning cannot be seen as assertions that belong to, or derivable from, the logical part or available evidence.[2] Rather, they should be viewed as *assumptions*, or hypotheses, we make about situations to explain evidences and predict further facts. This makes causal reasoning proper a special kind of assumption-based reasoning which in turn is a central part of general *nonmonotonic reasoning* that has been developed, mainly in artificial intelligence, since the end of the past century. This kind of reasoning will be described in chapter 3.

Chapters 2 and 3 will jointly provide a broad logical background for the main formalism of this book, the causal calculus, that will be described in chapter 4 while the latter will

---

2. A similar claim can be found in Pearl (2000).

serve as a basis for a logical theory of causality and its various applications that will be developed in subsequent chapters.

1.1 Aristotle's Apodeictic

At this point, as it often happens in philosophy, we suddenly realize that the path of inquiry we hoped to open is already marked by the footprints of Aristotle.
—Zeno Vendler, *Philosophy in Linguistics* (1967)

.

Knowledge is the object of our inquiry, and men do not think they know a thing till they have grasped the 'why' of it (which is to grasp its primary cause).
—Aristotle, *Physics* II.3 194b17–20.

With obvious adjustments concerning mainly modern understanding of logic and its language, Aristotle's theory of demonstration (*Apodeictic*) presented in his *Posterior Analytics* as well as its underlying understanding of knowledge come remarkably close to the view of causal reasoning that lies at the basis of the present study.

Logic has been viewed by Aristotle primarily as a tool for producing, or acquiring, knowledge. It has been designed by Aristotle as a replacement of dialectic that has reigned in Plato's Academy as the principal philosophical method in discovering the nature and essence of things and abstract concepts. This development of Aristotle's thoughts has actually reflected his transition from an orthodox Platonic view to a more measured one that has treated dialectic as a mere reasoning technique, both unessential and insufficient by itself to the pursuit of knowledge (see Hamblin 1970).

This transition has been completed in the *Analytics*. In contrast to Aristotle's previous dialectical works, such as the *Topics* and *Sophistical Refutations*, the exposition this time is carried out "monologically"; there is almost no mention of questioners and answerers, dialogues and sophists (except in a derogative way). However, in his search for more firm grounds of reasoning, Aristotle never loses sight of its purposes and, leaving dialectic as a method, he retains the main intent behind it and still holds that the principal aim of arguments and deductions should consist in the production of knowledge. In this sense, Aristotle's apodeictic could also be viewed as a theory of scientific knowledge (*episteme*) in its highest form.

According to Aristotle, we possess knowledge of a fact as opposed to knowing it in the accidental way "in which sophist knows," only when we know the *cause* on which the fact depends, and also that it is not possible for it to be otherwise (*APo*, I.2, 71b9–12).[3] Using a

---

3. In what follows we will freely blend two main English translations of Aristotle's *Analytics*, Ross (1949) and Barnes (1993).

modern terminology, knowledge can and even should be encoded in terms of cause–effect relations. This naturally leads Aristotle to the concept of *demonstration (apodeixis)*:

> By demonstration I mean a syllogism productive of scientific knowledge, a syllogism, that is, the grasp of which is *eo ipso* such knowledge .... Demonstrative knowledge in particular must proceed from items which are true and primitive and immediate and more familiar than and prior to and causes *(aition)* of the conclusions. (In this way the principles will also be appropriate to what is being proved.) There can be a syllogism even if these conditions are not met, but there cannot be a demonstration—for it will not bring about knowledge. (*APo*, I.2, 71b19–25)

A two-tier pattern of reasoning is clearly visible in the above description of demonstration. On Aristotle's account, the scientific knowledge of facts that are not "self-evident" is possible only through demonstration while the latter is achieved by a syllogism (deduction) that is not only logically valid but also, and most importantly, follows the natural order of things. This order is the order of priority, dependence and necessitation determined by causation. In this respect, demonstration is not just a preferred ordering of humanly constructed knowledge but a mapping of the structure of the real (Burnyeat 1981).

According to Aristotle, an important property of demonstrative knowledge is its universality and necessity. Knowledge and its object are different from opinion and its object because knowledge is universal and depends on necessary items and what is necessary cannot be otherwise. Such a knowledge can be used for predictions. Accordingly, there can be no knowledge of the things that can be otherwise (*APo*, 88b30–34).

Aristotle usually refers to his two *Analytics* as a single study, and the very first sentence of the *Prior Analytics* declares that the subject of this study is demonstration, or demonstrative knowledge (*APri*, I, 24a10–11). Still, syllogistic (which is described in the *Prior Analytics*) is an important prerequisite of this aim, so it should be discussed before demonstration because "syllogism is the more general; demonstration is a sort of syllogism, but not every syllogism is a demonstration" (*APri* I.4, 25b26–31). In other words, the causal reasoning of apodeictic stands firmly on the logic of the syllogism (see introduction to Barnes 1993). In addition, Aristotle provides a distinction between what is prior or better known *for us* and what is prior and better known "in the order of being." This distinction lies at the base of the difference between dialectic (and purely deductive reasoning in syllogistic) and demonstrative reasoning.

Aristotle has actually provided vivid examples of the distinction between deduction and demonstration. Thus, from the facts that the planets do not twinkle and that what does not twinkle is near, we can deduce that the planets are near (*APo*, I.13). However, this deduction is not demonstration since it gives not the reason why but only the fact: it is not because the planets do not twinkle that they are near—rather, it is because they are near they do not twinkle. It is only when we prove the latter through the former that the corresponding deduction will also be a demonstration because it will give the reason why.[4]

Some misconceptions accumulated in the history of interpretations of Aristotle's *Analytics* have been caused by the desire to make them more "understandable" for the modern logically and philosophically educated reader. Some of them have even influenced the very translation of Aristotle's texts. Thus, in the revised Oxford translation of the *Analytics* (see Barnes 1984), the term "cause" (used as a translation of *aitia* in all previous translations of the *Analytics*) has been replaced with "explanation." This has hidden to some extent the causal basis of the Aristotelian concept of demonstration. Similarly, Aristotle's term for knowledge (*episteme*) has been variously translated as "science" or "understanding," which has reflected general uncertainty about the proper place of the *Posterior Analytics* and its subject matter in the modern philosophical picture. In fact, the whole purpose of the larger *Posterior Analytics* has become unclear for the "logically educated" reader since the logic proper—that is, a theory of syllogism—has been described by Aristotle already in the relatively short *Prior Analytics*. An extreme expression of this uncertainty can be found in a (basically correct) claim that Aristotle's understanding of demonstration has not yet played any role in the modern understanding of demonstration, though accompanied with a conclusion that it only tends to deflect attention from the deep, clear, useful, and beautiful aspects of the *Analytics* (see Corcoran 2009).

The content of the *Posterior Analytics* has often been described as a particular theory of scientific method. Thus, Barnes (1993) has argued that Aristotle was the first to produce a full-fledged philosophical theory of axiomatized science. Indeed, Aristotle's apodeictic is not confined to exploring specific scientific domains; in this respect, it is a mere faculty of furnishing arguments, just as syllogistic or dialectic. Still, an axiomatic method (on a modern understanding of the latter) makes causal and even explanatory relations between particular assertions largely irrelevant; what really matters in the latter is only logical deduction of conclusions from the axioms. In contrast, the purpose of demonstrative proofs consists not in establishing facts that are previously unknown, but in providing a proper understanding (or true knowledge) of these facts. Accordingly, Aristotle has actually suggested an almost opposed direction in which sciences should be developed (for which the title *Analytics* is especially appropriate), one which starts with facts obtained, for example, from observations, and proceeds to their causal explanations because "without a demonstration you cannot become aware of what a thing is" (*APo*, II.8, 93b17–18). It is only at the end of this analytic process—namely, from successful demonstrations and whole theories—that we can get a grip of the principles (definitions and axioms) of a science (cf. Detel 2012). Actually, at this final stage of "complete" science, we can even

---

4. Numerous examples of this kind have reappeared in the past century as counterexamples to the regularity theory of causation, and especially to Hempel's deductive-nomological approach to explanation—see the next section.

reverse the order and deduce the relevant consequences, including new ones, in a purely logical way.[5]

The development of Aristotle's views is also visible in his new account of fallacies, which he gives in the *Analytics*. Here he is primarily interested not in the fallacies of argumentation but rather in the fallacies of demonstration. In other words, Aristotle objects here less to faults of the reasoning itself than he does to the fact that the reasoner goes the wrong way about getting knowledge. As a consequence, we see a drastic reduction in the number of fallacies mentioned as compared with the *Sophistical Refutations*, from thirteen to just four. However, what Aristotle retains is more significant. In the *Prior Analytics* we still find four fallacies: namely, *Begging the Question*, *Non-Cause*, *Misconception of Refutation*, and *Consequent*. As a matter of fact, all these fallacies are strongly related to (improper uses of) causal reasoning. However, the last two fallacies could also be viewed as formal or logical ones, but the first two fall beyond purely formal considerations and deductive validity. Thus, "begging the question" was originally intended by Aristotle to refer to a situation in the course of disputation in which the arguer "begs," or asks to be granted, the question at issue—that is, the very thesis he has set out to prove. However, in the *Prior Analytics* he treats it already as a species of failure of demonstration:

> Whenever a man tries to prove what is not self-evident by means of itself, then he begs the original question. This may be done by assuming what is in question at once; it is also possible to make a transition to other things which would *naturally* be proved through the thesis proposed, and demonstrate it through them, e.g. if A should be proved through B, and B through C, though it was *natural* that C should be proved through A: for it turns out that those who reason thus are proving A by means of itself. (*APri*, II.16, 64b29–65a9, emphasis added)

A logically educated reader should also notice that Aristotle targets here the well-known reflexivity postulate of logical inference: namely, that A implies A (cf. Duncombe 2014). In fact, irreflexivity is included in the very definition of a syllogism by Aristotle: a syllogism is defined as an argument in which, certain things being posited, something *different from the things posited* results by necessity (*APri*, I.1, 24b18–20). In this respect, a distinctive feature of the logical formalism of causal inference that will be described in chapter 4 will be the absence of the reflexivity postulate while preserving the rest of the properties of classical logical inference.

Similarly, the *Non-Cause* fallacy was described by Aristotle as a fallacy of "positing as cause what is not a cause" in deductive proofs, which clearly presupposes that premises of such deductions, or syllogisms, are not always causes of their conclusions. Again, a formal counterpart of this distinction will be described, in effect, in chapter 8 that will deal with the notion of actual causality.

---

5. A formal counterpart of this "logical reduction" will be described in chapter 7 on abductive reasoning.

In chapter 8, we will also describe the notion of causal inference with respect to an actual causal theory. This notion will be (tentatively) suggested as a formal counterpart of the Aristotelian notion of demonstration in the framework of our formalism of causal reasoning. Further important details about Aristotle's apodeictic will be recalled throughout our study.

To conclude this prohibitively short description of Aristotle's theory of demonstration, we should add that all three main systems of reasoning from the grand edifice of Aristotle's *Organon*—namely, dialectic, syllogistic, and apodeictic—will turn out to play their own distinctive roles in the logical framework of our study. Whereas the last two could be viewed as the origins of our two-tier system of causal reasoning, dialectic could be seen as the origin of a general theory of argumentation that will be described in chapter 3; it will provide a formal description of nonmonotonic reasoning that constitutes yet another essential ingredient of our logical approach to causality (see section 1.4 below).

1.2 Causation versus Logic: An Abridged History

It would be far beyond the scope of this book (and even the author's expertise) to provide a systematic history of relations between causation and logic. In this section we will make, however, a brief summary of the main "events" in the course of this history; they will serve as reference points in what follows. It will also show that the history of these relations is closely correlated with some major turning points in philosophical thought.

1.2.1 The Law of Causality

Briefly put, the widely known law of causality says that everything has a cause. It constitutes, in a sense, the main "maxim," or postulate, of causal reasoning. It is also the oldest such postulate; witness the following, more careful and dynamic formulation of the law, given by Plato:

> Everything that comes to be must of necessity come to be by the agency of some cause, for it is impossible for anything to come to be without a cause. (Plato, *Timaeus*, 28a4–5)

In his *Critique of Pure Reason* (Kant [1787]1998), Immanuel Kant has presented this law as an a priori principle of human understanding.

It is important to observe that, regardless of our education, the law of causality is indeed an essential part of how we view the world. If a plane crash has occurred and the investigation team returns with the conclusion that there are no visible causes of the crash, we will obviously be disappointed and probably will even send them back to complete their investigation. Similarly, if a physical particle that is previously at rest suddenly starts moving in some carefully crafted experimental setup, an expert physicist will assume that a clash with another particle was a probable cause. Note, however, that all such conclusions are logically ungrounded. Our best physical theories can describe precisely the consequences of a particular collision, but they cannot *logically infer* a collision from these consequences.

In chapter 4, the law of causality will be directly incorporated into the formalism of the causal calculus as a semantic constraint on adequate causal models.

The basic "internal" problem associated with the law of causality has also been discovered very early, presumably already by the skeptics. In a nutshell, it amounts to the *Agrippan trilemma* between the exhaustive three horns of (i) acceptance of an infinite regress of causation; (ii) acceptance of uncaused, brute facts; or (iii) acceptance of self-caused facts (circularity). Aristotle addresses this issue at length in the *Posterior Analytics* (*APo*, I.19–21), where he rejects the possibility of an infinite regress in demonstrations. Our logical theory of causality will also have to cope with this fundamental problem, and our proposed solution could be viewed as a formal version of the philosophical doctrine of *fallibilism* as put forth by Charles Sanders Peirce and Karl Popper, according to which all empirical knowledge outside mathematics and logic comprises defeasible *assumptions*, or hypotheses, and should be treated as such.

1.2.2   Leibniz's Principle of Sufficient Reason

There is nothing without a reason, or no effect without a cause.
—Gotfried W. Leibniz, *First Truths* (ca. 1680–1684)

From the standpoint of this book, the more objective law of causality and more epistemic principle of sufficient reason are just two sides of the same coin.

The principle of sufficient reason states that every fact has a reason why it holds. Accordingly, it implicitly asserts that the very process of human *reasoning* should proceed in terms of *reasons*. It would not be an exaggeration to claim that this principle is also as old as Western philosophy itself. Nearly all philosophers up until twentieth century employed it, explicitly or implicitly, in their thinking.

The term "principle of sufficient reason" was coined by Leibniz, though Spinoza has probably preceded him in appreciating the importance of the principle (see Melamed and Lin 2018). Like many rationalist philosophers, Leibniz hardly distinguished between reasons and causes. And as Aristotle long before him, he has considered sufficient reason as something that is provided by demonstration—a logical proof that reflects the causal order of things.

Leibniz has put *two* reasoning principles, the (logical) principle of contradiction and (causal) principle of sufficient reason, as a basis of his philosophy. Both these principles have become recognized laws of thought that held a place in European pedagogy of logic and reasoning in the eighteenth and nineteenth centuries. With the advance of formal logic, however, the principle of sufficient reason has disappeared from logic textbooks. The final move has been made by Ludwig Wittgenstein in his *Tractatus Logico-Philosophicus* (see Wittgenstein [1921]1961), where the principle of contradiction has been proclaimed as the only logical principle that governs the world. Moreover, Wittgenstein has already argued that there is no possible way of making an inference from the existence of one situation to the existence of another, entirely different situation and that there is no causal nexus to justify such an inference (Wittgenstein [1921]1961, 5.136), which has led him to the conclusion that belief in such connexions is simply a "superstition."

1.2.3 Hume on Causation and Its Inferential Source

The law of causality has become one of the targets of David Hume's criticism in his *Treatise of Human Nature* (see Hume [1739–1740]1978). He has considered several arguments that attempted to prove it and has found all of them wanting. However, just as for the theory developed in our book, Hume's main interest in causation has been in its connection with our *inferences* from causes to effects. Causal relation appears to be the foundation and immediate subject of all our empirical inferences. Most important, according to Hume, it is the only relation which enables us to infer from observed matters of fact to unobserved matters of fact (pp. 73–74). But what Hume himself has considered as his great discovery is that it is these inferences that constitute an ultimate "impression source" for the idea of causation, not the other way around (see Beebee 2016; Mackie 1974). Our very idea of the necessary connection between cause and effect arises from reification (or projection) of these inferences. Once the habit, or custom, of inference from one (kind of) event to another is established,[6] the events that a priori "seem entirely loose and separate" become necessarily connected, and we "then call the one object, *cause*; the other, *effect*" (Hume [1748]1975, p. 75). The purpose of forming such causal customs, or assumptions, is to extend our knowledge beyond what is currently present to the memory or senses, primarily in order to control and regulate future events by their causes. However, this kind of reasoning cannot be accomplished a priori, for example, by constructing purely logical deductions (that Hume has also called "demonstrations") solely on the basis of our knowledge of the causes. Pure reason is insufficient for inferring effects from causes. Nevertheless, we can succeed in using such a causal reasoning by tracking nature's regularities because nature has "implanted in us an instinct, which carries forward the thought in a correspondent course to that which she has established among external objects" (p. 55). Causal reasoning is a "just" reasoning already because it works (see Beebee 2006).

Hume's views of causation (occasionally reinterpreted), and especially of the necessary connection between cause and effect, have had profound impact on the subsequent relations between causation and logic. Thus, they have become the modern source of various positivist and analytic accounts that have rejected necessary connections between distinct

---

6. Founded on past experience of an appropriate "Humean" regularity that satisfies temporal priority, contiguity, and constant conjunction.

existences (cf. Wittgenstein's views above). However, Hume's views have also "provoked" Immanuel Kant to a systematic use of such connections as a priori concepts and principles of the understanding in his theory of experience (see, e.g., De Pierris and Friedman 2018). On the other hand, the ancient, Aristotelian views of natural necessity have also survived the passage of time and evolved to the use of forces, powers, and dispositions in contemporary realist approaches to causation.

1.2.4 Mill's Covering Law Account

John Stuart Mill's treatise *A System of Logic* is also an important contribution to a general theory of causal reasoning. To begin with, he has pointed out that not every Humean regularity determines a causal relation (for instance, the succession of day and night does not). According to Mill, only those regularities could serve this causal role that both invariably occur and are *unconditional* of any further circumstances. In particular, only such unconditional regularities could assign a sense of necessity to the causal relation.

Mill has identified such unconditional regularities with laws of nature and defined the cause of a phenomenon as "the antecedent, or the concurrence of antecedents, on which it is invariably and *unconditionally* consequent." In other words, a causal relation between facts or events can be established only when they are appropriately subsumed, or "covered," by some law.

Using a modern logical terminology, Mill's law could be viewed as a kind of a conditional, or inference rule, having a set of antecedents and a consequent. The precise logical form of such laws has emerged, however, as a key problem for what has become known as the regularity approach to causation.

To be unconditional, or circumstance free, a lawful conditional should include in its antecedent all the conditions on which the consequent depends. In principle, these should include not only positive conditions that support the corresponding lawful connection but also numerous negative conditions stating the absence of all interfering factors, and that is because, as Mill has remarked, "all laws of causation are liable to be counteracted or frustrated." Accordingly, the cause according to Mill is the "sum total of the conditions positive and negative taken together; the whole of the contingencies of every description, which being realised, the consequent invariably follows" (Mill 1872, Bk III, Ch V, S 3).

However, a common objection against such a notion of a law has become that no finite set of conditions could ever be truly sufficient for some consequent. Moreover, as has been aptly remarked by Russell (1912), once we have specified the full cause of a given event to render our regularity exceptionless, we'll find ourselves with a circumstance that is unlikely ever to be repeated and hence cannot be considered to be an instance of a "regularity" at all. On the other hand, if a lawlike conditional does not include all such negative conditions, it cannot be viewed as describing a logically invariable regularity (cf. Cartwright 1983).

Mill himself has addressed this issue as follows:

The negative conditions, however, of any phenomenon, a special enumeration of which would generally be very prolix, may be all summed up under one head, namely, the absence of preventing or countervailing causes.

This description has become known as Mill's omnibus negative condition. Note, however, that since it has been formulated in causal terms, it should be clear that if it were used as part of the definition of a law while the latter is used for defining causes, the whole approach would become hopelessly circular (cf. Moore 2009).

1.2.5 Hempel's Deductive-Nomological Approach

The problem of the logical representation of law sentences has actually transcended the boundaries of causal reasoning and has become a general problem for every theory of laws and lawlike generalizations, often in the form of a dispute about so-called ceteris paribus laws (see Reutlinger et al. 2019). Moreover, the problem has become especially acute when the covering law approach has been brought to its "logical end" by interpreting the underlying notion of a conditional as a (universally quantified) classical implication. This is what has been actually proposed by Carl Hempel in his deductive-nomological approach to scientific explanation (see Hempel 1965).

In the framework of classical logic, there is no principled distinction between conditional and nonconditional propositions. As a result, not only the causal but even the conditional component of the meaning of laws simply disappears in a purely logical representation, and we are left only with the property of universality as the basic characteristic feature of "lawlikeness." This elimination is fairly evident in the following passage from Hempel (1965, p. 266):

> A lawlike sentence usually is not only of universal, but also of conditional form; it makes an assertion to the effect that universally, if a set of conditions, C, is realized, then another specified set of conditions, E, is realized as well. The standard form for the symbolic expression of a lawlike sentence is therefore the universal conditional. However, since any conditional statement can be transformed into a non-conditional one, conditional form will not be considered as essential for a lawlike sentence, while universal character will be held indispensable.

The above logical reduction of the notion of a law makes the latter just a special instance of a *factual* assertion, which creates immediate problem already because it does not allow us to distinguish true laws from accidental generalizations like *All gold spheres are less than a mile in diameter* or even *All coins in my pocket are silver* (see Carroll 2016). Furthermore, the corresponding deductive representation loses the crucial directional feature of causal explanations that has been emphasized already by Aristotle in his apodeictic, so it readily "explains" the height of a flagpole on the basis of the length of its shadow (Bromberger 1966). As a by-product, this even creates the problem of epiphenomena, namely the problem that even different effects of a common cause can explain each other in this sense. And finally, even this highly regimented logical setting retains Mill's problem of negative conditions, this time as a problem of "provisos" in the description of scientific laws (see Hempel 1988). Summing up, numerous problems and counterexamples have been discovered in the literature that have revealed a wide gap between our commonsense understanding of (causal) explanations and laws, and their logical explications suggested by Hempel. From a historical perspective, these problems and discrepancies have led to a decline of the whole regularity approach to laws and causation and to multiple attempts of its replacement, most prominent being the counterfactual approach of David Lewis (see Lewis 1973a, as well as chapter 2 of this book). Most of these new approaches have disentangled, however, the connection between laws (regularities) and causation, and thereby even the very connection between causation and logic.

The problem of representing lawlike assertions has reemerged as a key practical problem for reasoning and representation in artificial intelligence, and it has become one of the main incentives for the development of nonmonotonic reasoning (see chapter 3). The solutions to this problem, suggested in the framework of nonmonotonic reasoning, will be used also in our representation of causal laws in this book.

1.2.6   A Sad Summary

The following joke (attributed to Kit Fine), taken from an unpublished book (Steedman 2005), provides also a remarkably concise description of the current relations between causation and logic:

> A man in need of trousers goes to a tailor, only to be told that tailors only make jackets, and that in fact only jackets are necessary, for it is easy to show that jackets are topologically equivalent to trousers. . . . Such is the authority of logicians that many otherwise decorous persons have found themselves in the position of trying to use jackets as trousers. When they have complained that jackets don't seem to work very well for the purpose—for example, that the pockets seem to be the wrong way up—the response has often been impatient.

> Sometimes the users have been led to give up on logic entirely and to go off and invent their own knowledge representations.[7] However serviceable these have been, they have often been derided by logicians as outlandish and even indecorous (perhaps the kilt is the metaphor). This is a shame, because in the end one's trousers are best made by tailors, and logicians are or ought to be the right people to make knowledge representations.

1.3   On Causal Relata

Natural language is tolerant, and even hospitable, to all kinds of entities that could serve as causal relata, such as facts, events, people, objects, properties and their aspects, and so on. On the other hand, contemporary philosophical theories of causation are far less tolerant in this respect and ordinarily restrict such relata to one primary kind. The choice is usually influenced by the level of "objectivity" such a theory tends to impose on the causal relation. This is in striking contrast with Aristotle's descriptions of causality, since causes

---

7. Such as structural equation models of causation.

have been treated by Aristotle both as objects or events in the world and as premises, and even middle terms, in syllogisms. This "obliteration" has persisted also in the scholastic and modern literature on causation—see Clatterbaugh (1999).

Various singularist theories, as well as traditional counterfactual theories of causation typically consider it to be a relation among events (see, e.g., Davidson 1980). Such theories have problems, however, with the fact that our causal claims often require more fine-grained distinctions than references to spatio-temporal events. This is especially evident in the legal discourse, and the following example from Moore (2009) can serve as an illustration.

Example 1.1 *An unlabeled can of rat poison is placed next to food in the kitchen of defendant's restaurant, and it is largely the risk of accidental poisoning that makes such act negligent. However, the unlabeled rat poison on a shelf next to a hot stove is not ingested but explode, as a result of which the plaintiff is injured.*

Clearly, the "naked" event of placing a can on a shelf cannot serve as an adequate and legally relevant causal relatum in this case.

Events as causal relata also create immediate problems for dealing with causation by absences, causation of absences, and causation via absences as intermediate steps. Absences, negative events or negative states of affairs are, using David Lewis's phrase, bogus entities; they are unsuitable relata for any sort of causal relation by reason of their nonexistence (see Lewis 2000). This also often leads to the conclusion that absences and omissions cannot be causes (see, e.g., Lombard and Hudson 2020).

This naturally brings us to facts as more suitable candidates on the role. A most appropriate natural language expression for this option is provided by causal claims of the form "E *because* C," where C and E are propositions that state facts and *because* is a sentential connective. A systematic philosophical defense of such a representation can be found in Bennett (1988) and Mellor (1995). For the purposes of the theory to be developed in the book, this representation has significant advantages over practically any other potential representation, first of all because it readily makes causation a subject of *reasoning*. With this formulation, we also acquire all the expressivity of a logical language, such as the use of propositional connectives like negation, conjunction, and disjunction. In particular, it allows causal claims that are true but not in virtue of the obtaining of a causal relation, such as causation by omission. Last but not least, this representation is also in accord with the contemporary structural approach to causation, in which assignments $X = x$ of values to variables can be viewed as basic causal relata.

The main objection against this propositional formulation has always been a metaphysical one, namely that propositions are not in the world; they are only *about* the world, so they cannot cause changes in the world (see, e.g., Moore 2009). Without entering this metaphysical dispute, however, we will show in the course of our exposition that this propositional formulation of causation will provide us with sufficient representation capabilities for all the purposes of this book.[8]

In the formal theory that will be presented in the book, causal claims will be represented as conditionals, or inference rules, of the form $C \Rightarrow E$ (*C causes E*) that connect classical propositions. This will provide a solid ground for the use of a logical language and corresponding tools in describing causation. In addition, our approach will preserve the close connection between causation and laws that constitutes an important ingredient of the traditional regularity account of causation but is absent in counterfactual and singularist theories. It will turn out, however, that this propositional formulation will not relieve us from all the problems involved in representing causal assertions, though they will become nontrivial internal problems of our formalism that will require a formal solution.

1.4   The Nonmonotonic Road to Causality

An important alternative source of the understanding of causation and its role in our reasoning has been provided by artificial intelligence. Causal reasoning has been employed first in theories of diagnosis and abduction in AI (see chapter 7), but the need in such a reasoning has become especially urgent in attempts to provide a proper—that is, succinct and efficient—representation for reasoning about actions and change in AI (we will discuss such theories in part IV of the book). In both these areas, causation has become a working concept in the corresponding formal theories, allowing us to single out intended models of commonsense descriptions. These novel applications have also confirmed that causal reasoning—that is, asking why and seeking explanations—can be used as a formal tool of knowledge representation in AI.

However, these AI applications have also made evident that, though causal reasoning includes an important logical part, it is *not reducible* to logical derivations in some ingenious causal logic. Instead, causal reasoning should be viewed as an instance of general nonmonotonic reasoning. The formalisms of nonmonotonic reasoning have provided a more adequate framework for representing the concept of causation itself.

1.4.1   Causal Assumptions and Defeasibility

The world, or rather that part of it with which we are acquainted, exhibits as we must all agree a good deal of regularity of succession. I contend that over and above that it exhibits no feature called causal necessity, but that *we make sentences called causal laws* from which (i.e. having made which) we proceed to actions and propositions connected with them in a certain way, and say that a fact asserted in a proposition which is an instance of causal law is a case of causal necessity. (Ramsey 1925, emphasis added)

8. See also Hitchcock (2012) about methodological advantages of using appropriate "means-ends metaphysics" in causal reasoning.

The above remark from the well-known paper of Frank Ramsey, "General Propositions and Causality," can be viewed as a particular blend of Hume's and Mill's views of causation. However, its constructive, nonmetaphysical content is largely in accord with the approach of this book. In the formal system of causal reasoning that will be described in what follows, causal rules will function as *assumptions*. More precisely, they will be viewed as *default* assumptions: namely, as (conditional) assertions that we accept and apply whenever there is no evidence to the contrary. This presumptive, normative understanding makes default assumptions polarly distinct from ordinary *factual* claims. In accordance with the principle of sufficient reason, factual assertions are accepted only when we have a reason why we should accept them. In causal reasoning, such reasons usually come in the form of the causes of this fact (in accordance with the law of causality). We do not need reasons for nonacceptance of factual assertions, only for their acceptance; in this sense, they are unaccepted "by default." In contrast, in our formal system of causal reasoning, we will not need reasons for acceptance of causal rules, only for their rejection or cancellation. This possibility of cancellation makes causal rules *defeasible*, which is a characteristic property of assumptions. Moreover, such cancellations will be represented as part of our system of causal reasoning using negative causal rules for *prevention*. It is important to note, however, that in contrast to outright refutations, possible preventions do not cancel the default status of assumptions, so we can continue to use them in all other situations that do not involve such preventions.

Deductive logical formalisms cannot serve as a basis for reasoning with default assumptions, primarily because such formalisms are monotonic: they preserve the obtained conclusions under additions of any further facts or assumptions, even those that contradict the assumptions that have been accepted earlier. For dealing with assumptions, we need nonmonotonic reasoning. In chapter 3, we will describe default logic as well as a modern theory of argumentation developed in AI that provide precise formal basis for this kind of assumption-based reasoning.

Remark. Instead of the distinction between assumptions and facts, Pearl (2000) stressed the importance of a related distinction between laws and facts, in particular a distinction between sentences describing mechanisms and those describing observations, because the former are presumed to be stable whereas the latter are transitory. Still, it is important to note that the key notion of Pearl's approach, intervention, provides a similar functionality of cancellation for structural equations ("laws") in terms of forming submodels. Indeed, a submodel is produced by canceling an equation for a particular endogenous variable and replacing it with a fixed value (see chapter 5).

The default status of causal rules creates immediate advantages for the representation of causal laws that has been a problem for the logic-based accounts we have mentioned in our abridged history above. To begin with, instead of unconditionality of Mill's laws, or universality of Hempel's laws, our representation makes causal rules defeasible from the outset. Still, given their default status, there is no need to include in the antecedents of such laws all the negative provisos that exclude possible interfering factors. Instead, such factors can be introduced modularly by adding appropriate causal rules of prevention. Specific details about the working of this mechanism will be provided in chapters 4 and 8 of the book.

This assumption-based approach to causality will actually make our representation of causal claims much similar to their commonsense language descriptions. Using an example from (Hart and Honoré 1985), it is perfectly legitimate to say that A's blow caused B's nose to bleed and to feel confidence in the truth of this statement, though we would find it difficult to formulate a general law purporting to specify conditions under which blows are invariably, or unconditionally, followed by bleeding from the nose. Moreover, even in this simple case, there is a *logical possibility* that just at the moment A struck, B independently ruptured a blood vessel! In other words, even here our causal claim is only a (defeasible) assumption, though a very plausible one.

We use causal rules and assumptions in general when we want to extend our knowledge beyond what is factually known. We need such an extended knowledge for guiding us in our decisions in ubiquitous situations where our factual knowledge, even coupled with logic, cannot provide us with sufficient information. The fact that nonmonotonic reasoning in general and causal reasoning in particular are not logically warranted make them a somewhat risky business. However, in most cases of interest, this is the best, if not the only, solution for our urgent, everyday problems.

Of course, the success of this endeavor of causal reasoning largely depends on the quality and adequacy of our causal assumptions. In this respect, significant progress has been achieved in the problem of systematic search of such assumptions (in the framework of causal graphs) using algorithms of *causal discovery*, which is one of the central topics in Spirtes, Glymour, and Scheines (2000) and Pearl (2000). The whole issue of discovery or of making causal assumptions, however, falls outside the scope of this book.

1.4.2 Locality and Closed World Assumption

The need in nonmonotonic reasoning in general and causal reasoning in particular become especially evident when we try to provide a formal description for *local situations*, because in such local descriptions a large part of the world should remain outside their scope. The problem, however, is that this outside world can influence the validity of most propositions that pertain to our situation. As a result, a purely logical description of such situations usually does not yield much information. It is not accidental that an extremely simple, standard semantics for classical logic is formulated in terms of complete valuations, or possible *worlds*, rather than local situations. However, the price of its simplicity is that it is much less suitable for providing a useful description of partial situations.[9] In chapter 2, we will describe an alternative, mereological semantics for classical logic that is based on situations. This latter semantics will provide an appropriate framework (as well as a necessary logical space) for introduction of nonmonotonic concepts and tools.

Remark. It has been argued in Pearl (2000, p. 420) that causality is not needed in descriptions of the entire universe, which is why it disappears in corresponding models, because interventions disappear—the manipulator and the manipulated lose their distinction. However, in most cases the scientist carves a piece from the universe while the rest of the universe is then considered out or background and is summarized by what we call boundary conditions. According to Pearl, it is this asymmetry that permits us to talk about outside intervention and hence about causality and cause–effect directionality.

Commonsense reasoning uses a lot of assumptions in describing local situations. In addition to causal assumptions, a large part of assumptions is based on the idea of *normality*: namely, on what should normally hold in a given situation. In chapter 9 on relative causality, we will consider in more detail the effect of such normality assumptions on causal reasoning.

Commonsense reasoning also often employs *formal* assumptions, or stipulations, that do not depend on particular situations. The most important such stipulation is *default negation*, or negation as failure, according to which negative assertions can be assumed to hold by default: that is, in the absence of reasons for accepting the corresponding positive assertion. Basically, this stipulation allows us to restrict our descriptions to positive propositions only and to use default negation for determining truth values of the rest. In other words, using this stipulation, also appropriately called the *closed world assumption*, a partial, positive description of a situation can be transformed into a complete valuation, or a "small" world, for all the relevant propositions of the language.

The first formal description of this closed world assumption has been provided by McCarthy's circumscription (see McCarthy 1980). The related notion of negation as failure plays a key role in modern logic programming (see, e.g., Baral 2003; Lifschitz 2019). In fact, we will show later that the formalism of logic programming can be viewed as a special case of the causal calculus that we will use throughout this book. This makes recent studies of human reasoning that are based on the logic programming framework and closed world assumption (see, e.g., Stenning and van Lambalgen 2008; Dietz Saldanha 2017) especially relevant for our approach to causal reasoning.

The concept of negation as default creates a complex problem when combined with the concept of causation. In the causal context, it boils down to the well-known problem of negative causation; its origins can be found already in the problem of negative causal

---

9. An illuminating discussion of the difficulties involved in providing classical models for realistic situations can be found in Lent and Thomason (2015).

conditions that has been discussed by Mill (see section 2 above), though it could actually be traced back even further to Aristotle's views of negation and negative demonstrations. This problem will occupy a decent place in what follows.

1.4.3 The Causal Calculus

Based on ideas from Geffner (1992), the causal calculus has been introduced in McCain and Turner (1997) as a nonmonotonic formalism purported to serve as a logical basis for reasoning about action and change in AI. A generalization of the causal calculus to the first-order classical language was described in Lifschitz (1997). This line of research has led to the action description language $\mathcal{C}+$, which is based on this calculus and serves for describing dynamic domains (Giunchiglia et al. 2004). A logical basis of the causal calculus has been described in Bochman (2003a) while Bochman (2004, 2007a) studied its possible uses as a general-purpose nonmonotonic formalism.

In the present study, we will use the causal calculus as a general logical formalism for causal reasoning. As such, it shares a common starting point with both an ancient view of causality (Aristotle) and its modern revival (Pearl) in that our knowledge can be stored in terms of cause–effect relationships (cf. Pearl 2017). In the language of the causal calculus, however, this relation is represented directly by causal rules of the form $A \Rightarrow B$ (*"A causes B"*), where A and B are classical propositions.[10] Structural equation models of Pearl turn out to be representable using such rules (see chapter 5), so our approach could be viewed as a (logical) generalization of the latter.

Causal rules $A \Rightarrow B$ will primarily represent *causal laws*—that is, general (type-level) causal claims—so they will naturally correspond to a host of notions from the literature, such as nomic or causal sufficiency, structural determination, and lawlike regularities. Just as for the latter, a causal rule will turn out to be an inherently *modal* concept (they will actually be assigned a simple possible worlds semantics; see chapter 4). Still, in contrast to the prevailing modern logical paradigm, this modal aspect of the notion will not play a significant role in our logical framework, whereas the status of causal rules as default assumptions will be held indispensable.

A plausible way of interpreting causal rules will consist in viewing them as representing (causal) *mechanisms* (cf. Pearl 2000). In fact, this interpretation will play an essential role in our approach to actual causation, though it will be based on a more fine-grained understanding of mechanisms than what is apparently assumed in structural equation models. On our account, a single structural equation may represent a number of different stable and autonomous causal mechanisms.

As one of its primary objectives, a causal theory should determine the set of situations (or worlds) that satisfy the rules of the theory. However, a distinctive feature of causal

---

10. Actually, such rules will allow us also to represent noncausal, factual knowledge using constraints of the form $A \Rightarrow \mathbf{f}$ (see chapter 4).

reasoning is that the relevant situations are determined not only by the rules that belong to the causal theory but also by what does *not* belong to it.[11] Accordingly, this principal semantic function is realized in the causal calculus by assigning a causal theory a particular *nonmonotonic* semantics. By the intended interpretation, situations that satisfy a causal theory should not only be closed with respect to the causal rules of the theory, but they should also satisfy the law of causality, according to which any proposition that holds in a model should have a cause in this model.

Causation is a notoriously difficult and complex notion. In our logical approach, this complexity will be reflected in the fact that the causal calculus is not a plain logical system with stipulated axioms but an essentially nonmonotonic formalism in which logic and nonmonotonic semantics are tightly intertwined.

1.5   Causation, Laws, and Counterfactuals

Traditionally, the trio of notions *causation, counterfactuals,* and *laws* has been at the heart of the philosophy of science, and the relations between them have been the focus of much discussion and controversy. An undeniable basis of these discussions, however, has always been the fact that these three notions are tightly correlated, or using David Lewis's phrase, they are rigidly fastened to one another, swaying together rather than separately. It seems reasonable to suppose that even the problems encountered in numerous attempts to provide precise analysis for each of these notions share the same conceptual source.

The above correlation has obviously required explanation, and many approaches have been suggested that tried to define practically each of these three concepts in terms of the rest. In particular, a reductive analysis of causation has been suggested, in effect, by Mill (in terms of laws) and David Lewis (in terms of counterfactuals). More recently, the framework of structural equation models has provided a rigorous basis for reasoning with these concepts, but it has not ended the controversy.

According to Pearl (2000), the basic building blocks of the structural account of causation are structural equations, which are functions that represent lawlike mechanisms. These equations can be naturally viewed as formal counterparts of (causal) *laws* since they describe generic (type-level) relations among variables that are applicable to every hypothetical scenario. That is why they are capable of determining corresponding counterfactuals (via the notions of intervention and submodel). Speaking more generally, structural equations provide all the information necessary for supporting various causal and counterfactual claims and, in particular, the claims of singular, or actual, causation.[12]

---

11. As has been repeatedly stressed by Pearl, in causal modeling, causal assumptions are encoded not in the links but rather in the missing links (that sanction, e.g., claims of zero covariance).

12. The initial definition of actual causation in Pearl (2000, chap. 10) used the notions of sustenance and causal beam that were directly defined in terms of structural equations.

Though in Pearl's formal approach structural equations were taken as primitive, an influential group of philosophers and researchers has continued David Lewis's legacy in arguing that counterfactuals, in one way or other, should still enjoy a principal status in causal reasoning. Pearl himself has argued that structural equations and counterfactuals are basically equivalent formalisms. Some researchers have viewed structural equations themselves as just a formal way of representing certain privileged counterfactuals.[13] This view has even been shared by the opponents of the structural approach.[14]

In a more elaborate approach of Woodward (2003), the main objective was to provide a manipulative (counterfactual) account of the causal notions that Pearl has taken as primitive. According to Woodward, facts about patterns of counterfactual dependence are more basic than facts about what causes what, and the essence of the manipulability account can be put in a slogan: "No causal difference without a difference in manipulability relations, and no difference in manipulability relations without a causal difference."

It seems that this line of thought has also influenced the bulk of recent counterfactual approaches to actual causation in the framework of structural equations, including counterfactual definitions of causality suggested in (Halpern and Pearl 2001, 2005) and (Halpern 2016a).

In the theory suggested in this book, causal laws of the form $A \Rightarrow B$ will be viewed as a basic concept while, as in the structural account, both counterfactuals and (actual) causality will be defined in terms of it. Moreover, using a translation of structural equations into the causal calculus, we will formally confirm that structural equations and causal counterfactuals are equivalent formalisms. However, our definition of actual causality, given in chapter 8, will not use counterfactuals; instead, it will be based on (a causal version of) the traditional regularity approach—the NESS test. Further, it will be shown that counterfactuals (and hence also structural equations) cannot be used for defining this notion of actual causality because they obliterate distinctions between certain causal descriptions that could produce different assertions of actual causality. We will show, in particular, that this shortcoming is intimately related to the well-known problem of structural equivalents in the structural account. By the same token, both counterfactuals and structural equations will be shown to be less expressive formalisms than the causal calculus, which means, in particular, that the latter theory cannot be translated neither to structural equations nor to counterfactuals.

In Maudlin (2004), Tim Maudlin has argued that causation and counterfactuals are analytically independent notions, whereas the correlations between them are due to the

---

13. Hitchcock (2007): "Counterfactuals are represented using equations among the variables, where each equation asserts several counterfactuals: one for each assignment of values to the variables that makes the equations true." See also Halpern (2016a) for a similar view.

14. Hall (2007): "Far from being indispensable, causal models merely provide a useful means for selectively representing aspects of an *antecedently understood* counterfactual structure."

common "third factor": namely, natural laws and lawlike regularities that provide an ultimate basis for both. This view is remarkably close to the suggested representation of causality in the causal calculus since in the latter causal rules will serve as causal laws that provide an ultimate basis for causal reasoning, including both counterfactuals and actual causation. Moreover, in full accordance with Maudlin's views, our respective definitions of these notions will not have direct analytical connections with each other, though both will be formulated in terms of causal rules.

The structural account of Pearl assigns a paramount role to interventions and counterfactuals in causal reasoning, and the results of this book should not be construed as an argument against this role. These results suggest, however, that the relations between counterfactuals and causation are less straightforward than what has been usually thought, especially for actual causation. In fact, our modified account of actual causality, described in chapter 9, will incorporate some important insights of the counterfactual approaches, especially the idea that our causal claims are not determined fully by the actual world but require comparison with alternative, "counterfactual" worlds.

1.6 Causality in Flux

We usually think of both causes and effects as changes, happenings, or events. This dynamic aspect of the concept of causation is visible already in the careful formulation of the law of causality, given by Plato, that restricts the scope of the law only to things that come to be (see section 1.2.1).

Taken in their general formulations, neither the structural account of causation nor our general causal calculus will reflect this dynamic aspect of causation. This does not necessarily mean that something is wrong with these formalisms. Rather, this means that they both have been designed to provide a most abstract, *atemporal* description of causality. Such a description constitutes a relatively stable and useful level of abstraction in causal reasoning in the same sense that propositional calculus constitutes an important abstract description of classical logical reasoning, despite the existence of first-order and even temporal logics. Still, the analogy immediately suggests a promising direction of development for a general theory of causality. In fact, in the course of our exposition, we will see many indications that taking this dynamic aspect of causality into account could help in resolving the problems we will encounter.

The necessity of considering causal dynamics has also been noted by researchers working in the framework of the structural approach.

Woodward (2003, pp. 112–113) has noted that changes in the values of variables, rather than the variables themselves, should be considered as causes and effects. Moreover, the very notion of an intervention will not be well defined if there is no well-defined notion of changing the values of a variable (as it happens, for instance, with a variable "animal" which takes the values {lizard, kitten, raven}). This implies, in particular, that such conditions as being a member of a particular race or being a certain age are problematic candidates for the role of a cause.

Glymour et al. (2010) have argued that intuitions about what causes what may vary because people implicitly make different assumptions about the prior states. So, to prevent equivocation, these assumptions should be made explicit (see also Glymour and Wimberly 2007).

Outside the structural approach, a dynamic model of causal reasoning has been suggested in Maudlin (2004). According to Maudlin, in judging causes, we try to carve up the situation into systems that can be assigned inertial behavior (behavior that can be expected if nothing interferes) along with at least a partial specification of the sorts of things that can disturb the inertial behavior, analogous to the Newtonian forces that disturb inertial motion. The latter can be viewed as causes of these disturbances.

In our approach to dynamic causation, we will distinguish two levels of "dynamic involvement." First, in chapter 9 we will introduce a relative generalization of the notion of actual causality that is semantically determined not only by the actual situation but also by an accompanying set of *background conditions*. A most natural interpretation of these background conditions will consist in viewing them as describing a preceding state or *preconditions* of the current situation. Given this modified semantic setting, we define the notion of *proximate causality* that will implement an approach to causation that has been suggested by Hart and Honoré (1985), according to which cause is essentially something that intervenes in the course of events that would normally take place and makes a difference in the way these events develop.

The modified semantic setting of chapter 9 will occupy an intermediate position between an atemporal description of causality and a fully dynamic setting that we will describe in part IV of the book. We will begin this part (in chapter 10) with a formal attempt to "reclaim" significant part of the territory currently occupied by the logical formalism of dynamic logic as properly belonging to causal reasoning. This alternative representation will implement, in particular, the idea that meaning of linguistic expressions could be viewed, in general, as a capability of context change, an idea that occupies today a prominent place in the natural language semantics. Then we will analyze (in chapter 11) the role of the dynamic Markov principle in current AI representations. Finally, in chapter 12 we will introduce a dynamic causal calculus and show its connections with high-level action description languages of Gelfond and Lifschitz (1998) that have been used in theories of action and change in AI.

2 Mereological Semantics for Classical Logic

As we argued in chapter 1, in order to provide a proper logical basis for an adequate theory of causality, we should abandon, or modify, some traditional assumptions surrounding the Logic itself. This does not mean that we have to change our (classical) logic. However, we need to change the ways logic is used, and as an important prerequisite for this change, we should upgrade its semantics to make it hospitable to this change in perspective. In accordance with that, in this chapter we are going to set a general *logical* background against which our theory of causality will be developed. We will make an effort to show that this alternative semantics has an independent interest and importance for general logical studies. Still, borrowing Pearl's metaphor, in the framework of this book it will serve only as a first, logical rung in our ladder of causation.

The generalization of the classical logical semantics that will be described below is based on a shift from possible worlds as points of evaluation to *situations*, or states of affairs, that will be viewed as natural parts of the world. We will show, however, that this shift creates a powerful representation framework that allows us to express some basic forms of dependence and independence among propositions, without changing the logic. Moreover, one of the important implications of these extended expressive capabilities will amount to a somewhat surprising result that counterfactuals on David Lewis's interpretation of the latter (see Lewis 1973b) are confined to this logical rung, and consequently they cannot serve as a basis for interpreting causality.

The plan of the chapter is as follows. We introduce first an important and relatively well-known class of partial semantics for classical logic and show that they correspond to a certain nonstandard generalization of classical Tarski consequence to multiple-conclusion sequents that will be called *classical* Scott consequence relation. Then we will demonstrate that the corresponding logical framework is actually far more expressive than what is strictly required for representing classical logic and its connectives. We will give a general description of the kinds of connectives definable in this semantic framework and show a special character of the classical connectives in this setting. Then we will single out a particular class of definable connectives that provide description of basic forms of dependence and independence among propositions, thereby creating an expressive framework for ana-lyzing such concepts as counterfactuals and related notions. Finally, it will be shown that the resulting semantics in the corresponding extended language is directly related to David Lewis's possible worlds semantics of comparative similarity, which was originally used for interpreting these concepts.

2.1 Two Perspectives of Classical Logic

The logical formalism that we are going to describe can be viewed as a product of juxtaposing two loosely related ideas, or problems, concerning classical logic and its semantics—one proof-theoretic, another a model-theoretic one. This juxtaposition will naturally lead us to a particular generalized formalism for classical logical reasoning.

It has been known at least since Carnap (1943) that ordinary formal systems of classical logic, such as classical Tarski consequence relations or systems of classical natural deduc-tion though complete, do not determine exactly the standard semantics of the classical connectives (i.e., familiar truth tables for connectives). More precisely, they always admit valuations that do not comply with this semantics.[1] It is also well known that this discrep-ancy disappears in the classical Gentzen sequent calculus but at the price of extending our representation, or inferential, capabilities with sequents: that is, multiple-conclusion infer-ence rules. However, this price has appeared to many as too high from a proof-theoretic point of view, since multiple-conclusion rules do not seem to be directly related to our actual inferential practices (see, e.g., Steinberger 2011). On the other hand, a purely proof-theoretic approach that is inherently based on ordinary, single-conclusion inference rules has found it difficult, or even impossible, to provide justified foundations for classical logic, which has also led to eschewing classical logic altogether in favor of intuitionistic logic (see, e.g., Dummett 2000).

There exists, however, a sound alternative conclusion that could be made from the above-mentioned discrepancy; it is that the ordinary classical semantics is not the only option for interpreting classical logic. Moreover, it could be taken to imply that the standard formalisms of classical reasoning permit a more general semantics for the latter.

It turns out that such a general semantics not only exists, it actually has quite a long history. Unfortunately, this history has not produced a single line of development and ter-minology, so the corresponding semantics has been rediscovered a number of times, often in different forms and guises. For reasons that will become clear shortly, we will use a generic term *classical partial semantics* to refer to these semantics. Its initial instantia-tions can be traced back to the possibility semantics of Humberstone (1981) and the partial semantics of van Benthem (1986). On a broader comparison scale, the classical partial semantics is closely related to the supervaluation semantics of van Fraassen (1971), Kit

---

1. A very detailed and instructive discussion of this issue can be found in Humberstone (2011).

Fine's semantics of vagueness (Fine 1975), and even to the forcing semantics in set theory (see, e.g., Burgess 1977). Among more recent variants of essentially the same semantics we could mention Cresswell (2004) and Peregrin (2006) as well as the inquisitive semantics (Ciardelli and Roelofsen 2011). Rumfitt (2015) has used such a semantics in his philosophical defense of classical logic. Finally, a special role of this semantics as the "natural semantics" of classical deductive formalisms has been pointed out in Garson (2001), and a very detailed description of the classical partial semantics and its properties, taken from this perspective, can be found in Garson (2013).

Despite the diversity, all the semantics mentioned above have basically the same structure, which is also (nonaccidentally) similar to the Kripke semantics for intuitionistic logic.[2] Namely, such a structure typically comprises a partially ordered set of states and a valuation function that interprets propositions with respect to these states. The semantic definitions of such connectives as negation, conjunction, and implication (though not disjunction) coincide with the corresponding definitions in intuitionistic semantics. Moreover, as in the latter, the valuation function is required to satisfy persistence with respect to the underlying partial order on states. However, a key difference from the intuitionistic semantics appears in the form of an additional constraint on the valuation function, called stability (van Benthem 1986) or refinability (Humberstone 1981); it is required to obtain a fully classical semantics.

We will provide a more detailed and formal description of the above semantics in the next section. Before doing this, however, we have to address yet another, model-theoretic aspect of the classical partial semantics.

2.1.1 A Model-Theoretic View

The above-mentioned studies have demonstrated that there exists a sound semantic framework for interpreting classical deductive formalisms, which is more general than the standard classical semantics. Taken by itself, however, this theoretical possibility still does not provide a compelling answer to a more interesting question: whether we *need* such a generalized classical semantics for some meaningful, nontechnical purposes. After all, the original classical semantics has obvious advantages in terms of simplicity and conceptual clarity over practically any other reasonable logical semantics. We contend, however, that an answer to this latter question can be obtained only if we will step out of the purely proof-theoretic view on the role of semantics in our reasoning.

On a more ambitious (though by no means uncontroversial) understanding, one of the important roles of logical semantics should consist in providing an adequate representation of the world, as well as of our knowledge about it. Whether this view is justified, however,

---

2. As was noted in Cresswell (2004), the difference between intuitionistic and classical logic is not a matter of the difference in the meaning of operators, but a matter of how the world presents itself for evaluation in terms of truth.

it is an undeniable fact that the standard classical semantics has actually had an overwhelming impact on the ways the world is formally represented in common logical formalisms. Indeed, the world according to these formalisms is still in the main the world of Wittgenstein's *Tractatus* in which "each item can be the case or not the case while everything else remains the same" (Wittgenstein [1921]1961, 1.21). This implies, in particular, that such a world is practically devoid of internal structure, being simply a set of facts, and the only restriction on its formation is based on the principle of contradiction.

There is, however, a number of concepts that have not found their natural place in such a picture of the world, most prominent among them being the notions of causation, law, and counterfactual that are central to this study. Speaking more generally, this picture of the world does not account for various dependencies that give the world its structure and thereby make it intelligible.

The above picture definitely looks unsatisfactory today. A modern logical way out, however, consists in preserving the traditional understanding of worlds, but considering a broader realm of possible worlds; here, the possible worlds semantics has found its place. Its viability has been based on a supposition that relations of our world with other possible worlds can be explanatory about what happens in our world (Lewis 1986a). By its very design, however, this semantic setting has invited also a rigid, bipartite division between *objective* facts that describe intrinsic features of particular worlds (and expressed by classical logical formulas) and *modal* facts that involve also other, related possible worlds.

Not all possible worlds may be involved in the evaluation of, say, counterfactual or law statements; only relevant worlds matter, and we may single out such worlds by imposing certain relations on the set of all possible worlds. However, the shortcomings of the standard classical semantics reappear at this point in the form of the fact that no nontrivial relations among Tractarian worlds could be determined by the worlds themselves. Of course, we can impose a relational structure on possible worlds, but the choice turns out to be completely unconstrained, so it is solely ours. Moreover, this conceptual difficulty becomes a severe practical problem when we try to determine the "right" relational structure on worlds that would produce intended results in particular formalizations.

The idea that the world has more structure than what is captured by its classical representation can also be found in the literature. Thus, the well-known situation semantics of Barwise and Perry (1983) can be viewed as a first systematic representation of the world as a structured entity that comprises situations as its essential parts (see Perry 1986). More implicitly, this idea has lingered in repeated attempts to construct alternative semantics for counterfactuals and causation that are not based on possible worlds (see, e.g., R. Turner 1981; Veltman 1986; Kratzer 1989; Sanford 1989). In fact, most of these alternative semantics have been based instead on a partially ordered structure of states much similar to the general classical partial semantics we discussed earlier.

A particular instantiation of the classical partial semantics, called a mereological semantics, has been suggested in Bochman (1992) with the objective to provide a unifying framework for analyzing the key concepts in the philosophy of science. Mereological semantics was based on a general mereological idea that the dependence structure of the world is determined by the structure of its parts (see Bochman 1980). Accordingly, the partially ordered structure of the classical partial semantics was interpreted as a partially ordered set of situations (viewed as proper world parts), whereas "ordinary" worlds were identified with the maximal elements of this structure.

By the account suggested in Bochman (1992), the world is not fully determined by the set of occurrent facts. Instead, the world is determined by the set of its parts. On a first approximation, it can be represented as a set of *well-formed* subsets of the set of occurrent facts: namely, the sets of facts that describe particular situations. This means that, in contrast to Wittgenstein's independence thesis, the world has "salient" subsets of facts (which correspond to situations), and not all omissions and additions to these sets are allowable. In particular, we may even have two different worlds sharing the same occurrent facts. In some sense such worlds would be descriptively indistinguishable. However, the part-whole structure of the world determines its qualitative character that in turn constitutes the ground for its nontrivial modal features. Two such worlds would have different dependence structures, meaning that they would have different laws, different causal relations among the facts, and different explanations and counterfactuals that would be appropriate with respect to them. And we would have to act differently in these two worlds.

Dependencies among the facts become apparent and relevant when we try to change, or modify, the current situation. They make themselves manifest in accompanying changes that must occur together with it (as well as in what remains intact). Consequently, to reveal the dependence structure of the world, we can consider accessible alternatives to our world, and the knowledge of these accessible possibilities could indeed provide us with a deeper understanding of the way the world is.

In Bochman (1992), an attempt has been made to provide specific explications for the key notions from the philosophy of science, namely laws, counterfactuals, causation, and explanation. Not all of these explications have survived the passage of time.[3] Still, all these explications were formulated in a certain extended language based on three new nonclassical connectives that are *definable* in the mereological semantics: necessity, dependence, and independence (see below). For our current objectives in this chapter, this has shown that the representation capabilities of the mereological semantics (and hence of the classical partial semantics in general) are significantly larger than what is properly required for representing classical logic and its connectives. Indeed, we will see in what follows that the

---

3. This pertains especially to the suggested reductive definition of causation. It has taken, however, almost thirty years and a radically different approach to rectify this shortcoming.

classical logical connectives form a very specific subclass of connectives that are definable in the mereological semantics.

The extended expressive capabilities of the mereological semantics will force us to reconsider the relations between the classical partial semantics and its proof-theoretic, inferential counterparts. The discrepancy between classical logical systems and the standard classical semantics we mentioned at the beginning of this chapter can now be reformulated as a more specific claim that, whatever proof-theoretic virtues these logical systems have, they are incapable of distinguishing among different classical semantics and, in particular, between the standard and partial classical semantics. We have mentioned also that such discrepancies disappear in the logical formalisms based on multiple-conclusion rules. Thus, the standard classical semantics is fully determined as the unique semantics of the classical sequent calculus.

Now, it turns out that the expressive capabilities of a multiple-conclusion formalism can be exploited also for a concise representation of the classical partial semantics. Thus, we will show that the mereological semantics has a natural syntactic counterpart in the framework of multiple-conclusion Scott consequence relations (Scott 1974). More specifically, we will show that it constitutes a canonical semantics of a natural class of such consequence relations that we will call classical Scott consequence relations. Moreover, the resulting deductive framework allows us to provide a relatively simple axiomatization for the nonclassical connectives used in Bochman (1992). And most importantly, it will be shown that the resulting semantics in this extended language is equivalent to a variant of Lewis's possible worlds semantics of comparative similarity (Lewis 1973b).

2.2 Mereological Semantics

We will start our explorations with a formal description of the mereological semantics. At the first stage our language will be restricted to that of pure classical logic, so at this stage there will be no essential difference between the mereological semantics and all other classical partial semantics we mentioned. Still, this will fix our terminology and will also pave the way to the extended language the will be used in subsequent sections.

The basic objects of mereological semantics are *situations*. They are ordered by the part-whole relation and serve as reference points for evaluating propositions.

Definition 2.1 Mereological semantics *is a triple* $\mathcal{M} = \langle S, \leqslant, \| \ \|_u \rangle$*, where S is a set of situations, partially ordered by the relation $\leqslant$, while for any $u \in S$, $\| \ \|_u$ is a valuation predicate on the set of propositional atoms.*

$\|p\|_u$ says that the atomic proposition p *holds* in (or is *supported* by) the situation u. The valuation predicate will be required to satisfy the following two conditions:

(Persistence) If $\|p\|_u$ and $u \leqslant v$, then $\|p\|_v$.
(Stability) If, for any $v \geqslant u$, there exists $w \geqslant v$ such that $\|p\|_w$, then $\|p\|_u$.

Persistence seems to be an obvious condition if we take into account the meaning of the relation of being part. The stability condition[4] is less obvious. Its contrapositive form says that if a proposition p is not supported by a situation, then this situation can be extended to a situation that is inconsistent with p (or, equivalently, supports $\neg p$, as we shall see below).[5] In other words, stability amounts to the "anti-intuitionist" thesis that absence of support for a given proposition implies a possibility of its rejection or refuting. It constitutes a semantic equivalent of the requirement of classicality; it secures the latter by imposing restrictions on the structure of possibilities open with respect to any given situation.

In fact, persistence and stability can be conjoined into a single condition:

$$\|p\|_u \;\Leftrightarrow\; (v)(u \leqslant v \to (\exists w)(v \leqslant w \ \& \ \|p\|_w)).$$

Now we will extend the valuation predicate to all classical formulas using the following familiar conditions for negation and conjunction:

(S¬) $\|\neg A\|_u$ iff no situation v such that $u \leqslant v$ supports A.

(S∧) $\|A \wedge B\|_u$ iff $\|A\|_u$ and $\|B\|_u$.

The condition (S¬) for negation says that $\neg A$ is supported in a given situation if it cannot be part of a situation that supports A. As can be seen, the above conditions coincide with the corresponding semantic definitions of these connectives in the intuitionistic logic.

Finally, using appropriate definitions of other classical connectives in terms of $\{\wedge, \neg\}$, we obtain the following semantic descriptions of classical disjunction and implication:

$$\|A \vee B\|_u \quad \text{iff} \quad (v)(u \leqslant v \to (\exists w)(v \leqslant w \ \& \ (\|A\|_w \vee \|B\|_w)))$$

$$\|A \to B\|_u \quad \text{iff} \quad (v)(u \leqslant v \to (\|A\|_v \to \|B\|_v)).$$

It is easy to show that persistence and stability are preserved by negation and conjunction, so they will hold for any classical propositional formula.

The mereological semantics is a thoroughly partial semantics. For instance, it may well be the case that neither A, nor $\neg A$ are supported by a given situation (although their disjunction, $A \vee \neg A$, always holds).

Remark. All the above conditions, except the condition for disjunction, coincide with corresponding definitions of these connectives in Kripke semantics for intuitionistic logic. An even stronger analogy exists, in fact, with *Beth semantics* for intuitionistic logic (Beth 1956), the only essential difference being that the principle of barring in Beth semantics is replaced with the principle of stability.

Although possible worlds are not basic objects of the above semantics, they find their proper place within it. Thus, we may identify possible worlds with *maximal situations*

---

4. Our terminology here follows van Benthem (1986).

5. Cf. the *refinability condition* in Humberstone (1981), also discussed in Humberstone (2011).

of the semantic structure (when it has such). Then the above conditions for connectives give us familiar classical definitions (in what follows we shall reserve the letters $i, j, \ldots$ as referring to worlds in this sense):

$$\|\neg A\|_i \Leftrightarrow \text{not } \|A\|_i$$

$$\|A \wedge B\|_i \Leftrightarrow \|A\|_i \text{ and } \|B\|_i.$$

Note also that under the supposition that any situation is included in some world, stability and persistence give us that for any situation u,

$$\|A\|_u \Leftrightarrow (i)(u \leqslant i \rightarrow \|A\|_i). \tag{PS}$$

In other words, a formula A holds in a situation if and only if it holds in all the worlds that include this situation. This will imply, in particular, that under some reasonable assumptions, the mereological semantics can be reformulated as a certain possible worlds semantics (see section 2.5 below).

Now we will turn to describing a syntactic counterpart of the mereological semantics.

2.3 Classical Scott Consequence

In a formal syntactic representation of the mereological semantics, we will make use of Scott consequence relations. Scott consequence relations, known also as multiple-conclusion consequence relations (Shoesmith and Smiley 1978; Segerberg 1982; Wojcicki 1988), take consequence relation to be a binary relation among sets of propositions. In contrast to ordinary single-conclusion, Tarski consequence relations, Scott sequents involve sets of formulas as conclusions; they are interpreted as implied alternatives. This multiple-conclusion character of Scott consequence relations will turn out to be essential for an adequate representation of the mereological semantics.

In the formal presentation of Scott consequence relations in this section we will mainly follow Gabbay (1981) and Bochman (2001).

$A, B, C, \ldots$ will denote propositions of some language $\mathcal{L}$, $a, b, c, \ldots$ finite sets of propositions while $u, v, w, \ldots$ will denote arbitrary such sets. Also, for a set u of propositions, $\overline{u}$ will denote the complement of u in $\mathcal{L}$.

Expressions of the form $a \Vdash b$ will be called *sequents*; they will be assumed to have the following informal interpretation:

If all propositions from a hold, then at least one proposition from b holds.

A binary relation $\Vdash$ on finite sets of propositions is called a *Scott consequence relation* if it satisfies the following conditions:

$$a \Vdash b, \text{ if } a \cap b \neq \emptyset \tag{Reflexivity}$$

$$\text{if } a \Vdash b, \text{ then } a \cup a' \Vdash b \cup b' \text{ for any } a', b' \tag{Monotonicity}$$

$$\text{if } a \Vdash \{A\} \cup b \text{ and } a \cup \{A\} \Vdash b, \text{ then } a \Vdash b. \tag{Cut}$$

In what follows we use ordinary terminological conventions and write, for example, $a, A \Vdash b, B$ instead of $a \cup \{A\} \Vdash b \cup \{B\}$ and $\Vdash a$ instead of $\emptyset \Vdash a$. Also, we will extend the source consequence relation to arbitrary sets of propositions by stipulating that for any sets u, v,

$$u \Vdash v \quad \equiv \quad a \Vdash b \qquad \text{for some finite } a \subseteq u, \ b \subseteq v. \qquad \text{(Compactness)}$$

Definition 2.2 *A set u of propositions is called a* theory *of a Scott consequence relation $\Vdash$ if $u \nVdash \bar{u}$.*

The following lemma shows that theories can be seen as sets of propositions that are closed with respect to the sequents of a consequence relation.

Lemma 2.1 *u is a theory of a Scott consequence relation $\Vdash$ iff $u \Vdash a$ implies $u \cap a \neq \emptyset$, for any set a of propositions.*

Theories in the above sense can be considered as multiple-conclusion analogues of theories of a Tarski consequence relation. Note, however, that such Scott theories do not have all the usual properties of Tarski theories. Most importantly, intersections of Scott theories are not in general theories. Still, they have a property of completeness with respect to directed sets, which implies that any theory of $\Vdash$ is included in a maximal theory and contains a minimal theory of $\Vdash$.

Any set of sets of propositions **T** determines a Scott consequence relation $\Vdash_T$, defined as follows:

$$a \Vdash_T b \quad \equiv \quad \text{For any } u \in \mathbf{T}, \text{ if } a \subseteq u, \text{ then } b \cap u \neq \emptyset.$$

The basic result about Scott consequence relations, called the Scott completeness theorem in Gabbay (1981), says that theories can serve in this sense as *canonical models* of the latter. Namely, if $\mathbf{T}_{\Vdash}$ is the set of all theories of a Scott consequence relation $\Vdash$, then we have the following theorem.

Theorem 2.2 (Representation Theorem) *If $\Vdash$ is a Scott consequence relation, then*

$$\Vdash = \Vdash_{\mathbf{T}_{\Vdash}} .$$

As a consequence, Scott consequence relations are uniquely determined by their theories.

Remark. Theories can also be seen as syntactic counterparts of *valuations*, that is, functions that assign truth or falsity to all propositions of the language (see, e.g., Humberstone 2011). More precisely, theories correspond to admissible valuations that comply with the rules of the source consequence relation.

A set u of propositions is *inconsistent* with respect to a Scott consequence relation if $u \Vdash$ holds; otherwise it will be called *consistent*. By the representation theorem, a set of propositions is consistent if and only if it is included in some theory. Consequently, maximal consistent sets coincide with maximal theories of a Scott consequence relation.

The set of all singular sequents $a \Vdash A$ belonging to a Scott consequence relation $\Vdash$ forms a usual Tarski consequence relation. We will call the latter the *Tarski subrelation of* $\Vdash$. This subrelation will play a key role in the subsequent development of this section.

Negativity and stability. General Scott consequence relations permit that the set of all propositions of the language can be consistent. Negative consequence relations exclude this possibility.

Definition 2.3 *A Scott consequence relation* $\Vdash$ *is* negative *if $a \Vdash$ for some set a.*

In the language of valuations, negativity amounts to the requirement that any admissible valuation should assign falsity to at least one proposition.

In stable Scott consequence relations, defined below, the Tarski subrelation is determined by the set of consistent sets.

Definition 2.4 $\Vdash$ *is said to be* stable *iff for any u and A, if $u \not\Vdash A$ then $v, A \Vdash$ for some consistent set v that includes u.*

The above syntactic stability condition is related to the semantic stability property of the mereological semantics, and it can be seen as a key property characterizing classical logical reasoning.[6]

The next lemma provides an important alternative characterization of stability. The proof is straightforward.

Lemma 2.3 *A Scott consequence relation* $\Vdash$ *is stable if and only if any theory u of $\Vdash$ is an intersection of maximal theories of $\Vdash$ that include it:*

$$u = \bigcap \{i \mid i \in W_{\Vdash} \ \& \ u \subseteq i\},$$

where $W_{\Vdash}$ is the set of all maximal theories of $\Vdash$.

Note, however, that we still do not have that all intersections of maximal theories correspond to theories of a consequence relation. Thus, even in the stable case, a Scott consequence relation is not determined uniquely by the set of its maximal theories.

2.3.1 Classicality

We will consider now Scott consequence relations in languages containing the usual classical connectives. The symbol $\vDash$ will denote the ordinary classical entailment.

---

6. A systematic development of this idea in the framework of ordinary Tarski consequence relations can be found in Peregrin (2015).

It turns out that the question of how classical inference (and the classical understanding of the associated connectives) can be represented in the framework of Scott consequence relations can be answered in a number of different ways. The strongest such way is provided by the well-known classical sequent calculus of Gentzen. Basically, a Scott sequent $a \Vdash b$ is reducible on this representation to a classical formula $\bigwedge a \to \bigvee b$.

However, we are going to describe now an alternative representation of classicality in the framework of Scott consequence, a representation under which Scott sequents are no longer reducible to classical formulas. As a guidance, we adopt a standard representation of classical inference in the framework of Tarski consequence relations.

A Tarski consequence relation $\vdash$ in a language containing the classical connectives is called *supraclassical* if it subsumes the classical entailment:

(Supraclassicality) If $a \vDash A$, then $a \vdash A$;

and *classical* if it satisfies, in addition,

(Deduction) If $a, A \vdash B$, then $a \vdash A \to B$.

Classical Tarski consequence relations provide a complete description of classical inference, and we will use this fact in suggesting the following extension of "classicality" to Scott consequence relations:

Definition 2.5 *A Scott consequence relation will be said to be* classical *if it is negative and its Tarski subrelation is classical.*

It is easy to verify that theories of a classical Scott consequence relation will be deductively closed sets. Negativity in this context amounts to exclusion of classically inconsistent theories; it can be replaced with the requirement:

(Consistency) $\bot \Vdash$,

where $\bot$ is a classical contradiction. Finally, it can be shown that Deduction makes the classical Scott consequence relation stable.

Classical Scott consequence relations allow for replacement of classically equivalent formulas in premises and conclusions of sequents. Also, they allow to replace sets of premises by their classical conjunctions: $a \Vdash b$ will be equivalent to $\bigwedge a \Vdash b$. Finally, Deduction makes any singular sequent $a \Vdash A$ equivalent to a provable formula $\Vdash \bigwedge a \to A$. Nevertheless, a classical Scott consequence relation still retains its multiple-conclusion character since conclusion sets of sequents are not replaceable, in general, by their classical disjunctions.

Another characteristic feature of classical Scott consequence relation is that its maximal theories are also maximal classically consistent sets. We will call such sets *(canonical) worlds* in what follows. In fact, the following theorem shows that classical Scott consequence relations are precisely stable consequence relations having this property. The proof can be found in the appendix to this chapter.

Theorem 2.4 *A Scott consequence relation is classical if and only if it is stable and all its maximal theories are worlds.*

The distinction between classical Scott and Tarski consequence relations can now be described as follows. In both cases a theory of a consequence relation is an intersection of theories that are worlds. In the case of a Scott consequence relation, however, not all intersections of worlds give rise to theories, so we have distinguished, or "well-formed," sets of worlds that correspond to such theories.

2.3.2 Axiomatization

In this section we will provide a direct logical description of classical Scott consequence relations in terms of the rules for the associated classical connectives. This description will highlight both similarities and differences with closely related logical formalisms.

As before, we will take the classical conjunction $\wedge$ and negation $\neg$ as basic, while the rest of the classical connectives will be viewed as definable.

Now, in describing these connectives we will make use of *bidirectional* (double-line) inference rules that have a history of their own. A systematic use of such rules for defining logical connectives has been advocated in Došen (1989), as well as in Sambin, Battilotti, and Faggian (2000), though their origins can be traced back to Kneale, Bernays, and Scott (see especially Scott 1974). In particular, the double-line notation for such rules has been introduced in Kneale (1956). A brief history of such rules can also be found in Humberstone (2011).

Using such rules, classical Scott consequence relations can be formalized precisely as Scott consequence relations in the classical language that satisfy the following rules:[7]

- Rules for negation:

$$\frac{\Gamma \Vdash A}{\Gamma, \neg A \Vdash} \; (\neg L) \qquad\qquad \frac{\Gamma, A \Vdash}{\Gamma \Vdash \neg A} \; (\neg R)$$

- Rule for conjunction:

$$\frac{\Gamma, A, B \Vdash \Delta}{\Gamma, A \wedge B \Vdash \Delta} \; (\wedge).$$

It can be shown that the above bidirectional rule for conjunction is equivalent to the usual, unrestricted introduction and elimination rules for the classical conjunction in the sequent calculus (see Sambin, Battilotti, and Faggian 2000). This cannot be said, however, about the above two rules for negation. Clearly, our rules subsume the corresponding (unidirectional) rules for *intuitionistic* negation. Still, they are stronger, and the following result shows that they determine, in effect, the classical negation in our setting.

---

7. We have changed our notation to highlight similarities with the corresponding sequent calculi.

Theorem 2.5 *A Scott consequence relation in a classical language is classical if and only if it satisfies the above rules for conjunction and negation.*

Proof. ($\rightarrow$) Due to Supraclassicality, theories of a classical Scott consequence relation are deductively closed sets, so the representation theorem will immediately justify the rule for conjunction (in both directions). Note also that Consistency makes $\Gamma \Vdash$ equivalent to $\Gamma \Vdash \bot$, and therefore the two rules for negation follow from the fact that the Tarski subrelation is classical.

($\leftarrow$) Note first that Reflexivity and the top-down direction of ($\neg L$) immediately give us $A, \neg A \Vdash$. Consequently, the bottom-up direction of ($\neg L$) implies stability. Now, the rule for conjunction implies that, for any theory u, $A \wedge B \in u$ if and only if $A \in u$ and $B \in u$. In addition, the rules for negation immediately imply that, for any maximal theory i, $\neg A \in i$ iff $A \notin i$. Consequently, any maximal theory is a world, and the result follows now from theorem 2.4. $\qquad\Box$

2.3.3 Completeness

Now we will turn to the correspondence between the mereological semantics and classical Scott consequence relations.

Definition 2.6 *A sequent $a \Vdash b$ will be said to be* valid *in a mereological semantics $\mathcal{M}$ iff, for any situation s, all propositions from a hold in s only if at least one proposition from b also holds in s.*

We will denote by $\Vdash_{\mathcal{M}}$ the set of all sequents that are valid in a mereological semantics $\mathcal{M}$. This notation is justified by the following

Lemma 2.6 *For any mereological semantics $\mathcal{M}$, $\Vdash_{\mathcal{M}}$ is a classical Scott consequence relation.*

Proof. Straightforward check of the bidirectional rules for conjunction and negation. $\qquad\Box$

Definition 2.7 *A consequence relation $\Vdash$ will be said to be* complete *for a mereological semantics $\mathcal{M}$ if $\Vdash = \Vdash_{\mathcal{M}}$.*

In order to show that any classical Scott consequence relation has a mereological semantics, we will introduce the following canonical semantics.

Definition 2.8 *The* canonical semantics *of a Scott consequence relation $\Vdash$ is a mereological semantics $\mathcal{M}_{\Vdash} = \langle \mathcal{T}, \leqslant, \| \ \|_u \rangle$, where $\mathcal{T}$ is the set of all theories of $\Vdash$, $u \leqslant v \equiv u \subseteq v$, and the valuation predicate is defined as $\|p\|_u \equiv p \in u$.*

Using the above canonical semantics, we obtain the following key result (see appendix 2.A for the proof):

Theorem 2.7 (Classical Completeness Theorem) *A Scott consequence relation is classical if and only if it has a mereological semantics.*

2.4 Standard Connectives

Since in this chapter we are considering the mereological semantics as a semantic representation of the world that has an independent interest, we are naturally interested in the kinds of information that are embodied in it.[8] In order to make this information an integral part of our reasoning, however, we should have syntactic means and connectives that would allow us to express this information in the associated deductive framework of Scott consequence relations.

In this section we will consider a certain class of connectives that are definable in the mereological semantics. Since the latter semantics is determined by a partial order on situations, any such connective can be uniformly described using an appropriate formula in a monadic first-order language of a partial order having exactly one free variable. It will be shown, however, that the classical connectives constitute in this respect only a small part of the possibilities.

2.4.1 What Is a Classical Connective?

We consider first what sort of information about situations is expressed by the classical connectives. On a closer look, classical connectives form a very special class that is characterized by the following lemma (see its proof in the appendix to the chapter).

Below by *cuts* of a partially ordered set we mean sets of the form $\{v \mid u \leqslant v\}$ for some element u.

Lemma 2.8 *Let $\mathcal{L}$ be a monadic first-order language of a partial order. Then a formula in $\mathcal{L}$ with one free variable determines a classical propositional connective if and only if (i) it is persistent and stable when its elementary predicates are and (ii) it is preserved under formation of cuts.*

The above lemma characterizes the classical connectives as those pertaining only to the extensions of any given situation. From the syntactic point of view this means that classical connectives characterize only possible extensions of theories and are largely insensitive, for example, to their subtheories or to theories that are consistent with them. In other words, classical connectives reflect only a very small part of dependencies generated by the mereological semantics.

To obtain a more expressive representation of objective situations, we must have means for describing a broader "neighborhood" of a given situation in the partial ordering of the mereological frame. As a step in this direction we shall consider below three non-classical propositional connectives that have been used in Bochman (1992) for describing

---

8. Cf. a similar approach to intuitionistic semantics in van Benthem (2009).

counterfactuals, laws, and related concepts. These connectives are also definable in the mereological semantics, though they belong to a broader class of connectives that can be characterized as follows:

Definition 2.9 *An $\mathcal{L}$-formula with one free variable will be said to determine a* standard *propositional connective if it is persistent and stable when its elementary predicates are.*

As we will show later, any persistent and stable propositional formula is determined by its behavior on worlds. Consequently, it is easy to show that there exists a one-to-one correspondence between standard connectives in the above sense and arbitrary formulas of $\mathcal{L}$ with one free variable ranging over *worlds*. We will use this correspondence in what follows.

2.4.2　Necessity, Dependence, and Independence

We will introduce first the following notion of compatibility for situations.

Definition 2.10 *Two situations s and t will be said to be* compatible *(notation $C(s,t)$) if there exists a situation u such that $s, t \leqslant u$.*

For mereological semantics in which every situation is a part of some world, the above condition amounts to the requirement that both situations are parts of at least one common world.

Necessity.　Let us define the following necessity operator:

$$\| \boxdot A \|_u \quad \text{iff} \quad (v)(C(u,v) \to \|A\|_v). \tag{SN}$$

The above operator describes a natural kind of necessity of A in a situation u by which any situation where A does not hold must be incompatible with u. The dual form of this modality, interpreted as "A is *contingent*," has been extensively used in Bochman (1992) for distinguishing propositions that can change their truth-values from situation to situation.

It is easy to show that $\boxdot$ is a standard connective. Hence it is uniquely determined by its behavior on worlds. An equivalent "possible-worlds" description of this connective could be construed as follows. Let us define the following accessibility relation on the set of worlds of a mereological semantics :

$$iRj \quad \text{iff} \quad (\exists u)(u \leqslant i \,\&\, u \leqslant j),$$

where u is some situation. Then it can be shown that for any world i,

$$\| \boxdot A \|_i \quad \text{iff} \quad (j)(iRj \to \|A\|_j).$$

Dependence.　We will say that A *depends* on B in a situation u if any compatible situation that supports A, supports also B:

$$\|A \succ B\|_u \quad \text{iff} \quad (v)(C(u,v) \,\&\, \|A\|_v \to \|B\|_v). \tag{SD}$$

If the above dependence holds, then any situation that supports A without supporting B must contain something that is inconsistent with u. This connective is stronger than classical implication $A \to B$ since it involves not only extensions of a given situation but also all situations that are compatible with respect to it. Note, in particular, that the above necessity operator $\boxdot$ is definable as $\top \succ A$, where $\top$ is an arbitrary tautology.

The dependence connective can be viewed as covering the phenomenon of "dependent," or derived, propositions and facts. The relation of dependence is similar to the relation of *involving* used in the situation semantics (see Barwise and Perry 1983, p. 101). It naturally corresponds also to the relation of *lumping* in Kratzer (1989).

Since $\succ$ is stable and persistent, it is also determined by its behavior on worlds. The corresponding condition for worlds is

$$\|A \succ B\|_i \quad \text{iff} \quad (u)(u \leqslant i \ \& \ \|A\|_u \to \|B\|_u).$$

Independence.   Finally, we will introduce a connective that is purported to capture strong independence among propositions. Namely, we will require that B is *independent* of A (notation $A \dashv B$) in a situation when the support of B in this situation can be always chosen to be neutral with respect to the support of A.

To simplify the description, we give the truth-conditions of the above connective only with respect to worlds. We shall say that $A \dashv B$ holds in a world i if and only if any situation in i which does not support A can be extended to a situation in i which supports B, but still does not support A:

$$\|A \dashv B\|_i \quad \text{iff} \quad (u)(u \leqslant i \ \& \ \neg\|A\|_u \to (\exists v)(u \leqslant v \leqslant i \ \& \ \neg\|A\|_v \ \& \ \|B\|_v)).$$

Using the condition (PS) from section 2.2, this definition can be extended to a definition for arbitrary situations.

The above defined connective satisfies the rule of replacement of provable equivalents and has the following basic properties which are common to connectives of a modal kind:

$$\text{If } A \leftrightarrow B \text{ is provable, then so is } (A \dashv C) \leftrightarrow (B \dashv C). \tag{I1}$$

$$\text{If } B \to C \text{ is provable, then so is } (A \dashv B) \to (A \dashv C). \tag{I2}$$

$$(A \dashv B) \ \& \ (A \dashv C) \to (A \dashv B \wedge C). \tag{I3}$$

These properties show, in effect, that the independence connective can be considered as a kind of relative necessity operator. These properties, however, almost exhaust the area of agreement with familiar modal connectives.

Counterfactuals.   The above independence connective has been used in Bochman (1992) for defining counterfactuals. A formal explication of the latter has been based on an old idea from Goodman (1947), according to which a counterfactual $A > B$ holds if B is a logical consequence of A and some occurrent facts and laws that are "cotenable" with A. More precisely, the latter have been identified with propositions that are independent of $\neg A$, which has led, ultimately, to the following definition of a counterfactual:

$$A > B \equiv \neg A \dashv (A \rightarrow B).$$

On this account, a counterfactual has been taken to mean something like

¬A notwithstanding, if A then B.

If we unfold the above definition using the semantic characterization of independence, we will obtain a description that coincides, in effect, with the definition suggested by Angelika Kratzer in her premise-based semantics for counterfactuals (Kratzer 1981). Moreover, we will see later that, under the general correspondence between the mereological semantics and comparative similarity semantics of Lewis, the above definition of counterfactuals will turn out to be equivalent to the now standard definition of Lewis.

2.5 Possible Worlds Semantics

As a final step in our development, we are going to show that classical Scott consequence relations admit also some possible worlds semantics.

Definition 2.11 *A* mereological possible worlds semantics *is a triple* $\mathcal{M}_W = \langle W, S, (\ ) \rangle$, *where W is a set of* possible worlds, *S a set of subsets of W such that for every* $i \in W$, $\{i\} \in S$, *and for any* $i \in W$, *( )$_i$ is a 'classical' valuation predicate satisfying the conditions:*

$$(\neg A)_i \ \text{iff} \ \text{not-}(A)_i \tag{W$\neg$}$$

$$(A \wedge B)_i \ \text{iff} \ (A)_i \ \text{and} \ (B)_i. \tag{W$\wedge$}$$

A mereological possible worlds semantics is an ordinary possible worlds semantics with some distinguished sets of worlds. However, the real difference with the traditional possible worlds semantics will lie in its associated semantic entailment.

For any formula A and any set w of worlds we will use $(A)_w$ to denote the fact that $(A)_i$ holds for every world $i \in w$.

Definition 2.12 *A sequent* $a \Vdash b$ *will be said to be* valid *in a mereological possible worlds semantics* $\mathcal{M}_W$ *iff, for any* $w \in S$, *if* $(A)_w$ *for all* $A \in a$, *then* $(B)_w$, *for some* $B \in b$.

Under this definition of validity, the mereological possible worlds semantics turns out to correspond to a particular kind of mereological semantics that will be called *regular* one. Moreover, since the canonical mereological semantics is also regular in this sense, the mereological possible worlds semantics will also turn out to be complete for classical Scott consequence relations (see the appendix for the relevant definitions and proofs):

Theorem 2.9 *A Scott consequence relation is classical if and only if it has a mereological possible worlds semantics.*

It is interesting to note that "Tarski" sequents of the form $a \Vdash A$ are valid in a mereological possible worlds semantics $\mathcal{M}_W$ just in the case they are valid in all worlds of

$\mathcal{M}_W$. Consequently, mereological possible worlds semantics, coupled with the standard definition of validity, is already sufficient for representing classical Tarski rules. Still, this standard classical semantics is insufficient for representing arbitrary Scott sequents.

Remark. To end this section, we should note an obvious similarity of the above mereological possible worlds semantics with a recent *inquisitive semantics* (Ciardelli, Groenendijk, and Roelofsen 2018). In fact, from a purely formal point of view, the latter semantics (and its corresponding logic) can be seen as a result of extending our formalism with a new disjunction connective that "reifies" the disjunctive conclusions of Scott sequents as formulas of the language. From a semantic point of view, however, this move amounts to adding the well-known "intuitionistic" disjunction to the language having the following deceptively simple definition:

($S^{\veebar}$) $\|A \veebar B\|_u$ iff $\|A\|_u$ or $\|B\|_u$.

Unfortunately, this disjunction connective is not standard in our sense; though persistent, it does not preserve stability, which means, in effect, that it violates the classicality property of our formalism. In practice, this implies that stable and nonstable parts of the resulting language should receive separate treatment as is actually done in the studies of inquisitive semantics. A more systematic discussion of these differences, however, is well beyond the scope and aims of this chapter.

2.6 Comparative Similarity Semantics

In this section we will make our last step and show the correspondence between the mereological possible worlds semantics and Lewis's possible worlds semantics of comparative similarity. As a matter of fact, this final reduction can be seen as a systematic generalization of the correspondence between the premise-based semantics for counterfactuals (Kratzer 1981) and Lewis's semantics that has been established already in Lewis (1981).

We will begin with the following technical notion.

Definition 2.13 *A partially ordered set $(S, \leqslant)$ will be said to be convex if any subset of S, having an upper bound, has the least upper bound.*

Convex partially ordered sets are also known under the name *conditionally complete upper semilattices*. Now, it is easy to show that mereological possible worlds semantics corresponds to mereological semantics on convex frames precisely when the distinguished sets of worlds are closed with respect to arbitrary nonempty intersections. This justifies our terminology because closure with respect to arbitrary intersections is a characteristic property of convex sets. Hence, we shall also call convex those mereological worlds frames which satisfy this property.

The following definition stems from Lewis (1973b).

Definition 2.14 *A* system of spheres *is a pair* $\langle W, \$ \rangle$, *where W is a set of worlds, while* $\$$ *is a function that assigns any* $i \in W$ *a set* $\$_i$ *of nonempty subsets of W (the set of spheres around i) subject to the following conditions:*

1. $\{i\} \in \$_i$;
2. $i \in u$, *for any* $u \in \$_i$;
3. $\$_i$ *is closed with respect to arbitrary intersections.*

A system of spheres will be said to be symmetric *if, for any* $i, j \in W$ *and any* $u \in \$_i$, *if* $j \in u$ *then* $u \in \$_j$.

Remark. The above definition of a system of spheres is more general than the original Lewis's definition, since Lewis has assumed also that a system of spheres around a world must be *nested*, that is, totally ordered by inclusion. Note also that the symmetry requirement has not been considered by Lewis.

The following lemma shows, in effect, that convex mereological semantics is strongly equivalent to symmetric system-of-spheres semantics. The proof is immediate and will be omitted.

Lemma 2.10

(a) *Let* $\langle W, S \rangle$ *be a convex mereological worlds frame. For any* $i \in W$ *define* $\$_i = \{u \mid u \in S \ \& \ i \in u\}$. *Then* $\langle W, \$ \rangle$ *is a symmetric system of spheres.*

(b) *Let* $\langle W, \$ \rangle$ *be a symmetric system of spheres and let S be the union of all* $\$_i$ *for* $i \in W$. *Then* $\langle W, S \rangle$ *is a convex worlds frame.*

(c) $\langle W, \$ \rangle$ *is a system of spheres derived from a convex frame* $\langle W, S \rangle$ *by the definition from (a) iff* $\langle W, S \rangle$ *is a convex frame derived from* $\langle W, \$ \rangle$ *through the definition from (b).*

Lewis has also proposed a somewhat different description of his semantics based on a three-place relation $j \leqslant_i^\circ k$ of comparative similarity on worlds. The corresponding generalization to partial comparative similarity orders has been described in Burgess (1981).

Definition 2.15 *A* comparative similarity frame *is a pair* $\langle W, \leqslant^\circ \rangle$, *where W is a set of worlds, while* $\leqslant^\circ$ *is a three-place comparative similarity relation on W, which satisfies the following conditions:*

1. $i \leqslant_i^\circ i$,
2. $j \leqslant_i^\circ i \rightarrow i = j$,
3. $j \leqslant_i^\circ k \rightarrow i \leqslant_i^\circ j$,
4. $j \leqslant_i^\circ k \ \& \ k \leqslant_i^\circ l \rightarrow j \leqslant_i^\circ l$.

A comparative similarity frame will be said to be symmetric *if* $k \leqslant_i^\circ j$ *implies* $k \leqslant_j^\circ i$ *for any* $i, j, k \in W$.

The following results show the equivalence of this semantic representation to the two preceding ones. They directly follow from the corresponding definitions.

Lemma 2.11 *Let* $\langle W, \$ \rangle$ *be a system of spheres. Define*

$$k \leqslant_i^\circ j \quad iff \quad (\exists U)(U \in \$_i \ \& \ j \in U) \ \& \ (U)(U \in \$_i \ \& \ j \in U \to k \in U).$$

Then $\langle W, \leqslant^\circ \rangle$ *is a comparative similarity frame. Moreover, if* $\langle W, \$ \rangle$ *is symmetric, then* $\langle W, \leqslant^\circ \rangle$ *is also symmetric.*

Lemma 2.12 *Let* $\langle W, \leqslant^\circ \rangle$ *be a comparative similarity frame. Define*

$$U \in \$_i \quad iff \quad U \neq \emptyset \ \& \ (j, k)(j \in U \to i \leqslant_i^\circ j \ \& \ (k \leqslant_i^\circ j \to k \in U)).$$

Then $\langle W, \$ \rangle$ *is a system of spheres.*

The definition of spheres in the above lemma (which can also be found in Lewis 1973b) does not preserve the symmetry property. Below we give a modified definition that is more suitable for our purposes.

Lemma 2.13 *Let* $\langle W, \leqslant^\circ \rangle$ *be a symmetric comparative similarity frame. Define the following set S of subsets of W:*

$$U \in S \quad iff \quad U \neq \emptyset \ \& \ (i, j, k)(i \in U \ \& \ j \in U \to i \leqslant_i^\circ j \ \& \ (k \leqslant_i^\circ j \to k \in U)).$$

Then $\langle W, S \rangle$ *is a convex mereological worlds frame.*

Our final result in this section will establish, in effect, that though the comparative similarity semantics corresponds only to a special (convex) kind of general mereological semantics, it is still adequate for the restricted language containing the necessity and two dependency connectives discussed earlier.

Let $\mathcal{L}_d$ denote the classical language extended with these three connectives. Then we have

Theorem 2.14 *Convex mereological semantics is adequate for the language* $\mathcal{L}_d$.

The proof is rather involved, and it can be found in the appendix to this chapter.

2.6.1 Language Correspondences

As has been shown above, convex mereological semantics is equivalent to symmetric system-of-spheres semantics. This equivalence generates a correspondence between the three standard connectives described in section 2.4.2 and certain connectives that have been actually defined by Lewis in his semantics.

To begin with, the following lemma describes the counterparts of the dependency connectives in the comparative similarity semantics. The proof is straightforward.

Lemma 2.15 *Let* $\langle W, \leqslant^\circ, (\ ) \rangle$ *be a comparative similarity semantics generated by a symmetric system-of-spheres semantics for the language* $\mathcal{L}_d$ *(see lemma 2.11). Then*

$$(A \succ B)_i \quad \textit{iff} \quad (j)(i \leqslant_i^\circ j \ \& \ \neg(B)_j \to (\exists k)(k \leqslant_i^\circ j \ \& \ \neg(A)_k)).$$

$$(A \dashv B)_i \quad \textit{iff} \quad (j)(i \leqslant_i^\circ j \ \& \ \neg(A)_j \to (\exists k)(k \leqslant_i^\circ j \ \& \ \neg(A)_k \ \& \ (l)(l \leqslant_i^\circ k \to (B)_l))).$$

As a result, the following correspondences can be immediately established:

- The necessity operator corresponds to Lewis's connective of *outer necessity* in Lewis (1973b, section 1.5);
- The dependence connective is a definitional variant of Lewis's connective of *comparative possibility* $\preccurlyeq$: $A \succ B$ is equivalent to $\neg A \preccurlyeq \neg B$;
- The independence connective corresponds to a variant of Lewis's definition of *cotenability* in Lewis (1973b, section 2.6): $A \dashv B$ is equivalent to cotenability of B with $\neg A$.

Finally, it has been proved in Lewis (1973b, section 2.6) that the above notion of cotenability can be used to provide an equivalent characterization of Lewis's counterfactuals $A \, \square\!\!\rightarrow B$, which can be immediately exploited for showing the equivalence between the latter and the definition of $A > B$ given in section 2.4.2 above.

Theorem 2.16 *Under the correspondence between the mereological and comparative similarity semantics, the definition of* $A > B$ *is equivalent to Lewis's definition of* $A \, \square\!\!\rightarrow B$.

2.6.2 On Lewis's Definition of Causation

Counterfactuals have been used by Lewis as a basis for his definition of (actual) causality between particular events in (Lewis 1973a). Briefly put, causation has been defined by Lewis as a transitive closure of *counterfactual dependence*, where the latter has been described using counterfactuals of the form $\neg C \, \square\!\!\rightarrow \neg E$ for some propositions C and E that are true in the actual world.

We will omit a detailed analysis of Lewis's definition, primarily because nobody believes today that it is correct. We will only point out that his defintion can actually be simplified to a plain counterfactual description by using the following result.

Lemma 2.17 $A \vee B \, \square\!\!\rightarrow B$ *holds if and only if there exists a chain of valid counterfactuals* $A \, \square\!\!\rightarrow C_1, C_1 \, \square\!\!\rightarrow C_2, \ldots, C_n \, \square\!\!\rightarrow B$.

A proof sketch. Let us use $A < B$ as a shorthand for $A \vee B \, \square\!\!\rightarrow B$. Then the direction from right to left follows from the fact that $<$ is a transitive relation (this fact can be verified semantically) and that $A \, \square\!\!\rightarrow B$ implies $A < B$. The reverse direction follows from the fact that if $A \vee B \, \square\!\!\rightarrow B$, then we have $A \, \square\!\!\rightarrow C$ and $C \, \square\!\!\rightarrow B$, where C is $A \vee B$. $\square$

As a result, Lewis's definition of causality (among occurred events) could be expressed directly as follows:

$$C \Rightarrow E \ \equiv \ \neg C \vee \neg E \ \Box\!\!\rightarrow \neg E.$$

This formulation allows us to see Lewis's definition as a straightforward reduction of causation to counterfactuals.

2.7 Conclusions

From the general perspective of this book, the classical formalism described in this chapter constitutes a purely logical formalism that forms only the first, *logical* level in our theory of causation. It corresponds roughly to the data level in Pearl's ladder of causation, though in our account it covers also Lewis' counterfactuals and, more generally, basic dependence and independence relations among facts. Just as for Pearl's approach, however, we have argued in chapter 1 that causation cannot be expressed on this logical level because some essential, nonmonotonic ingredients necessary for causal reasoning are still missing.

Taken from this perspective, it may appear surprising that the whole Lewis approach to counterfactuals, and thereby his counterfactual definition of causation, are confined to this logical level. The apparent surprise can be somewhat moderated, however, if we recall that Lewis's accounts of counterfactuals and causation peacefully coexisted with his principle of *Humean supervenience* (which is not too distant from Wittgenstein's reduction thesis) according to which "if two worlds match perfectly in all matters of particular fact, they match perfectly in all other ways too—in modal properties, laws, causal connections, chances." (see Lewis 1986b). In other words, by this principle the world is fully determined by its "Humean mosaic"—the distribution of perfectly natural properties and relations throughout space-time. This conceptual basis allows us to see the whole Lewis approach to causality as a sophisticated version of a reductive "Humean" regularity theory (cf. Beebee 2014).

It has been shown in this chapter that the mereological semantics subsumes Lewis's possible worlds semantics of comparative similarity. Speaking more generally, this means that the mereological semantics obviates the need in the "external geometry" of possible worlds (as in the general geometric approach of Lewis 1986a) for describing the properties of our world. Instead, it considers the world itself to be a *structured* entity that is determined by the structure of its parts—situations. Still, despite being a purely logical framework on our interpretation, the mereological semantics constitutes an important step toward our goal of describing causation in that it shifts, in effect, the primary reasoning and representation tasks from possible worlds to local situations. This situational setting creates a new dimension of reasoning problems that invites the use of nonmonotonic methods and ultimately provides the room for causal reasoning.

2.A Appendix: Proofs of the Main Results

Theorem (2.4) *A Scott consequence relation is classical if and only if it is stable and all its maximal theories are worlds.*

Proof. Let ⊩ be a classical Scott consequence relation. If $u \not\Vdash A$, for some u and A, then $u \not\Vdash \neg A \to \bot$ by supraclassicality and hence $u, \neg A \not\Vdash$ by Deduction. Since $u, A, \neg A \Vdash$ we have that $v = u \cup \{\neg A\}$ is consistent, includes u, and is inconsistent with A. Thus ⊩ is stable.

Since v above is consistent, applying the above argument to the case when u is a maximal theory, we obtain that v should then coincide with u and hence $\neg A \in u$. Thus, for any proposition A, either A or $\neg A$ should belong to u, and therefore u is a world.

In the other direction, if ⊩ is stable and all its maximal theories are worlds, then it is negative (since *Fm* is not a theory) and satisfies Supraclassicality, since any theory is deductively closed, being an intersection of worlds. Hence, we need only to check the validity of Deduction. Assume that $a \not\Vdash A \to B$ for some a, A, and B. Then there exists a theory u including a and such that $A \to B \notin u$. Since u is an intersection of maximal theories (by stability), there should exist a world i such that $u \subseteq i$ and $A \to B \notin i$, that is, $A \in i$ and $B \notin i$. But i is a theory of ⊩, and hence $a, A \not\Vdash B$. Thus ⊩ is a classical consequence relation. □

Theorem (2.7, Classical Completeness Theorem) *A Scott consequence relation is classical iff it has a mereological semantics.*

Proof. As we have shown, any mereological semantics determines a classical Scott consequence relation. Now, let $\mathcal{M}_{\Vdash}$ be a canonical semantics of ⊩. Due to the general representation theorem for Scott consequence relation, $a \Vdash b$ iff $b \cap u \neq \emptyset$ for any theory u that includes a. Consequently, to demonstrate that ⊩ coincides with the set of sequents that are valid in $\mathcal{M}_{\Vdash}$, it is sufficient to show that, for any theory u and any proposition A, $\|A\|_u$ iff $A \in u$. The latter can be shown by induction on the complexity of formulas of the language. The case of propositional atoms follows from the definition of the valuation predicate. The case of conjunction is immediate. Finally, by $S\neg$ and the inductive assumption, $\|\neg A\|_u$ iff no theory of ⊩ that extends u is consistent with A. Since ⊩ is stable, this is equivalent to $u \Vdash A$. But u is a theory, and hence the latter amounts to $A \in u$. This concludes the proof of the theorem. □

Lemma (2.8) *Let $\mathcal{L}$ be a monadic first-order language of a partial order. Then a formula in $\mathcal{L}$ with one free variable determines a classical propositional connective if and only if (i) it is persistent and stable when its elementary predicates are and (ii) it is preserved under formation of cuts.*

Proof. It is easy to check that all classical connectives satisfy the above properties. On the other hand, if some formula $F(x)$ in $\mathcal{L}$ satisfies these properties, then persistence and stability imply that the corresponding semantic definition in any regular mereological semantics is uniquely determined by its behavior on worlds, whereas the above preservation condition implies that for any world i, $F(i)$ is equivalent to some classical propositional formula built from elementary propositions of the form $P(i)$. $\qquad\Box$

From Mereological to Possible Worlds Semantics

In this section we establish a one-to-one correspondence between mereological possible worlds semantics and a particular kind of general mereological semantics that will be called regular.

Theorem 2.18 *Let* $\langle W, S, (\,) \rangle$ *be a mereological possible worlds semantics. For any* $u, v \in S$ *and any formula A define*

$$u \leqslant v \ \textit{ iff } \ v \subseteq u;$$

$$\|A\|_u \ \textit{ iff } \ (i)(i \in u \to (A)_i).$$

Then $\langle S, \leqslant, \| \ \| \rangle$ *is a mereological semantics with* $W_0 = \{ \{i\} \mid i \in W \}$ *as the set of worlds, which satisfies the following additional property:*

$$(i)(i \in W_0 \ \& \ v \leqslant i \to u \leqslant i) \ \to \ u \leqslant v. \tag{R}$$

Proof. Straightforward checking of the rules characterizing classical mereological semantics, and of the condition (R). $\qquad\Box$

A mereological semantics will be called *regular* if it satisfies the above condition (R). Then the following results will demonstrate, in effect, that the regular mereological semantics is equivalent to the possible worlds semantics.

Lemma 2.19 *Let* $\mathcal{M} = \langle S, \leqslant, \| \ \| \rangle$ *be a regular mereological semantics with W as a set of its worlds. Then for any* $u \in S$:

$$\|A\|_u \ \leftrightarrow \ (i)(i \in W \ \& \ u \leqslant i \to \|A\|_i).$$

Proof. The regularity requirement (R) implies that any state is included in some world. But then the above condition follows directly from persistence and stability (in fact, they are jointly equivalent to it). $\qquad\Box$

Theorem 2.20 *Let* $\mathcal{M} = \langle S, \leqslant, \| \ \| \rangle$ *be a regular mereological semantics with W as a set of its worlds. Define the set* S_0 *of subsets of W as follows:*

$$V \in S_0 \textit{ iff for some } v \in S, \ V = \{ i \mid i \in W \ \& \ v \leqslant i \}.$$

Define also the valuation $(\,)$ *as the restriction of* $\| \ \|$ *to W. Then* $\langle W, S_0, (\,) \rangle$ *is a mereological possible worlds semantics and its generated mereological semantics is isomorphic to* $\mathcal{M}$.

Proof. It is easy to see that the characteristic conditions $(S\neg)$ and $(S\wedge)$ for the classical connectives give us $(W\neg)$–$(W\wedge)$ when applied to worlds. Hence $\langle W, S_0, (\,) \rangle$ is a mereological worlds semantics. Now the isomorphism between the generated mereological semantics and the source worlds semantics follows from the regularity condition and the preceding lemma. □

Finally, we can show that mereological possible worlds semantics is also an adequate semantics for classical Scott consequence relations (theorem 2.9).

Corollary 2.21 *A Scott consequence relation is classical if and only if it has a mereological possible worlds semantics.*

Proof. Since any classical Scott consequence relation is stable, lemma 2.3 implies that its canonical semantics is regular. Hence, classical Scott consequence relations are complete with respect to regular classical mereological semantics and the result follows from the above theorem. □

From Mereological to Comparative Similarity Semantics

In this section we are going to show that the comparative similarity semantics is adequate for the language containing the dependency connectives (theorem 2.14). Due to the fact that the necessity operator $\boxdot$ is definable in terms of $\succ$, this will also cover the latter.

Due to the results established earlier, it is sufficient to show that convex mereological semantics is adequate for this restricted language. Actually, we will show that the relevant partial order can even be chosen to be a quasi-tree as defined below.

Definition 2.16 *A partially ordered set $(S, \leqslant)$ will be said to be a quasi-tree if any pair of incomparable elements has no more than one upper bound.*

As a preparation, we will prove the following

Lemma 2.22 *A quasi-tree is convex iff any its bounded chain has the least upper bound.*

Proof. Let J be a set of elements of a quasi-tree which is bounded by some element j. If J contains a pair of incomparable elements then j is the least upper bound of the pair (since there cannot be others) and hence also of J. On the other hand, if J does not contain incomparable elements, then it is a bounded chain, and hence it must also have the least upper bound. □

Recall that $\mathcal{L}_d$ denotes a classical language extended with the two dependency connectives. The following basic theorem shows that quasi-trees form adequate mereological models for this language. In fact, this theorem has a more general character; it could be seen as a "classical" counterpart of the well-known theorem that tree Kripke structures are adequate models for intuitionistic logic (see, e.g., Gabbay 1981, theorem 23).

Theorem 2.23 *Any adequate regular mereological semantics for $\mathcal{L}_d$ is equivalent to some regular quasi-tree mereological semantics.*

To prove this theorem, we introduce first some terminological conventions. For any finite non-empty sequence $s = (s_0, s_1, ..., s_n)$ we denote by $A(s)$ the sequence $(s_0, s_1, ..., s_{n-1})$ and by $l(s)$ its last element, s_n. Hence if $*$ denotes the concatenation of sequences, then $s = A(s) * (l(s))$. Now let $\langle S, \leqslant, \| \ \| \rangle$ be some regular mereological semantics, and let S° denote the set of all finite nonempty $\leqslant$-increasing sequences of situations while $\hat{S}$ is a set of all finite nonempty sequences of elements of S° satisfying the following conditions: if $\hat{u} = (u_0, u_1, ..., u_n) \in \hat{S}$, then for any $k < n$,

1. $l(u_k)$ is a world;
2. any situation from u_{k+1}, which is not a world, is a part of $l(u_k)$.

Thus, the situations $l(u_k)$ must be worlds which include all the elements from u_{k+1}, except possibly the last, if it also happens to be a world. For any $\hat{u} \in \hat{S}$ let $|\hat{u}|$ denote the situation $l(l(\hat{u}))$. On the set $\hat{S}$ we define the following partial order: for any $\hat{u}, \hat{v} \in \hat{S}$, $\hat{u} \leqslant \hat{v}$ iff either $A(\hat{u}) = A(\hat{v})$ and $l(\hat{u})$ is an initial segment of $l(\hat{v})$, or $A(\hat{u}) = \hat{v}$ and $|\hat{u}|$ is not a world. Then we have

Lemma 2.24 $(\hat{S}, \leqslant)$ *is a regular quasi-tree.*

Proof. Let $\hat{u} \leqslant \hat{w}$ and $\hat{v} \leqslant \hat{w}$ for some incomparable $\hat{u}$ and $\hat{v}$. Suppose that $A(\hat{u}) \neq \hat{w}$ and $A(\hat{v}) = \hat{w}$. Then $A(\hat{u}) = A(\hat{v}) = A(\hat{w})$ and both $l(\hat{u})$ and $l(\hat{v})$ are initial segments of $l(\hat{w})$. But in this case either $l(\hat{u})$ is an initial segment of $l(\hat{v})$ or $l(\hat{v})$ is an initial segment of $l(\hat{u})$. In both these cases $\hat{u}$ and $\hat{v}$ are comparable, contrary to our assumption. Hence, either $A(\hat{u}) = \hat{w}$ or $A(\hat{v}) = \hat{w}$. Now if $\hat{u} \leqslant \hat{x}$ and $\hat{v} \leqslant \hat{x}$ for some $\hat{x} \neq \hat{w}$, then we also have $A(\hat{u}) = \hat{x}$ or $A(\hat{v}) = \hat{x}$. But this is possible only if either $A(\hat{u}) = \hat{w}$ and $A(\hat{v}) = \hat{x}$ or if $A(\hat{u}) = \hat{x}$ and $A(\hat{v}) = \hat{w}$. Since these cases are similar, we shall consider only the first one. We have $A(\hat{u}) \neq \hat{x}$ and $A(\hat{v}) \neq \hat{w}$ (since $\hat{w} \neq \hat{x}$), and therefore $A(\hat{u}) = A(\hat{x})$ and $A(\hat{v}) = A(\hat{w})$. But then $\hat{w} = A(\hat{x})$ and $\hat{x} = A(\hat{w})$, which is impossible. Thus, $\hat{w} = \hat{x}$, and therefore $(S, \leqslant)$ is a quasi-tree. We shall show now that this quasi-tree is also regular, that is, satisfies the condition (R) from theorem 2.18. Note that $\hat{u}$ is a world in $\hat{S}$ iff $|\hat{u}|$ is a world in S. Let $\hat{u} \not\leqslant \hat{v}$ for some $\hat{u}, \hat{v} \in \hat{S}$. If $\hat{v}$ is not a world by itself then $|\hat{v}|$ is not a world in S and hence by regularity it must be a part of at least two different worlds i, j from S. Define $\hat{i} = A(\hat{v}) * (l(\hat{v}) * (i)), \hat{j} = A(\hat{v}) * (l(\hat{v}) * (j))$. It is easy to see that $\hat{i}$ and $\hat{j}$ are worlds in $\hat{S}$ that include $\hat{v}$. If one of them does not include $\hat{u}$, then we are done. Otherwise, $\hat{u}$ and $\hat{v}$ must be comparable and hence $\hat{v} \leqslant \hat{u} \leqslant \hat{i}, \hat{j}$, which is impossible because there are no elements of $\hat{S}$ which lie strictly between $\hat{v}$ and $\hat{i}$. $\square$

Define the following valuation predicate on $\hat{S}$: $\|p\|_{\hat{u}}^\circ \equiv \|p\|_{|\hat{u}|}$, for any $\hat{u} \in \hat{S}$ and any elementary proposition p. This valuation predicate can be extended to all formulas of $\mathcal{L}_d$, and we obtain:

Lemma 2.25 $\langle \hat{S}, \leqslant, \| \ \|^\circ \rangle$ *is a regular mereological semantics for* $\mathcal{L}_d$ *and for any proposition A and any* $\hat{u} \in \hat{S}$,

$$\|A\|_{\hat{u}}^\circ \ iff \ \|A\|_{|\hat{u}|} \tag{*}$$

Proof. We shall show first that $\langle \hat{S}, \leqslant, \| \ \|^{\circ} \rangle$ is indeed a mereological semantics. Note that $\hat{u} \leqslant \hat{v}$ always implies $|\hat{u}| \leqslant |\hat{v}|$. Now, if $\|p\|_{\hat{u}}^{\circ}$ and $\hat{u} \leqslant \hat{v}$, then $\|p\|_{|\hat{u}|}$ and therefore $\|p\|_{|\hat{v}|}$. Hence, persistence holds. On the other hand, if $(\hat{v})(\hat{u} \leqslant \hat{v} \to (\exists \hat{w})(\hat{v} \leqslant \hat{w} \ \& \ \|p\|_{\hat{w}}^{\circ}))$ then obviously $(v)(|\hat{u}| \leqslant v \to (\exists w)(v \leqslant w \ \& \ \|p\|_{w}))$ and hence $\|p\|_{|\hat{v}|}$. Therefore, stability also holds.

Now we prove (*) by induction on the complexity of formulas. Since any classical connective is expressible in terms of conjunction and negation, we may restrict ourselves to only these latter classical connectives. The case of conjunction is obvious. For negation, $\neg \| \neg A\|_{\hat{u}}^{\circ}$ iff for some $\hat{v}$, $\hat{u} \leqslant \hat{v}$ and $\|A\|_{\hat{v}}^{\circ}$ iff for some v, $|\hat{u}| \leqslant v$ and $\|A\|_{v}$. This holds because if $\hat{v}$ is given, we take $v = |\hat{v}|$, while if v is given, we take $\hat{v} = A(\hat{u}) * (l(\hat{u}) * (v))$.

Consider now the dependency connectives. Since the above semantics is regular, we may restrict ourselves only to the case of worlds. Consider first the dependence connective. For any world $\hat{i}$, $\neg \|A \succ B\|_{\hat{i}}^{\circ}$ iff for some $\hat{u} \leqslant \hat{i}$, $\|A\|_{\hat{u}}^{\circ}$ and not $\|B\|_{\hat{u}}^{\circ}$ iff for some $u \leqslant |\hat{i}|$, $\|A\|_{u}$ and not $\|B\|_{u}$. This holds because if $\hat{u}$ is given, take $u = |\hat{u}|$, while if u is given, take $\hat{u} = \hat{i} * ((u))$. Hence (*) holds for the dependence connective. Consider finally $\dashv$. If $\|A \dashv B\|_{\hat{i}}^{\circ}$ then for any u, such that $u \leqslant |\hat{i}|$ and not $\|A\|_{u}$, let $\hat{u} = \hat{i} * ((u))$. We have $\hat{u} \leqslant \hat{i}$ and not $\|A\|_{\hat{u}}^{\circ}$. Therefore, for some $\hat{v}$ such that $\hat{u} \leqslant \hat{v}$ and $\hat{v} \leqslant \hat{i}$, $\neg \|A\|_{\hat{v}}^{\circ}$ and $\|B\|_{\hat{v}}^{\circ}$. But then for $v = |\hat{v}|$ we have $u \leqslant v \leqslant |\hat{i}|$, $\neg \|A\|_{v}$ and $\|B\|_{v}$ and hence $\|A \dashv B\|_{|\hat{i}|}$ holds. On the other hand, if $\|A \dashv B\|_{|\hat{i}|}$ then for any $\hat{u}$ such that $\hat{u} \leqslant \hat{i}$ and not $\|A\|_{\hat{u}}^{\circ}$ we have $|\hat{u}| \leqslant |\hat{i}|$ and not $\|A\|_{|\hat{u}|}$. Therefore, for some v, $|\hat{u}| \leqslant v \leqslant |\hat{i}|$, $\neg \|A\|_{v}$ and $\|B\|_{v}$. Take $\hat{v} = A(\hat{u}) * (l(\hat{u}) * (v))$. Then it is easy to see that $\hat{u} \leqslant \hat{v} \leqslant \hat{i}$, $|\hat{v}| = v$ and hence $\|A \dashv B\|_{\hat{i}}^{\circ}$ holds. $\qquad \square$

Theorem 2.23 follows now from the above two lemmas.

If a mereological semantics is chain-complete, then it can be verified that the construction given in the proof of the above theorem also gives a chain-complete frame which is thereby convex. Now, since any canonical semantics is chain-complete, we finally obtain:

Corollary 2.26 *Convex mereological semantics is an adequate semantics for the language* $\mathcal{L}_d$.

3 Assumption-Based Nonmonotonic Reasoning

It is logically impossible to reason successfully about the world around us using only deductive reasoning. All interesting reasoning outside mathematics involves defeasible steps.
—Pollock (Cognitive Carpentry 1995, p. 41)

There is an instructive correspondence between the above remark of John L. Pollock and David Hume's observation in his *Enquiry* (that we have used as an epigraph in chapter 1) that all reasonings concerning matter of fact are founded on the relation of Cause and Effect. Hume has also argued that pure (deductive) reasoning is insufficient for capturing causal reasoning. The analogy suggests, however, that defeasible, and more generally nonmonotonic reasoning, could be highly relevant for causal reasoning.

The field of nonmonotonic reasoning today is an essential part of the logical approach to artificial intelligence (AI). There exists a vast literature on the topic, including some books, as well as a number of general overviews of the field, in particular Reiter (1987b), Minker (1993), Brewka, Dix, and Konolige (1997), Thomason (2003), and Bochman (2007b).

AI has practical purposes that give rise to problems and solutions of a new kind, apparently different from the questions relevant, say, for philosophers. The authors of first nonmonotonic theories have tried, of course, to express their formalisms using available logical means ranging from classical first order language to modal logics. Still, from its very beginning, logical AI has created formalisms and approaches that had no counterparts in existing logical theories. It is even advantageous to see nonmonotonic reasoning as a brand-new approach to reasoning; this would save from hasty attempts to subsume such a reasoning by existing logical formalisms at the price of losing precious new content. Nonmonotonic reasoning is not yet another application of logic but a relatively independent field of broadly logical research that has a great potential in informing, in turn, future logical theory as well as many areas of philosophical inquiry (Thomason 2003).

The origins of nonmonotonic reasoning within the broader area of logical AI lay in dissatisfaction with the traditional logical methods in representing and handling the problems posed by AI. Basically, the problem was that reasoning necessary for an intelligent behavior and decision making in realistic situations has turned out to be difficult, even impossible, to represent as deductive inferences in some logical system.

An influential technical report (Minsky 1974) has included an appendix, *Criticism of the Logistic Approach*, in which Marvin Minsky explained why he thinks that logical approaches will not work. To begin with, he directly questioned the suitability of representing commonsense knowledge in a form of an axiomatic deductive system:

> There have been serious attempts, from as far back as Aristotle, to represent common sense reasoning by a "logistic" system. . . . No one has been able successfully to confront such a system with a realistically large set of propositions. I think such attempts will continue to fail, because of the character of logistic in general rather than from defects of particular formalisms.

Traditional formal logic cannot discuss what ought to be deduced under ordinary circumstances, and its monotonicity could be viewed as a source of the problem:

> MONOTONICITY: . . . In any logistic system, all the axioms are necessarily "permissive"—they all help to permit new inferences to be drawn. Each added axiom means more theorems, none can disappear. There simply is no direct way to add information to tell such the system about kinds of conclusions that should not be drawn! To put it simply: if we adopt enough axioms to deduce what we need, we deduce far too many other things.

More generally, Minsky argued that the requirement of consistency (alias the principle of contradiction) makes the corresponding logical systems too weak:

> I cannot state strongly enough my conviction that the preoccupation with Consistency, so valuable for Mathematical Logic, has been incredibly destructive to those working on models of mind. At the popular level it has produced a weird conception of the potential capabilities of machines in general. At the "logical" level it has blocked efforts to represent ordinary knowledge, by presenting an unreachable image of a corpus of context-free "truths" that can stand separately by themselves.

As a solution, Minsky proposed the notion of a frame, a complex data structure meant to represent also stereotyped and default assumptions.

In commonsense reasoning, we usually have just partial information about a given situation, and we make a lot of assumptions about how things normally are in order to carry out further reasoning. If we learn that Tweety is a bird, we usually assume that it can fly. Without such assumptions, it would be almost impossible to carry out the simplest commonsense reasoning tasks. Human reasoning is not reducible to collecting facts and deriving their consequences; it embodies an active epistemic attitude that involves making assumptions about particular situations and wholesale theories about the world and acting in accordance with them. We do not only perceive the world, we also give it structure to make it intelligible and controllable. Commonsense reasoning in this sense is just a rudimentary form of a general scientific methodology.

In this chapter, however, we are going to provide an overview of only those theories and parts of nonmonotonic reasoning that either subsume the main formalism of this book, the causal calculus (that will be described in chapter 4), or else contain nonmonotonic tools and constructions that will be used in what follows. The purpose of this chapter is also to demonstrate that the causal calculus is not a stand-alone formalism but an organic and important part of general nonmonotonic reasoning, connected to many other nonmonotonic theories and concepts. Still, as far as nonmonotonic reasoning itself is concerned, this chapter is obviously not self-contained, and the interested reader is invited to consult the above mentioned overviews of nonmonotonic reasoning or even Bochman (2005) for missing details and a more systematic picture.

3.1 Default Logic

Default logic has been born as just one of a number of alternative formalisms for nonmonotonic reasoning. In the course of its development, however, it has become increasingly clear that it occupies a central place in nonmonotonic reasoning, both with respect to its representation capabilities and in its relations with other nonmonotonic formalisms.

Originally (see Reiter 1980) default theory was defined as a pair (W, D), where W is a set of classical propositions (the axioms) and D a set of *default rules* of the form $A : b/C$, where A and C are propositions and b a finite set of propositions. Very informally, a rule $A : b/C$ was intended to state something like:

"If A is believed, and each $B \in b$ can be consistently assumed, then C should be believed."

Default rules were intended to act as meta-rules for extending the initial knowledge base W beyond what is strictly known. Accordingly, the nonmonotonic semantics of default logic was defined by determining a set of *extensions* of a default theory. An extension was defined by a fixed point construction: for a set u of propositions, let $\Gamma(u)$ be the least deductively closed set that includes W and satisfies the following condition:

If $A : b/C \in D$, $A \in \Gamma(u)$ and $\neg B \notin u$, for every $B \in b$, then $C \in \Gamma(u)$.

Then a set s is an *extension* of the default theory if $\Gamma(s) = s$.

As can be seen, default claims were represented in default logic as inference *rules* affecting our beliefs. In this respect, Reiter's default logic has been largely inspired by the need to provide logical foundations for the procedural approach to nonmonotonicity found in deductive databases, logic programming, and Doyle's truth maintenance (Doyle 1979).

Marek and Truszczyński (1989) have suggested a more abstract description of default logic using the notion of a context-dependent proof as a way of formalizing Reiter's operator Γ. This representation has been developed in Marek, Nerode, and Remmel (1990) to a general theory of nonmonotonic rule systems (see also Marek and Truszczyński 1993).

Given a set s of propositions (the "context"), let us consider the set $\mathcal{D}(s)$ of all propositions that are derivable from W using classical entailment and the following ordinary inference rules that are allowed by the context:

$$\{A \vdash C \mid A : b/C \in D \ \& \ \neg B \notin s, \text{ for every } B \in b\}.$$

Then s is an extension of the default theory if and only if $s = \mathcal{D}(s)$.

The above representation has made it vivid that a large part of reasoning in default logic involves ordinary rule-based inference, the only distinction from traditional inference systems being that the very set of rules allowed in the inference process is determined by the (assumptions made in the) context. In particular, an extension of a default theory can be viewed as a set of propositions that are logically provable (= justified) on the basis of taking itself as an assumption context.

In what follows, instead of a pair (W, D) of propositional axioms and default inference rules, we will represent a default theory simply as a set of rules $A : b/C$, where b may be empty. In the latter case, the default rule $A : /C$ represents an ordinary inference rule $A \vdash C$, and therefore the axioms W are representable via rules of the form $: /A$, for $A \in W$.

Remark. The above reformulation has also suggested a generalization based on extending the logical background from ordinary Tarski inference rules to disjunctive, multiple-conclusion sequents (see chapter 2). Such a generalization of default logic based on disjunctive inference rules has been proposed in Gelfond et al. (1991), guided by the need to provide a logical basis for disjunctive logic programming as well as more perspicuous ways of handling disjunctive epistemic information. An even more powerful generalization of this kind has been suggested in (Lin and Shoham 1992; Lifschitz 1994) as a unified formalism for nonmonotonic reasoning and logic programming.

Nonmonotonic formalisms are usually defined similarly to ordinary logical formalisms, namely by describing the syntax of their formulas or rules and supplying corresponding theories with an associated nonmonotonic semantics. It should be stressed, however, that unlike the ordinary logical semantics, the nonmonotonic semantics does *not* provide a semantic interpretation for the syntax. In particular, nonmonotonic theories that are different with respect to their informational content may "accidentally" have the same nonmonotonic semantics. Moreover, such differences cannot be viewed as apparent since it may well happen that by adding the same additional default rules to each of these theories, we obtain new default theories that already have different nonmonotonic semantics (namely different extensions).

The best way to handle this situation amounts to a clear separation between logical and nonmonotonic components of nonmonotonic formalisms. For instance, the *logic* of default rules and its associated logical (monotonic) semantics should be responsible for providing the meaning of default rules while the nonmonotonic semantics of extensions will provide us with nonmonotonic consequences of a given default theory. Many nonmonotonic formalisms have actually been based on this understanding all the time. Thus, McCarthy's circumscription (McCarthy 1980) was explicitly based on classical logic. Similarly, D. McDermott (1982) has considered the influence of the underlying modal logic on the nonmonotonic semantics. Subsequent studies in modal nonmonotonic logics have made this connection explicit (see Marek, Schwarz, and Truszchinski 1993). A general discussion of the relations between nonmonotonic reasoning and logic can be found in Bochman (2011).

Fortunately, there is a systematic way of "restoring" the underlying logic of default rules on the basis of the nonmonotonic semantics. The key notion here is the following concept of strong equivalence among default theories.

Definition 3.1 *Default theories D_1 and D_2 are strongly equivalent if, for any set D of default rules, $D_1 \cup D$ has the same extensions as $D_2 \cup D$.*

Originally, the above notion of strong equivalence has been suggested in the framework of logic programming (see Lifschitz, Pearce, and Valverde 2001), but it has turned out to have a general significance. Note that strong equivalence is already a monotonic, logical notion since strongly equivalent theories are interchangeable in any larger theory without changing the associated nonmonotonic semantics. This suggests that there might exist a logical system of default rules such that default theories are strongly equivalent if and only if they are logically equivalent in this formalism. When this is the case, the relevant logical system can be viewed as the underlying logic of default rules that is in accord with the nonmonotonic semantics of extensions (though it does not determine the latter).

An explicit description of such a logic for default theories, formulated in terms of *bisequents* (or bi-context sequents), can be found in Bochman (2005). It turns out that a large class of nonmonotonic formalisms, including the causal calculus that will be described in chapter 4, can be expressed in such a generalized default logic. More precisely, they can be obtained from the basic formalism just by strengthening its underlying logic (see also Bochman 2008, for further details).

3.1.1 Default Logic Simplified

It turns out that default logic is reducible to a rather simple formalism that contains only ordinary (monotonic) inference rules and default assumptions. This reduction will display default logic as a principal instantiation of the general assumption-based framework for nonmonotonic reasoning.

The reduction. Let us extend the source propositional language $\mathcal{L}$ with new propositional atoms A° for every classical proposition A in $\mathcal{L}$. For a set u of propositions from $\mathcal{L}$, u° will denote the set of new atoms $\{A^\circ \mid A \in u\}$.

Next, if D is a default theory in $\mathcal{L}$, then D° will denote the following set of plain inference rules in the extended language:

$$\{A, b^\circ \vdash C \mid A : b/C \in D\},$$

plus the following two rules for every formula A from $\mathcal{L}$ that appears as a justification in the rules from D:

$$\neg A \vdash \neg A^\circ \qquad \text{and} \qquad : A^\circ/A^\circ.$$

The following theorem shows, in effect, that the above translation is a polynomial, faithful, and modular translation (PFM) in the sense of Janhunen (1999). The proof can be found in Bochman (2008).

Theorem 3.1 *A set u is an extension of a default theory D if and only if there is a unique extension u_0 of D° such that $u = u_0 \cap \mathcal{L}$.*

The above translation reduces an arbitrary default theory to a default theory containing only plain inference rules and *supernormal* default rules of the form $: A/A$. We will call such default theories *simple* in what follows. The above reduction allows us to provide a natural alternative description of default reasoning as deductive reasoning in the presence of assumptions.

Simple default theories.   Supernormal default rule $:A/A$ asserts, in effect, that proposition A can and should be accepted whenever it is consistent with the rest of the facts and rules of a default theory. Such a proposition can be viewed as a *default assumption* of the theory, which allows us to reformulate any simple default theory as a formalism that contain only plain inference rules and assumptions.

For an arbitrary set D of (Tarski) inference rules, let Cn_D denote the provability operator associated with the least supraclassical consequence relation containing D. In other words, for any set u of propositions, $\mathrm{Cn}_D(u)$ is the set of propositions that are derivable from u using the rules from D and classical entailment.

Definition 3.2 *A simple default theory is a pair $(D, \mathcal{A})$, where D is a set of inference rules, and $\mathcal{A}$ a distinguished set of propositions called* assumptions *(or defaults).*

Default reasoning in this setting amounts to deriving plausible conclusions of a default theory by using its rules and assumptions. However, in the case when the set of all defaults $\mathcal{A}$ is jointly incompatible with the background theory D, we must make a reasoned choice. Here default reasoning requires that a reasonable set of assumptions that can be actually used in this context not only should be consistent and maximal but also should explain why the rest of assumptions should be rejected. An important prerequisite of such explanations is that the underlying inference system D contains *cancellation rules* by which some sets of assumptions refute others (given the known facts). The appropriate choices of assumptions will determine then the nonmonotonic semantics of a default theory.

Definition 3.3 *Given a simple default theory* $(D, \mathcal{A})$,

- *A set* $\mathcal{A}_0$ *of assumptions will be called* stable *in* $(D, \mathcal{A})$ *if it is consistent and refutes any assumption outside the set:*

$$\neg A \in \text{Cn}_D(\mathcal{A}_0), \text{ for any } A \in \mathcal{A} \setminus \mathcal{A}_0.$$

- *A set* s *of propositions is an* extension *of a simple default theory if* $s = \text{Cn}_D(\mathcal{A}_0)$ *for some stable set of assumptions* $\mathcal{A}_0$. *The set of extensions determines the* nonmonotonic semantics *of the default theory.*

Combining the above definitions of a stable set and that of extension, we obtain the following description of the nonmonotonic semantics.

Proposition 3.2 *A set* s *of propositions is an extension of a simple default theory* $(D, \mathcal{A})$ *if and only if it satisfies the following two conditions:*

- s *is the deductive closure of the set of its defaults:* $s = \text{Cn}_D(\mathcal{A} \cap s)$;
- s *decides the default set: for any* $A \in \mathcal{A}$, *either* $A \in s$, *or* $\neg A \in s$.

Simple default theories and their nonmonotonic semantics provide presumably the most transparent description of assumption-based nonmonotonic reasoning.

3.2 Argumentation Theory

The argumentation theory is even older than logic, and its systematic description can be found already in Aristotle's dialectical studies, such as *Topics* and *Sophistical Refutations*. Since then, however, it has also undergone significant development, especially in the past century, and has become an integral part of AI.

Dung's argumentation frameworks (Dung 1995b) are viewed today as a general formal basis of the argumentation theory in AI, but they have deep roots in nonmonotonic reasoning, and it is due to these roots that they constitute a new stage in the development of the theory of argumentation.

Traditional formal argumentation theory (see, e.g., Hamblin 1971) has been based, implicitly or explicitly, on the standard deductive paradigm, according to which our corpus of knowledge and beliefs comprises a set of propositional assertions, coupled with a set of strict, universal deductive rules (= the logic) that govern their acceptance. These rules allow us, in particular, to derive (support) further propositional claims as well as to reveal possible inconsistencies among them. This underlying logic can also be given a precise argumentative (dialectical) formulation in the form of allowable attack and defense moves in argumentation games (Lorenzen and Lorenz 1978). In this way, the traditional formal argumentation theory could be largely viewed as a "human-friendly" instantiation of standard deductive reasoning.

The above deductive paradigm has been challenged, however, with the advent of non-monotonic reasoning in AI. Studies in the latter, as well as contemporaneous studies of defeasible reasoning in philosophical logic (see Pollock 1987), have shown that epistemic states underlying our reasoning are more complex and structured than plain sets of beliefs governed by logic. They have shown, in particular, an important role of *default assumptions* in our reasoning, assumptions that we normally accept in the absence of evidence to the contrary. These assumptions usually appear in the conditional form *"If A, then normally B,"* and it can even be argued that such normality conditionals constitute one of the central ingredients of our commonsense epistemic states.

Despite their presumptive acceptability status, default assumptions are *defeasible*, that is, they can be attacked, and even eventually refuted due to other assumptions and available evidence. The corresponding adjudication process, however, already cannot be represented as a deductive inference or proof in some logic, primarily because it is in general *nonlocal* and *nonmonotonic*. The eventual acceptability of such assumptions depends on other assumptions present, and it can change from acceptance to rejection and vice versa with addition of new assumptions or facts. This reasoning process displays, however, distinctive features of genuine argumentation.

As a matter of fact, the intimate connections between argumentation, assumptions and nonmonotonic reasoning has been noticed at the very beginning of the studies in nonmonotonic reasoning. The starting point of this understanding can be found already in the Truth Maintenance System (TMS) of Doyle (1979) and assumption-based truth maintenance (ATMS) of de Kleer (1986). This understanding has even led to a general view of nonmonotonic reasoning as a theory of the reasoned use of assumptions in Doyle (1994). In fact, even before Dung, a significant argumentation-based representation of default logic and other nonmonotonic formalisms has been suggested in Lin and Shoham (1989).

Developing further this line of research, Dung has shown a fundamental and unifying role of argumentation in logic programming and general nonmonotonic formalisms such as default and modal nonmonotonic logics. More precisely, he has shown that all these formalisms can be viewed as particular instantiations of a uniform argumentation scheme that implements the principle of default acceptability for arguments in an abstract framework based solely on a single relation of attack among them.

Dung's argumentation frameworks have had two crucial novel features. First, they implemented the basic principle of default acceptance for arguments.[1] In Dung's vision, however, this principle admits a number of different interpretations, which lead to different possible *nonmonotonic semantics* for the argumentation frameworks.

The second, more formal, novel feature of the Dung's formalism was the asymmetric (directional) character of the attack relation; it was this "degree of freedom" that has allowed to provide an adequate representation of the above-mentioned nonmonotonic

---

1. Dung himself has called it the basic principle of argumentation and described it in Dung (1995b) as the principle *"The one who has the last word laughs best."*

formalisms. This feature has also marked an important formal difference with the traditional, deductive argumentation that has been based primarily on symmetric inconsistency relations.

It could even be argued that the main contribution of Dung's theory has consisted in incorporating these two novel features as central conceptual ingredients of argumentation. It is this conceptual advancement that has given the argumentation theory its current impetus.

3.2.1 Logic in Argumentation

One of the fundamental questions that has been reopened, however, with the advent of Dung's argumentation frameworks has become the question of the relations between argumentation and logic. Indeed, on the face of it, Dung's abstract frameworks do not include, or even require, any explicit logical components. On the other hand, it has been shown in subsequent argumentation literature that arbitrary, unrestricted combinations of argumentation frameworks with deductive rules may lead to patently inappropriate results, so such compositions should be constrained by some (more or less) reasonable "rationality postulates"—see, for example, Caminada and Amgoud (2007), Amgoud and Besnard (2013), and Dung and Thang (2014).

Though the relations between argumentation and logic have irrevocably changed, logic still plays (or, better, should play) an important role in argumentation. However, a crucial prerequisite for a proper understanding of this role amounts to a clear separation of the logical and nonmonotonic aspects of argumentation. In fact, the latter objective is not specific to argumentation theory but pertains to all nonmonotonic formalisms.

In nonmonotonic formalisms, logic no longer "pervades the world" (using the famous Wittgenstein phrase). Namely, the logic, taken by itself, cannot provide the final "output" of these formalisms; this latter task is relegated to the associated nonmonotonic semantics. Nevertheless, logic still plays a distinctive and even crucial role in these formalisms. First of all, logic and its associated (monotonic) semantics should still provide a formal interpretation and meaning for the very syntax of a nonmonotonic formalism. Note that a nonmonotonic semantics is usually defined as a distinguished *subset* of the corresponding logical semantics, so it cannot be used for interpreting the source language. In addition, logic provides deductive inferences that are "safe" with respect to the nonmonotonic semantics, so it can be used to facilitate proofs and computations of the latter.

However, an even more profound benefit of the separation between logical and nonmonotonic aspects of a reasoning formalism emerges from the fact that, once the separation is made, many of these formalisms can be reconstructed as instantiations of the same nonmonotonic semantics in different logical languages.

The logic appropriate for Dung's argumentation frameworks can be constructed on the basis of a four-valued logical semantics that can be found, in effect, already in Jakobovits and Vermeir (1999). In that paper, the authors described a general semantic framework based on acceptance and rejection of arguments. This semantics was essentially four-valued because assignments of acceptance and rejection to arguments were primarily viewed as mutually independent, permitting valuations in which arguments can be both accepted and rejected or neither accepted nor rejected. The semantics suggested in Jakobovits and Vermeir (1999) were designed, however, to be generalizations of existing *nonmonotonic* argumentation semantics, so they incorporated also some nonlogical, nonmonotonic features (see below). Still, a properly generalized four-valued semantics can be used as a *logical* basis of Dung's argumentation frameworks. It will allow us, in particular, to augment the underlying language with appropriate logical connectives that will transform this abstract argumentation to real argumentation reasoning with full-fledged logical capabilities.

The assumption-based frameworks of Bondarenko et al. (1997) can be viewed as a focal point of this logical development. On our reconstruction of the latter that we will describe below, assumption-based frameworks can be obtained from abstract Dung's frameworks just by adding a particular negation connective to the underlying language of arguments. This connective will also allow us to establish straightforward relations between attack and inference as well as between (nonpropositional) arguments and (propositional) assumptions that can be viewed as their reified counterparts in the object language.

Further stages of this logical development allow us to provide a more systematic description of many general nonmonotonic formalisms, including logic programming, default logic, abstract dialectical frameworks, and (last, but not least) the causal calculus.

The general picture that will emerge from this formal development is not only that argumentation is important for nonmonotonic reasoning but also the other way round: namely, that the main nonmonotonic formalisms and argumentation systems constitute actually primary instantiations of Dung's abstract argumentation in appropriately extended logical languages.

3.2.2   Collective Argumentation

As a general formal basis of argumentation theory, we will use the formalism of collective argumentation suggested in Bochman (2003b) as a generalization of Dung's argumentation theory. In this formalism, a primitive attack relation holds between *sets* of arguments: in the notation introduced below, $a \hookrightarrow b$ says that a set a of arguments attacks a set of arguments b. This fact implies, of course, that these two sets of arguments are incompatible. However, $a \hookrightarrow b$ says more than that, namely that the set a of arguments, being accepted, provides a *reason* or explanation for rejection of the set b of arguments (taken as a whole). Accordingly, the attack relation will not in general be symmetric since in this situation acceptance of b need not give reasons for rejection of a. In addition, the attack relation is not reducible to attacks among individual arguments. For instance, we can disprove some conclusion jointly supported by a disputed set of arguments, though no particular argument in the set, taken alone, could be held responsible for this.

In what follows, $a, b, c, \ldots$ will denote finite sets of arguments while $u, v, w, \ldots$ will denote arbitrary such sets. We will use the same agreements for the attack relation as for usual consequence relations. Thus, $a, a_1 \hookrightarrow b, B$ will have the same meaning as $a \cup a_1 \hookrightarrow b \cup \{B\}$, and so on.

Proofs of all the claims in this section can be found in Bochman (2003b, 2005).

Definition 3.4 *Let $\mathcal{A}$ be a set of arguments. A (collective) attack relation is a relation $\hookrightarrow$ on finite sets of arguments satisfying the following postulate:*

(Monotonicity)   *If $a \hookrightarrow b$, then $a, a_1 \hookrightarrow b, b_1$.*

The attack relation can be extended to arbitrary sets of arguments by imposing the compactness requirement: for any $u, v \subseteq \mathcal{A}$,

(Compactness) $u \hookrightarrow v$ iff there exist finite $a \subseteq u$ and $b \subseteq v$ such that $a \hookrightarrow b$.

The Dung's original argumentation frameworks (Dung 1995a) can be seen as a special case of collective argumentation in which all attacks are generated by attacks $A \hookrightarrow B$ between single arguments.

By an *argument theory* we will mean an arbitrary set of attacks $a \hookrightarrow b$ between finite argument sets. Any argument theory Δ generates a unique least attack relation that we will denote by $\hookrightarrow_\Delta$. The latter is obtained from Δ just by closing it with respect to the monotonicity rule. Accordingly, $\hookrightarrow_\Delta$ can be described directly as follows:

$$u \hookrightarrow_\Delta v \text{ iff } a \hookrightarrow b \in \Delta, \text{ for some } a \subseteq u, b \subseteq v.$$

Four-valued logical semantics.   Collective argumentation can be given a four-valued semantics that can be seen as describing the (abstract) *meaning* of the attack relation. This formal meaning stems from the following understanding of an attack $a \hookrightarrow b$:

If all arguments in a are accepted, then at least one of the arguments in b is rejected.

The argumentation theory does not impose, however, the classical constraints on acceptance and rejection of arguments, so an argument can be both accepted and rejected, or neither accepted nor rejected. Such an understanding can be captured formally by assigning any argument A a *subset* $\nu(A) \subseteq \{t, f\}$, where t denotes acceptance (truth) while f denotes rejection (falsity). This is nothing other than the well-known *Belnap's interpretation* of four-valued logic (see Belnap 1977). On this understanding, $t \in \nu(A)$ means that an argument A is accepted while $f \in \nu(A)$ means that A is rejected. In accordance with this, collective argumentation acquires a four-valued logical semantics described below.

Definition 3.5

- *An attack $a \hookrightarrow b$ will be said to* hold *in a four-valued interpretation ν of arguments, if either $t \notin \nu(A)$, for some $A \in a$, or $f \in \nu(B)$, for some $B \in b$.*

- *An interpretation ν is a* model *of an argument theory Δ if every attack from Δ holds in ν.*

Since an attack relation can be seen as a special kind of an argument theory, the above definition determines also the notion of a model for an attack relation.

For a set I of four-valued interpretations, $\hookrightarrow_I$ will denote the set of all attacks that hold in each interpretation from I. Then the following result is actually a basic representation theorem showing that the four-valued semantics is adequate for collective argumentation.

Theorem 3.3 $\hookrightarrow$ *is an attack relation iff it coincides with $\hookrightarrow_I$, for some set of four-valued interpretations I.*

3.2.3 Nonmonotonic Semantics

In the preceding section we have described a structural logical basis of argumentation. The argumentation theory should be viewed, however, as a two-layered formalism which has both logical and nonmonotonic components. This means that, in addition to the logical semantics, an argumentation formalism should be assigned also a *nonmonotonic* semantics that will determine the actual acceptance and rejection of arguments in each reasoning context. As one of its main desiderata, the latter semantics should incorporate and thoroughly implement the basic principle of default acceptance for arguments.

Partly due to historical reasons (primarily, the logic programming origins), there is a bewildering number of nonmonotonic semantics that are actively investigated in the current argumentation literature. There have been a number of attempts to systematize these semantics (see, e.g., Baroni and Giacomin 2007), though no uniform picture has emerged.

We will provide below a rough sketch of the basic principles and desiderata for constructing the nonmonotonic semantics of argumentation. As a starting point, we will formulate the main *principle of argumentation* as the claim that arguments (in sharp distinction with factual assertions) bear with them presumption of acceptance:

An argument is accepted unless there is a reason for its rejection.

In the framework of the formal argumentation theory, the reasons for rejection of arguments come only in the form of attacks by other arguments. Thus, our logical interpretation of the attack relation immediately sanctions that if an argument A attacks an argument B and A is a accepted, then B should be rejected. In what follows, we will say that an argument is *refuted* if it is attacked by an accepted argument set. Then our main principle of argumentation implies that an argument should be accepted whenever all its attacking arguments are not accepted. In other words, it evolves to:

An argument is accepted if and only if it is not refuted.

Now, if we combine the above principle with the natural "classical" requirement that any argument should be either accepted or rejected, but not both, we will immediately obtain the primary nonmonotonic semantics of argumentation, the *stable semantics*.[2] According to this semantics, acceptable sets of arguments are conflict-free sets that attack any argument outside them (cf. definition 3.3 of a stable set of default assumptions in section 3.1).

In the general correspondence between Dung's argumentation theory and other non-monotonic formalisms, the stable semantics corresponds to the main nonmonotonic semantics of the latter. This makes it a proper candidate on the role of the *standard* non-monotonic semantics for argumentation, much in the same sense as the classical logic can be viewed as the standard logic for our reasoning (whatever the objections one could possibly have against this logic).

Despite its naturalness and simplicity, however, there are quite simple argumentation frameworks where the stable semantics fails to determine an acceptable set of arguments,[3] which creates an obvious incentive for trying alternative, more tolerant nonmonotonic semantics. It turns out that the general four-valued logical semantics of acceptance and rejection of arguments provides all the necessary degrees of freedom for defining such alternative nonmonotonic semantics, and the way to do this amounts to adopting different "partial" generalizations of the main argumentation principle in the four-valued setting.

Retaining our earlier definition of refutation, a first such relaxed argumentation principle can be formulated as follows:

An argument is rejected if and only if it is refuted.

Note that the above principle is not equivalent to our original main argumentation principle since the assignments of acceptance and rejection are logically independent. Instead, combined with our logical characterization of the attack relation, this principle will give us precisely the notion of labeling from Jakobovits and Vermeir (1999).

An even stronger general constraint on nonmonotonic semantics can be obtained by adding the following alternative generalization of the main argumentation principle:

An argument is accepted if and only if all its attackers are rejected.

Now, if we will restrict the set of valuations to consistent ones (that exclude simultaneous acceptance and rejection of arguments), we will obtain exactly the Caminada labelings (see Caminada and Gabbay 2009). These labelings have been shown to encompass the main nonmonotonic semantics of Dung's argumentation frameworks.

3.2.4 Negation, Deduction, and Assumptions

The notion of an argument has been taken as primitive in Dung's argumentation theory, which allows for a possibility of considering arguments that are nonpropositional in

---

2. Pollock (1987) was apparently the first to suggest this argumentation semantics.
3. The simplest such framework comprises a single argument that attacks itself.

character (e.g., arguments as inference rules or derivations). Still, there exists a natural, direct connection between abstract argumentation frameworks and traditional deductive argumentation; it has been established, in effect, already in Bondarenko et al. (1997). In this formalism of *assumption-based argumentation*, arguments were constructed as plain deductive arguments that may involve, however, auxiliary propositional *assumptions* (cf. our description of simple default theories in section 3.1). Moreover, the attack relation can already be defined in this framework, so the assumption-based argumentation can be viewed as a special case of Dung's abstract argumentation. Nevertheless, it has been shown in Bondarenko et al. (1997) that this special kind of argumentation still provides a natural and powerful generalization of the main nonmonotonic formalisms and various semantics for logic programming.

It turns out that the entire formalism of assumption-based argumentation can be obtained just by adding a single negation connective to the logical system of abstract argumentation, a connective that is actually implicit in the formalism of Bondarenko et al. (1997) in the form of the contrary mapping on assumptions. This move will also constitute a first, and most important, step toward a full-fledged theory of *propositional argumentation*.

Let us extend our underlying language with a negation connective, $\sim$, having the following precise (four-valued) semantic definition:[4]

$$\sim\!A \text{ is accepted iff } A \text{ is rejected}$$

$$\sim\!A \text{ is rejected iff } A \text{ is accepted.}$$

The above definition makes $\sim$ a particular four-valued connective; it is called a *global negation* since it switches the evaluation contexts between acceptance and rejection.

An axiomatization of this negation in abstract argumentation theory can be obtained by imposing the following rules on the attack relation:

$$A \hookrightarrow \sim\!A \qquad \sim\!A \hookrightarrow A$$

$$\text{If } a \hookrightarrow A, b \text{ and } a, \sim\!A \hookrightarrow b, \text{ then } a \hookrightarrow b \qquad\qquad \text{(AN)}$$

$$\text{If } a, A \hookrightarrow b \text{ and } a \hookrightarrow b, \sim\!A, \text{ then } a \hookrightarrow b.$$

Attack relations satisfying the above postulates will be called *N-attack relations*. It turns out that the latter are interdefinable with a particular kind of consequence relations.

Definition 3.6 *A Belnap consequence relation* in a propositional language with a global negation $\sim$ is a Scott consequence relation[5] *satisfying the following two double negation rules for the global negation:*

$$A \Vdash \sim\!\sim\!A \qquad \sim\!\sim\!A \Vdash A.$$

---

4. This negation connective played a prominent role in information lattices of Belnap (1977).
5. See chapter 2 for a definition.

For any set u of propositions, we will denote by $\sim u$ the set $\{\sim A \mid A \in u\}$. Now, for a given N-attack relation, we can define the following consequence relation:

$$a \Vdash b \equiv a \hookrightarrow \sim b. \tag{CA}$$

Similarly, for any Belnap consequence relation we can define the corresponding attack relation as follows:

$$a \hookrightarrow b \equiv a \Vdash \sim b. \tag{AC}$$

As has been shown in Bochman (2003b), the above definitions establish an exact equivalence between N-attack relations and Belnap consequence relations. This correspondence allows us to represent an assumption-based argumentation framework from Bondarenko et al. (1997) entirely in the framework of attack relations.

Assumptions versus factual propositions. Though the global negation $\sim$ is a logically well-defined connective, it implicitly interferes with the main principle of argumentation that presupposes an asymmetric treatment of acceptance and rejection for arguments. Indeed, if A is an argument, then $\sim A$ cannot already be viewed as an argument since otherwise presumptive acceptance of $\sim A$ would directly imply presumptive *rejection* of A itself!

The emerging problem immediately reminds us, however, that our commonsense epistemic states are not homogeneous: in addition to assumptions (that can be viewed as primitive arguments), they contain also ordinary factual claims, and the latter have an opposite nature as compared to arguments. Namely, according to Leibniz's principle of sufficient reason, they are presumably rejected unless we have reasons for their acceptance.

A simplest and perhaps the most natural way of resolving the above issues consists in a clear separation between assumptions and factual propositions. In this setting, the global negation will switch also between assumptions and facts. This bipartite scheme of things has been actually implemented in the assumption-based approach to argumentation presented in Bondarenko et al. (1997).

Assumption-based argumentation (ABA). Slightly changing the formulation of Bondarenko et al. (1997), an assumption-based argumentation framework can be defined as a triple consisting of an underlying deductive system, a distinguished subset of propositions Ab called *assumptions*, and a mapping from Ab to the set of all propositions of the language that determines the *contrary* $\overline{A}$ of any assumption A.

Now, the underlying deductive system can be expressed directly in the framework of N-attack relations by identifying deductive rules $a \vdash A$ with attacks of the form $a \hookrightarrow \sim A$. Moreover, the global negation $\sim$ can also serve as a faithful logical formalization of the operation of taking the contrary. More precisely, given an arbitrary underlying (factual) language $\mathcal{L}$ that does not contain $\sim$, we can *define* assumptions as propositions of the form $\sim A$, where $A \in \mathcal{L}$. Then, since $\sim$ satisfies double negation, a negation of an assumption will be a proposition from $\mathcal{L}$. Accordingly, N-attack relations can be seen as a proper generalization of the assumption-based framework.

In Bondarenko et al. (1997), the connection between (assumption-based) argumentation and main nonmonotonic formalisms has been established by showing that these nonmonotonic systems can be viewed as assumption-based frameworks just by defining assumptions and their contraries. As a partial converse of these results, it can be shown that many of these formalisms constitute actually primary instantiations of propositional argumentation in appropriately chosen logical languages.

3.2.5 Default Argumentation

Taking seriously the idea of propositional argumentation, it is only natural to make further steps toward extending the underlying language of arguments to the usual classical propositional language. These steps should be coordinated, however, with the inherently four-valued nature of the attack relation. And the way to do this amounts to requiring that the relevant classical connectives should behave in a usual classical way with respect to both acceptance and rejection of arguments.

As a first such connective, we introduce the *conjunction* $\wedge$ of arguments that is determined by the following familiar semantic conditions:

$A \wedge B$ is accepted iff A is accepted and B is accepted

$A \wedge B$ is rejected iff A is rejected or B is rejected.

As can be seen, $\wedge$ behaves as an ordinary classical conjunction with respect to acceptance and rejection of arguments. On the other hand, it is a four-valued connective since the above conditions determine a four-valued truth table for conjunction in the Belnap's interpretation of four-valued logic (see Belnap 1977). The following postulates provide a simple syntactic characterization of this connective for attack relations:

$$a, A \wedge B \hookrightarrow b \quad \text{iff} \quad a, A, B \hookrightarrow b$$
$$a \hookrightarrow A \wedge B, b \quad \text{iff} \quad a \hookrightarrow A, B, b. \qquad (A_\wedge)$$

Collective attack relations satisfying these postulates will be called *conjunctive*. The next result shows that they give a complete description of the four-valued conjunction.

Corollary 3.4 *An attack relation is conjunctive if and only if it coincides with $\hookrightarrow_I$, for some set of four-valued interpretations I in a language with the four-valued conjunction $\wedge$.*

An immediate benefit of introducing conjunction into the language of argumentation is that any finite set of arguments a becomes reducible to a single argument $\bigwedge a$:

$$a \hookrightarrow b \quad \text{iff} \quad \bigwedge a \hookrightarrow \bigwedge b.$$

As a result, the collective attack relation in this language is reducible to an attack relation among individual arguments, just as it has been assumed in Dung (1995b).

Having a conjunction at our disposal, we only have to add a classical negation ¬ to obtain a full classical language. Moreover, since sets of arguments are reducible to their conjunctions, we can represent the resulting argumentation theory using just a binary attack relation on classical formulas.

As a basic condition on argumentation in the classical propositional language, we will require only that the attack relation should respect the classical entailment $\models$ in the precise sense of being monotonic with respect to $\models$ on both sides.

Definition 3.7 *A propositional attack relation is a relation* $\hookrightarrow$ *on the set of classical propositions satisfying the following postulates:*

(Left Strengthening) *If $A \models B$ and $B \hookrightarrow C$, then $A \hookrightarrow C$;*
(Right Strengthening) *If $A \hookrightarrow B$ and $C \models B$, then $A \hookrightarrow C$;*
(Truth) $\mathbf{t} \hookrightarrow \mathbf{f}$;
(Falsity) $\mathbf{f} \hookrightarrow \mathbf{t}$.

Left Strengthening says that logically stronger arguments should attack any argument that is attacked already by a logically weaker argument, and similarly for Right Strengthening. Truth and Falsity postulates characterize the limit cases of argumentation by stipulating that any tautological argument attacks any contradictory one, and vice versa.

There exists a simple definitional way of extending the above attack relation to a collective attack relation among arbitrary sets of propositions. Namely, for any sets u, v of propositions, we can define $u \hookrightarrow v$ as follows:

$$u \hookrightarrow v \ \equiv \ \text{there exist finite } a \subseteq u, b \subseteq v \text{ such that } \bigwedge a \hookrightarrow \bigwedge b.$$

The resulting attack relation will satisfy the properties of collective argumentation as well as the postulates $(A_\wedge)$ for conjunction.

Finally, to acquire full expressive capabilities of the argumentation theory, we can add the global negation $\sim$ to the language. Actually, a rather simple characterization of the resulting collective argumentation theory can be obtained by accepting the basic postulates AN for $\sim$, plus the following rule that permits the use of classical entailment in attacks:

(Classicality) If $a \models A$, then $a \hookrightarrow \sim A$ and $\sim A \hookrightarrow a$.

It can be verified that the resulting system satisfies all the postulates for propositional argumentation. The system will be used below for a direct representation of default logic.

Logical semantics. A semantic interpretation of propositional attack relations can be obtained by generalizing four-valued interpretations to pairs (u, v) of deductively closed theories, where u is the set of accepted propositions while v the set of propositions that are not rejected. Such pairs will be called (classical) *bimodels* while a set of bimodels will be called a *classical binary semantics*.

Definition 3.8 *An attack* $A \hookrightarrow B$ *will be said to be* valid *in a binary semantics* $\mathcal{B}$ *if there is no bimodel* (u, v) *from* $\mathcal{B}$ *such that* $A \in u$ *and* $B \in v$.

We will denote by $\hookrightarrow_{\mathcal{B}}$ the set of attacks that are valid in a semantics $\mathcal{B}$. This set forms a propositional attack relation. Moreover, the following result shows that propositional attack relations are actually complete for the binary semantics.

Theorem 3.5 $\hookrightarrow$ *is a propositional attack relation if and only if it coincides with* $\hookrightarrow_{\mathcal{B}}$, *for some binary semantics* $\mathcal{B}$.

Default logic. It turns out that propositional argumentation allows us to provide a direct representation of Reiter's default logic.

Given a system of propositional argumentation in the classical language augmented with the global negation $\sim$, we will interpret Reiter's default rule $a{:}b/A$ as an attack[6]

$$a, \sim\neg b \hookrightarrow \sim A,$$

or equivalently as a rule $a, \sim\neg b \Vdash A$ of the associated Belnap consequence relation. Similarly, an axiom A of a default theory will be interpreted as an attack $\mathbf{t} \hookrightarrow \sim A$. For a default theory Δ, we will denote by $tr(\Delta)$ the corresponding argument theory obtained by this translation.

By our general agreement, by *assumptions* we will mean propositions of the form $\sim A$, where A is a classical proposition. For a set u of classical propositions, we will denote by $\tilde{u}$ the set of assumptions $\{\sim A \mid A \notin u\}$. Finally, a set w of assumptions will be called *stable* in an argument theory Δ if, for any assumption A, $A \in w$ iff $w \not\hookrightarrow_{\Delta} A$, where $\hookrightarrow_{\Delta}$ is the least propositional attack relation containing Δ. Then we have the following theorem.

Theorem 3.6 *A set* u *of classical propositions is an extension of a default theory* Δ *if and only if* $\tilde{u}$ *is a stable set of assumptions in* $tr(\Delta)$.

The above result is similar to the corresponding result in Bondarenko et al. (1997, theorem 3.10), but it is much simpler and is formulated entirely in the framework of propositional attack relations. The simpler representation was made possible due to the fact that propositional attack relations already embody the deductive capabilities treated as an additional component in assumption-based frameworks.

3.2.6 Logic Programming

Finally, the formalism of logic programming, which could be seen as one of the main sources of Dung's argumentation theory, can also be viewed as a special kind of propositional argumentation.

---

6. As before, we use set notation according to which $\neg b$ denotes the set $\{\neg B \mid B \in b\}$.

A *general logic program* Π is a set of rules of the form:[7]

$$\mathbf{not}\, d, c \leftarrow a, \mathbf{not}\, b, \tag{*}$$

where a, b, c, d are finite sets of propositional atoms. These are program rules of a most general kind that contain disjunctions and negations as failure **not** in their heads.

Now, any such rule can be translated as an attack

$$a, \neg c \hookrightarrow \neg b, d. \tag{AL}$$

This allows us to translate any logic program into an argument theory. In order to capture the nonmonotonic semantics of logic programming, however, we need to accept the following argumentative counterpart of the closed world assumption:

(Default Assumption) $p \hookrightarrow \neg p$, for any atom p.

Let $tr(\Pi)$ denote the argument theory obtained by this translation from a logic program Π. Then we obtain the following theorem.[8]

Theorem 3.7 *A set u of propositional atoms is a stable model of a logic program Π if and only if $\tilde{u}$ is a stable set of assumptions in $tr(\Pi)$.*

It is interesting to note that, due to the reduction rules $(R_\sim)$ for the global negation $\sim$, described earlier, the above representation (AL) of the program rules is equivalent to $a, \sim b \hookrightarrow \sim c, d$ and therefore to the inference rules

$$a, \sim b \Vdash c, \sim d$$

of the associated Belnap consequence relation. For normal logic programs (single atoms in heads), this latter representation coincides with that given in Bondarenko et al. (1997).

3.3 The Problem of Defeasible Entailment

The problem of defeasible inference can be seen as one of the main objectives as well as one of the main problems of the general theory of nonmonotonic reasoning. An impressive success has been achieved in our understanding of it, in realizing how complex it is, and most importantly, how many forms it may have. Many formalisms have been developed that capture significant aspects of defeasible inference, though a unified picture has not yet emerged.

The problem of defeasible inference can be viewed as a problem of formalizing, or representing, commonsense rules, or conditionals, of the form "*A normally implies B*" (they will be written as $A \vdash B$ in what follows). Viewed in this way, it can be safely argued that the problem of defeasible inference was born precisely with the birth of nonmonotonic reasoning. Indeed, one of the main objectives of the first nonmonotonic formalisms,

---

7. Again, **not** a denotes the set $\{\mathbf{not}\, A \mid A \in a\}$.
8. As in the preceding section, $\tilde{u}$ denotes the set $\{\sim A \mid A \notin u\}$.

namely, circumscription, default logic, and modal nonmonotonic logics, was a faithful formal representation of such commonsense rules.

The sentence "Birds (normally) fly" is weaker than "All birds fly"; there is a seemingly open-ended list of exceptions—ostriches, penguins, Peking ducks, and so on and so forth. So, if we would try to use classical logic for representing "Birds fly," the first problem[9] would be that it is practically impossible to enumerate all exceptions to flight with an axiom of the form

$$(\forall x).Bird(x)\&\neg Penguin(x)\&\neg Emu(x)\&\neg Dead(x)\&... \rightarrow Fly(x).$$

The second crucial problem is that, even if we could enumerate all such exceptions, we still could not derive *Fly(Tweety)* from *Bird(Tweety)* alone. This is so since we are not given that Tweety is not a penguin, or dead, and so on. The antecedent of the above implication cannot be derived, in which case there is no way of deriving the consequent. Nevertheless, if told only about a particular bird, say Tweety, without being told anything else about it, we would be justified in assuming that Tweety can fly without knowing that it is not one of the exceptional birds. This fact indicates that our commonsense assumptions about the world are often global in character, saying something like "the world is as normal as possible, given the known facts." So the problem is how we can actually *make* such assumptions in the absence of information to the contrary.

This suppositional character of commonsense reasoning conflicts with the monotonic character of logical derivations. Monotonicity is just a characteristic property of deductive inferences arising from the very notion of a proof being a sequence of steps starting with accepted axioms and proceeding by inference rules that remain valid in any context of its use. Consequently, if a set *a* of formulas implies a consequence *C*, then a larger set $a \cup \{A\}$ will also imply *C*. Commonsense reasoning is nonmonotonic in this sense because adding new facts may invalidate some assumptions made earlier.

All three initial formal systems of nonmonotonic reasoning have proposed a translational approach to this problem. Thus, Reiter (1980) has suggested to represent $A \mathrel{|\!\sim} B$ in default logic using "normal" default rules,

$$A : B/B.$$

The main effect of this representation consisted in securing two important properties of the corresponding commonsense rule $A \mathrel{|\!\sim} B$: (i) the rule is applicable whenever *A* holds and $\neg B$ is not known to hold, yet (ii) a possible factual refutation $A \wedge \neg B$ does not create contradiction in the system; it only cancels the rule. A similar modal translation of defeasible rules, $A \wedge \mathbf{M}B \supset B$, was suggested in the modal nonmonotonic logics of McDermott and Doyle (1980).

---

9. Which is precisely Mill's problem of covering laws—see section 1.2 in chapter 1.

Initially, Reiter has mentioned in (Reiter 1980) that he knows of no naturally occurring default which cannot be represented in this form. His views have changed, however, already in Reiter and Criscuolo (1981) where it has already been argued that normal default rules are insufficient to deal with interactions of different defaults. For example, in combining two defaults, *Birds fly* and *Penguins can't fly*, the specificity principle naturally suggests that the second, more specific default should be preferred, so *Birds fly* shouldn't be applied to penguins. The authors argued that to capture this outcome in default logic, we need at least *semi*-normal defaults of the form $A : B \wedge C/C$, so *Birds fly* could now be represented roughly as "Birds normally fly, unless they are penguins," encoded as $B : F \wedge \neg P/F$.

Unfortunately, this complication has significantly undermined the transparency and modularity of the initial default representation: according to the modified proposal, the very representation of the claim "*A* normally implies *B*" has become dependent on other defaults and constraints present in the description. Worse still, no systematic method of constructing the resulting default representation for a given set of rules has emerged since then. The problem of defeasible inference has shown its real complexity.

In contrast to the above representations, John McCarthy (1980) has suggested a purely classical translation of normality conditionals $A \mathrel{\vert\!\sim} B$ as implications of the form

$$A \wedge \neg ab \supset B,$$

where *ab* is a new "abnormality" proposition serving to accumulate the conditions for violation of the source rule.[10] Thus, *Birds fly* was translated into something like "Birds fly if they are not abnormal."

In fact, viewed as a formalism for nonmonotonic reasoning, the central concept of McCarthy's circumscriptive method was not circumscription itself, but his notion of an abnormality theory—a set of classical conditionals containing the abnormality predicate *ab* that provides a representation for defeasible information. The default character of commonsense rules $A \mathrel{\vert\!\sim} B$ was captured in McCarthy's theory by a circumscription policy that minimized abnormality (and thereby maximized the acceptance of the corresponding normality claims $\neg ab$). Since then, this representation of defeasible rules using auxiliary (ab)normality propositions has been employed both in applications of circumscription and in many other theories of nonmonotonic inference in AI, sometimes in alternative logical frameworks. Some major examples are inheritance theories (Etherington and Reiter 1983), logic-based diagnosis (Reiter 1987a), general representation of defaults in Konolige and Myers (1989), and reasoning about time and action. Note also that naming of defaults (as in Poole 1988) can also be viewed as a species of this idea. In fact, this idea will also be used in our representation of defeasible causal rules in subsequent chapters.

---

10. Note again the similarity of this representation with Mill's omnibus negative condition—see section 1.2.4.

Classical abnormality theories have brought out, however, much the same problems of representing conflicting commonsense rules that plagued the default representation. A general approach to handle such problems in circumscription, suggested in Lifschitz (1985) and endorsed in McCarthy (1986), was to impose priorities among minimized predicates and abnormalities. The corresponding variant of circumscription has been called prioritized circumscription. The problem of defeasible inference was reduced in this way to the problem of finding the "right" priority order for a given set of defaults. Unfortunately, in this case also no systematic understanding or principles behind constructing such a priority order have emerged.

A more specific problem of the abnormality representation has arisen from the need to cope with the possibility of having two independent commonsense rules $A \mathrel{\vert\!\sim} B$ and $A \mathrel{\vert\!\sim} C$ with the same antecedent: for example, *Birds fly* and *Birds have wings*. In order to preserve one of these rules when the other is refuted, McCarthy has been forced to relativize the abnormality claims with respect to particular *aspects* so that some aspects can be abnormal without affecting others (see McCarthy 1986). For our example, birds might be abnormal with respect to flying without being abnormal with respect to having wings, and vice versa. McCarthy has not been fully satisfied with this representation, however, since he thought that this compels us to introduce (aspects of) abnormalities as new entities into our ontology, the things that exist. He mentioned that the aspects are abstract entities and their unintuitiveness is somewhat a blemish on the theory.

The difficulties encountered in representing defeasible inference in the first nonmonotonic formalisms have become a strong incentive for developing *direct*, non-translational theories of defeasible entailment. One of the first approaches of this kind has been based on an attempt of describing the very notion of defeasible inference, or derivation, that would take into account the impact and interaction of conflicting rules. Defeasible logic of Nute (see, e.g., Nute 1994) could be seen as a representative example of this approach.

In fact, the same approach lay at the basis of the theory of nonmonotonic inheritance nets (Touretzky 1986); see also an overview in (Horty 1994). Reasoning in inheritance hierarchies was represented in this theory as a process of constructing acceptable derivations ("paths") that formed a basis of making conclusions. Inheritance reasoning has dealt with a quite restricted class of conditionals built on literals. Nevertheless, in this restricted domain it has achieved a remarkably close correspondence between what is derived and what is expected intuitively. Accordingly, inheritance reasoning has emerged as an important test bed for adjudicating proposed theories.

The next stage in the development of the direct approach to defeasible inference was undoubtedly a theory of preferential entailment (Kraus, Lehmann, and Magidor 1990). In this theory, the normality conditionals $A \mathrel{\vert\!\sim} B$ have acquired a precise semantic interpretation in terms of a preference relation on worlds and a complete syntactic characterization in a set of natural postulates for defeasible inference. Moreover, it has been shown that the resulting theory preserves the main intuitive features of normality claims and, in particular, properly handles the specificity principle.

Unfortunately, it was realized quite early that preferential entailment also cannot serve as the ultimate theory of defeasible inference. Preferential inference is severely sub-classical and does not allow us, for example, to infer *Red birds fly* from *Birds fly*. Clearly, there are good reasons for not accepting such a derivation as a *logical* rule for defeasible inference; otherwise *Birds fly* would imply also *Birds with broken wings fly* and even *Penguins fly*. Still, this should not prevent us from accepting *Red birds fly* on the basis of *Birds fly* as a reasonable *defeasible* conclusion; namely, a conclusion made in the absence of information against it. By doing this, we would just follow the general strategy of nonmonotonic reasoning that involves making reasonable assumptions on the basis of available information. In other words, it was realized that preferential inference should be augmented at least with a mechanism of making nonmonotonic conclusions that goes beyond the latter. This has led to various theories of *defeasible entailment* built upon preferential inference.

Actually, the literature on nonmonotonic reasoning is abundant with theories of defeasible entailment. A history of studies on this subject could be briefly summarized as follows. Initial formal systems, namely Lehmann's rational closure (Lehmann and Magidor 1992) and Pearl's system Z (Pearl 1990), have turned out to be equivalent. This encouraging development was followed by a realization that both theories are still insufficient for representing defeasible inference since they do not allow to make certain intended conclusions. Hence, they have been refined in a number of ways, giving such systems as lexicographic inference (Benferhat et al. 1993; Lehmann 1995) and similar modifications of Pearl's system. Unfortunately, these refined systems have encountered an opposite problem: namely, together with some desirable properties, they invariably produced some unwanted conclusions. All these systems have been based on a supposition that defeasible entailment should form a rational inference relation. A more general approach in the framework of preferential inference has been suggested in Geffner (1992).

Conditional entailment of Geffner (1992) was based on a systematic prioritization of conditional bases securing that $A \vdash B$ will hold for every default conditional that belongs to the base. Then the intended models of a conditional base were defined as models that are generated by all such admissible priority orders.

It is interesting to note that, in the framework of conditional entailment, the conditionals $A \vdash B$ admitted a representation much similar to their representation in McCarthy's circumscription. Namely, $A \vdash B$ can be represented as a classical implication

$$A \wedge \delta \supset B,$$

where δ is a new "normality" proposition unique to each default conditional, plus a new, conceptually simpler conditional $A \vdash \delta$. This alternative representation has been actually implemented in Geffner and Pearl (1992).

Conditional entailment has shown itself as a serious candidate on the role of a general theory of defeasible entailment. Still, it has preserved strong connections between defeasible conditionals and corresponding material implications; as a by-product, it has allowed for a good deal of backward (contrapositive) reasoning. As a result, Geffner himself has shown in Geffner (1992) that conditional entailment still does not capture some important derivations that are based on a more directional, rule-based view of conditionals. Accordingly, in the last chapter of his book he has suggested extending conditional entailment with an explicit representation of a *causal order* among propositions. It is this latter causal representation of defeasible conditionals that has been one of the important sources for the development of the causal calculus in McCain and Turner (1997).

II THE BASICS

4 Causal Calculus

After all the preparations that have been made, we will introduce in this chapter the basic formalism of causal reasoning that will be used in this book, the causal calculus.

The causal calculus can be seen as a most natural and immediate generalization of classical logic that allows for causal reasoning. From a purely logical point of view, this generalization amounts to dropping the reflexivity postulate of classical inference. However, the associated inference systems will be assigned both a normal logical semantics (that will give a semantic interpretation to causal rules) and a natural nonmonotonic semantics which will provide a representation framework for causal reasoning.

The initial constructions and results of this chapter have been taken from Bochman (2004) and chapter 8 of Bochman (2005). There the reader could also find all the proofs of these initial results that have been omitted in the text.

As before, our basic language will be a classical propositional language with the usual classical connectives and constants $\{\wedge, \vee, \neg, \rightarrow, \mathbf{t}, \mathbf{f}\}$. The symbol $\models$ will stand for the classical entailment while Th will denote the associated classical provability operator. In this chapter, $p, g, r, \ldots$ will usually denote propositional atoms while $A, B, C, \ldots$ will denote arbitrary classical propositions.

4.1 Production Inference

The basic informational units of the causal calculus are conditionals of the form $A \Rightarrow B$ (saying that A *causes* B) that hold among classical propositions. They will determine our causal language, which is built on top of the underlying language of classical logic. Moreover, the main role of the postulates in the definition below amounts to securing that the corresponding causal reasoning respects this underlying classical logic of propositions.

Definition 4.1 *A production inference relation is a binary relation $\Rightarrow$ on the set of classical propositions satisfying the following postulates:*

(Strengthening) *If $A \models B$ and $B \Rightarrow C$, then $A \Rightarrow C$;*
(Weakening) *If $A \Rightarrow B$ and $B \models C$, then $A \Rightarrow C$;*

(And)   *If $A \Rightarrow B$ and $A \Rightarrow C$, then $A \Rightarrow B \wedge C$;*
(Truth) $\mathbf{t} \Rightarrow \mathbf{t}$;
(Falsity) $\mathbf{f} \Rightarrow \mathbf{f}$.

Remark. Though driven by entirely different considerations and objectives, production and causal inference relations described in this and subsequent sections have originated in input-output logics of Makinson and van der Torre (2000), and our description of different kinds of causal inference in this chapter will closely follow their classification.

From a logical point of view, all the postulates above are obviously valid for ordinary, deductive inference rules. However, the most significant "omission" of the above set of postulates is the absence of the reflexivity postulate $A \Rightarrow A$. It is precisely this feature of causal rules that will create a possibility of nonmonotonic reasoning in this framework.

We will extend causal rules to rules having arbitrary sets of propositions as premises using the compactness recipe: for any set u of propositions, we define $u \Rightarrow A$ as follows:

$$u \Rightarrow A \equiv \bigwedge a \Rightarrow A, \text{ for some finite } a \subseteq u.$$

For a set u of propositions, $\mathcal{C}(u)$ will denote the set of propositions caused by u, that is,

$$\mathcal{C}(u) = \{A \mid u \Rightarrow A\}.$$

The causal operator $\mathcal{C}$ plays much the same role as the usual derivability operator for consequence relations. Note, in particular, that it is monotonic.

(Monotonicity) If $u \subseteq v$, then $\mathcal{C}(u) \subseteq \mathcal{C}(v)$.

Note also that $\mathcal{C}(u)$ is always a deductively closed set: for any set u,

$$\mathcal{C}(u) = \mathrm{Th}(\mathcal{C}(u)).$$

Still, $\mathcal{C}$ is not inclusive: that is, $u \subseteq \mathcal{C}(u)$ does not always hold. Also, it is not idempotent: that is, $\mathcal{C}(\mathcal{C}(u))$ can be distinct from $\mathcal{C}(u)$.

Remark. Though we consider the classical entailment $\vDash$ as our official background logic in this book, we contemplate also a more liberal understanding of the underlying logic that could include *meaning postulates* and rules, including ordinary *definitions*. In accordance with our general two-tier design for causal reasoning, this logical level is purported to encode not only plain data but also all correlations and constraints that are not causal, such as correlations between my whereabouts and whereabouts of my wallet. It could include, in particular, all constraints and correlations among events and their parts that are excluded by definition from being causal in most accounts of causation.

Technically, this more liberal understanding of the underlying logic could be achieved by replacing classical entailment $\vDash$ with an arbitrary *supraclassical consequence relation* $\vdash$ (see chapter 2) in constructions and applications of the causal calculus. It should be noted, however, that *axioms* of this underlying logic, that is, rules of the form $\vdash A$, could also be expressed directly in the original causal calculus by accepting special causal rules of the form $\mathbf{t} \Rightarrow A$. A few examples of the use of such an extended logic will be given later in this chapter as well as in chapter 8.

Causal theories. Throughout this book, by a *causal theory* we will mean an arbitrary set of causal rules. For any causal theory Δ, there exists a least production relation that includes Δ. We will denote it by $\Rightarrow_\Delta$ while $\mathcal{C}_\Delta$ will denote the corresponding causal derivability operator. Clearly, $\Rightarrow_\Delta$ is the set of all causal rules that can be derived from Δ using the postulates for production inference relations.

For any set u of propositions and a causal theory Δ, we will denote by $\Delta(u)$ the set of all propositions that are directly caused by u in Δ, that is,

$$\Delta(u) = \{A \mid B \Rightarrow A \in \Delta, \text{ for some } B \in u\}.$$

Using this notion, the following explicit description of $\Rightarrow_\Delta$ has been given already in Makinson and van der Torre (2000):

Proposition 4.1 $\mathcal{C}_\Delta(u) = \text{Th}(\Delta(\text{Th}(u)))$.

4.1.1 Logical Semantics of Causal Rules

The semantic framework for production relations will be built from pairs of deductively closed theories called bimodels. In accordance with the "input-output" understanding of productions, a bimodel will represent an initial state (input) and a possible final state (output) of a derivation based on a given set of causal rules. The set of such bimodels will give a semantic description for these causal rules.

Definition 4.2 *A pair of consistent deductively closed sets will be called a* classical bimodel. *A set of classical bimodels will be called a* classical binary semantics.

A classical binary semantics can also be viewed as a *binary relation* on the set of deductive theories. Accordingly, given a set of bimodels (semantics) $\mathcal{B}$, we will write $u\mathcal{B}v$ to denote the fact that the a bimodel (u, v) belongs to $\mathcal{B}$. These descriptions will be used interchangeably in what follows.

Classical bimodels have been defined above in a syntactic fashion, namely, as pairs of theories of the language. This formulation will make subsequent constructions simpler and more transparent. Still, any such bimodel (u, v) can be safely equated with a pair of situations in the mereological semantics of chapter 2. Accordingly, a classical binary semantics can be viewed as an arbitrary binary relation on situations. All our subsequent constructions will permit such a purely semantic reformulation.

Now we will define the notion of validity of causal rules with respect to a classical binary semantics.

Definition 4.3 *A rule $A \Rightarrow B$ will be said to be* valid *in a classical binary semantics $\mathcal{B}$ if, for any bimodel (u, v) from $\mathcal{B}$, $A \in v$ only if $B \in u$.*

We will denote by $\Rightarrow_B$ the set of all causal rules that are valid in a semantics B. It can be easily verified that this set forms a production relation.

To prove completeness, for any production relation $\Rightarrow$ we can construct its *canonical semantics* $B_\Rightarrow$ as the set of all classical bimodels of the form $(\mathcal{C}(w), w)$, where w is an arbitrary consistent and deductively closed theory. Then the following basic completeness result can be obtained.

Theorem 4.2 *A binary relation $\Rightarrow$ on the set of propositions is a production inference relation if and only if it is determined by a classical binary semantics.*

4.1.2 Regular Inference

A production inference relation will be called *regular* if it satisfies:

(Cut)   If $A \Rightarrow B$ and $A \wedge B \Rightarrow C$, then $A \Rightarrow C$.

Cut is one of the basic rules for ordinary consequence relations. In the context of causal inference it plays the same role: namely, it allows for a reuse of caused propositions as premises in further derivations.[1] It corresponds to the following characteristic property of the causal operator:

$$\mathcal{C}(u \cup \mathcal{C}(u)) \subseteq \mathcal{C}(u).$$

Regular production relations have a number of additional properties. Thus, any such relation will already be transitive: that is, it will satisfy

(Transitivity)   If $A \Rightarrow B$ and $B \Rightarrow C$, then $A \Rightarrow C$.

Transitivity corresponds to the following familiar property of the production operator:

$$\mathcal{C}(\mathcal{C}(u)) \subseteq \mathcal{C}(u).$$

Note, however, that Transitivity is a weaker postulate than Cut since it does not imply the latter (see below).

A causal rule of the form $A \Rightarrow \mathbf{f}$ will be called a *constraint*. Such rules can be used for incorporating a purely factual information into causal theories: a rule $A \Rightarrow \mathbf{f}$ says, in a sense, that A should not hold in any intended model.

Now, an important property of regular production relations is that any causal rule implies the corresponding constraint:

(Constraint)   If $A \Rightarrow B$, then $A \wedge \neg B \Rightarrow \mathbf{f}$.

(Indeed, if $A \Rightarrow B$, then $A \wedge \neg B \Rightarrow B$ by Strengthening and $A \wedge B \wedge \neg B \Rightarrow \mathbf{f}$ by Falsity. Hence $A \wedge \neg B \Rightarrow \mathbf{f}$ by Cut.)

In particular, we have that $\mathbf{t} \Rightarrow A$ implies $\neg A \Rightarrow \mathbf{f}$. Note, however, that the reverse entailment does not hold even in the latter special case. More generally, though ordinary causal

---

1. They correspond to input-output logics with reusable output in Makinson and van der Torre (2000).

rules imply corresponding factual constraints, the latter cannot produce nontrivial causal rules (see below).

As a special case of Constraint, we have also the rule

(Coherence)   If $A \Rightarrow \neg A$, then $A \Rightarrow \mathbf{f}$.

It says that if a proposition causes propositions that are incompatible with it, then it is causally inconsistent. Actually, Coherence turns out to be equivalent to Constraint for all production inference relations.

Remark. Regular inference sanctions, in effect, an *atemporal* understanding of the notion of production inference. For example, the rule $p \wedge q \Rightarrow \neg q$ cannot be understood as saying that p and q jointly cause $\neg q$ (afterwards) in a temporal sense; instead, by Coherence it implies $p \wedge q \Rightarrow \mathbf{f}$, which means, in effect, that $p \wedge q$ cannot hold. Speaking generally, causally consistent propositions cannot cause a result incompatible with them. Just as in classical logic, however, a representation of temporal domains in this formalism can still be obtained by adding explicit temporal arguments to propositions (see chapter 12).

Regular inference relations can already be described in terms of a usual notion of a propositional theory.

Definition 4.4 *A set u of propositions will be called a* theory *of a production relation if it is deductively closed and $\mathcal{C}(u) \subseteq u$. A set u will be called a (propositional)* theory *of a causal theory Δ if it is a theory of $\Rightarrow_\Delta$.*

A theory of a production relation is a set of propositions that is closed both deductively and with respect to its causal rules: namely, if $A \in u$ and $A \Rightarrow B$, then $B \in u$. Accordingly, such theories have much the same properties as ordinary theories of consequence relations. Note, in particular, that the set of theories of a production inference relation is closed with respect to arbitrary intersections, and consequently any set of propositions is included in the least such theory. In addition, theories that are worlds (maximal deductively closed sets) have a very simple characterization.

Lemma 4.3 *A world α is a theory of a regular inference relation if and only if $\alpha \not\Rightarrow \mathbf{f}$.*

As could be expected, propositional theories of a causal theory Δ are sets of propositions that are closed with respect to the rules of Δ. Thus, the following description follows immediately from proposition 4.1.

Lemma 4.4 *A deductively closed set u is a propositional theory of a causal theory Δ if and only if $\Delta(u) \subseteq u$.*

The semantic characterization of regular inference relations can be obtained by considering only classical bimodels (u, v) such that $u \subseteq v$. We will call such bimodels (and the corresponding semantics) *consistent* ones.

Theorem 4.5 $\Rightarrow$ *is a regular inference relation if and only if it is generated by a consistent classical binary semantics.*

The proof of the above theorem (given in Bochman 2005) contains, in effect, the following characterization of regular inference relations in terms of their theories.

Corollary 4.6 $\Rightarrow$ *is a regular inference relation if and only if, for any set v and any A,*

$$v \Rightarrow A \text{ iff } A \in \mathcal{C}(u), \text{ for any theory } u \text{ of } \Rightarrow \text{ that contains } v.$$

We will denote by $\Rightarrow_\Delta^r$ the least regular production relation containing a causal theory Δ. As a consequence of the above characterization, we obtain a constructive characterization of $\Rightarrow_\Delta^r$, given in Makinson and van der Torre (2000).

Proposition 4.7 $v \Rightarrow_\Delta^r A$ *iff* $A \in \text{Th}(\Delta(u))$, *for any theory u of Δ that contains v.*

Regular inference relations allow us, in particular, to explicitly define an appropriate notion of equivalence among propositions such that equivalent propositions would be substitutable in any causal rule. Namely, let us assume that propositions A and B are equivalent with respect to the underlying logic $\vdash$, that is, $\vdash A \leftrightarrow B$ holds. As we have mentioned earlier, this equivalence can be expressed directly in the causal calculus by accepting the rule $\mathbf{t} \Rightarrow (A \leftrightarrow B)$. Accordingly, we will say that A and B are *causally equivalent* with respect to a production inference relation if the latter rule belongs to it. Then we have:

Lemma 4.8 *Propositions A and B are causally equivalent with respect to a regular production relation $\Rightarrow$ iff any occurrence of A can be replaced with B in the rules of $\Rightarrow$.*

Due to the above result, causal equivalence can be used for describing definitional extensions of the underlying language with new propositions (cf. H. Turner 1999).

Production versus consequence.   A further insight into the properties of regular inference can be obtained by comparing it with associated consequence relations.

Note that any causal theory, and hence any production inference relation, can also be considered as an ordinary conditional theory (a set of inference rules), so it determines the corresponding supraclassical Tarski consequence relation (see chapter 2). In order to construct such a consequence relation, we need only to "restore" reflexivity of the corresponding inference. The following construction gives a direct description of this consequence relation in terms of the source production relation. Namely, for a (regular) production relation $\Rightarrow$, we can define the following *consequence* relation:

$$A \vdash_\Rightarrow B \equiv A \Rightarrow (A \to B).$$

Theorem 4.9 *If $\Rightarrow$ is a regular production relation, then $\vdash_\Rightarrow$ is a least supraclassical consequence relation containing $\Rightarrow$.*

Let $\mathrm{Cn}_{\Rightarrow}$ denote the provability operator corresponding to $\vdash_{\Rightarrow}$. Then the above description can be extended to the following equality, for any set u of propositions:

$$\mathrm{Cn}_{\Rightarrow}(u) = \mathrm{Th}(u \cup \mathcal{C}(u)).$$

Remark. Incidentally, the above equality shows that regular causal inference and its associated logical consequence are indeed close relatives since it shows that $\mathcal{C}(u)$ captures all nontrivial consequences included in $\mathrm{Cn}_{\Rightarrow}(u)$, save for u itself. This means, in particular, that in many cases the causal consequences of a set of propositions could be obtained by using ordinary deductive tools. An even stronger connection can be established for the special case of literal causal theories that will be described in the next section.

As a first consequence of the above correspondence, we obtain:

Lemma 4.10 *Theories of* $\Rightarrow$ *coincide with the theories of* $\vdash_{\Rightarrow}$.

Since theories of $\vdash_{\Rightarrow}$ are exactly sets of propositions of the form $\mathrm{Cn}_{\Rightarrow}(u)$, the above result implies that such sets are precisely theories of $\Rightarrow$.

As yet another consequence of the above equation, we obtain the following alternative characterization of regular inference:

$$\mathcal{C}(u) = \mathcal{C}(\mathrm{Cn}_{\Rightarrow}(u)).$$

Note that we have also $\mathcal{C}(u) = \mathrm{Cn}_{\Rightarrow}(\mathcal{C}(u))$, so the causal operator absorbs $\mathrm{Cn}_{\Rightarrow}$ on both sides:

$$\mathrm{Cn}_{\Rightarrow} \circ \mathcal{C} = \mathcal{C} \circ \mathrm{Cn}_{\Rightarrow} = \mathcal{C}.$$

Due to these results, the supraclassical consequence relation $\mathrm{Cn}_{\Rightarrow}$ can be viewed as a *maximal underlying logic* that is suitable for a given regular production relation.

The above description of associated consequence relations can be immediately generalized to arbitrary causal (= conditional) theories. Thus, if Cn_{Δ} denotes the least supraclassical consequence relation containing a causal theory Δ while $\mathcal{C}_{\Delta}^r$ is the production operator of the least regular production relation containing Δ, then we have:

$$\mathrm{Cn}_{\Delta}(u) = \mathrm{Th}(u \cup \mathcal{C}_{\Delta}^r(u)).$$

Moreover, proposition 4.7 will immediately imply the following simplified characterization of $\mathcal{C}_{\Delta}^r$:

Corollary 4.11 $\mathcal{C}_{\Delta}^r(u) = \mathrm{Th}(\Delta(\mathrm{Cn}_{\Delta}(u)))$.

4.2 Nonmonotonic Semantics of Causal Theories

In the preceding sections, we have given a formalization and a (binary) logical semantics for the logical system of production inference. However, production inference relations determine also a natural nonmonotonic semantics, which is an ordinary *propositional* semantics that singles out a set of models that conform to the causal rules of the inference relation. This means, of course, that such models should be closed with respect to these causal rules. Causal reasoning imposes, however, a further constraint on appropriate models, a constraint that stems from the law of causality, or principle of sufficient reason. Speaking generally, the constraint says that adequate models of a causal theory should also be "explanatory closed" in the sense that any proposition that holds in them should have a reason (i.e., cause) why it holds in the model.

Formally, the fact that the causal operator $\mathcal{C}$ is not reflexive creates an important distinction among theories of a production relation.

Definition 4.5 *For a production inference relation* $\Rightarrow$,

- *A set u of propositions will be called* explanatory closed, *if* $u \subseteq \mathcal{C}(u)$.
- *A theory u of* $\Rightarrow$ *will be called* exact, *if it is consistent and explanatory closed, that is,*

$$u = \mathcal{C}(u).$$

- *A set u is an* exact theory of a causal theory Δ, *if it is an exact theory of* $\Rightarrow_\Delta$.

An exact theory describes a situation that not only satisfies the relevant causal rules, but also is such that every proposition that holds in it is caused by other propositions in accordance with these rules. In this sense, it directly implements the law of causation, or Leibniz's principle of sufficient reason (see section 1.2), as part of its definition. This leads us to the following notion of a nonmonotonic semantics:

Definition 4.6 *A nonmonotonic semantics of a production inference relation or a causal theory is the set of all its exact theories.*

All the information that can be discerned from the nonmonotonic semantics of a causal theory can be seen as nonmonotonically implied by the latter. This includes, for instance, so-called skeptical conclusions (what holds for all exact theories) or credulous conclusions (what holds in at least one exact theory).

Remark. It is important to observe that an exact theory (being just a set of classical propositions) and the nonmonotonic semantics in general contain only purely propositional, *factual* information. In this respect, they provide only a factual output (a "factual shadow," if you like) of the rich causal information embodied in the source causal theory or an inference relation. Unlike the case of an ordinary, logical semantics, this factual output is insufficient for determining, or capturing back, the source causal information.

The above nonmonotonic semantics is indeed nonmonotonic in the sense that adding new rules to a causal theory may lead to not only removal of some models but also occasionally create *new* exact models. Consequently, it can produce nonmonotonic changes in the derived information, not only its growth (a number of instances of such a behavior will be presented below). This happens even though causal rules themselves are monotonic since they satisfy Strengthening (the antecedent)—see definition 4.1.

Exact theories are precisely fixed points of the causal operator $\mathcal{C}$. Since the latter operator is monotonic and continuous, exact theories (and hence the nonmonotonic semantics) always exist. Also, the general properties of monotonic operators imply the following properties of exact theories:

Lemma 4.12

1. *Any production inference relation has a least exact theory.*
2. *Any theory of a production relation contains a greatest exact theory.*
3. *The union of any chain of exact theories (with respect to set inclusion) is an exact theory.*
4. *Any exact theory is included in a maximal exact theory.*

However, the following example shows that exact theories are not closed with respect to arbitrary intersections.

Example 4.1 *Let us consider a causal theory* $\Delta = \{p_i \Rightarrow p_i, p_i \Rightarrow q \mid i \geq 0\}$. *Then, for any natural n, the set* $u_n = \text{Th}(q, p_n)$ *is an exact theory of* $\Rightarrow_\Delta$. *However,* $\bigcap u_n = \text{Th}(q)$ *is not an exact theory of* $\Rightarrow_\Delta$ *since it does not include causes of q.*

As a consequence, a least exact theory containing a given set of propositions does not always exist. Still, the least exact theory of a regular inference relation always exists; it coincides with the set of propositions that are caused by truth **t**.

Lemma 4.13 *The least exact theory of a causal theory* Δ *coincides with the set of propositions that are provable from* Δ *using the postulates of regular production inference.*

The following simple lemma gives a constructive description of the nonmonotonic semantics of a causal theory. The proof is immediate by proposition 4.1.

Lemma 4.14 *u is an exact theory of a causal theory* Δ *iff* $u = \text{Th}(\Delta(u))$.

We are going to show now that regular inference provides an adequate and maximal logical framework for reasoning with exact theories.

Definition 4.7 *Causal theories* Γ *and* Δ *will be called* g-equivalent *if they have the same nonmonotonic semantics.*

Note that production inference relations can also be considered as (rather big) causal theories. As before, let $\Rightarrow_\Delta^r$ denote the least regular production relation that contains a causal theory Δ. Then we have:

Lemma 4.15 *A causal theory* Δ *is g-equivalent to* $\Rightarrow_\Delta^r$.

The above lemma implies that the postulates of regular inference are adequate for reasoning with exact theories since they preserve the latter. Moreover, regular inference relations constitute a maximal logic suitable for the above nonmonotonic semantics.

Let us say that causal theories Δ and Γ are *regularly equivalent* if each can be obtained from the other using the postulates of regular inference or, equivalently, when $\Rightarrow^r_\Delta = \Rightarrow^r_\Gamma$. Now, as an immediate consequence of theorem 4.15, we obtain:

Corollary 4.16 *Regularly equivalent theories are g-equivalent.*

The reverse implication in the above corollary does not hold, and a deep reason for this is that the regular equivalence is a monotonic (logical) notion; and hence, unlike the nonmonotonic g-equivalence, it is preserved under addition of new causal rules. What we need, therefore, is a stronger, monotonic counterpart of the notion of g-equivalence that would be preserved under addition of new causal rules. This immediately suggests the following definition.

Definition 4.8 *Two causal theories Δ and Γ are* strongly g-equivalent *if, for any set Φ of causal rules, $\Delta \cup \Phi$ is g-equivalent to $\Gamma \cup \Phi$.*

The above notion of strong equivalence is actually a causal counterpart of the notion of strong equivalence for default theories we mentioned in chapter 3 (see definition 3.1).

Strongly equivalent theories are "equivalent forever"—that is, they are interchangeable in any larger causal theory without changing the general nonmonotonic semantics. Consequently, strong equivalence can be seen as an equivalence with respect to some background logic of causal theories. And the next result shows that this logic is precisely the logic of regular production relations.

Theorem 4.17 *Two causal theories are strongly g-equivalent if and only if they are regularly equivalent.*

The above result implies, in effect, that regular production relations are maximal inference relations that are adequate for general nonmonotonic reasoning in causal theories: any postulate that is not valid for regular production relations can be "falsified" by finding a suitable extension of two causal theories that would determine different nonmonotonic semantics and hence would produce different nonmonotonic conclusions.

It turns out that production inference relations also satisfy a certain restricted monotonicity property with respect to the nonmonotonic semantics.

Definition 4.9 *Causal theories Γ and Δ will be called* supraclassically equivalent *if they determine the same supraclassical consequence relation.*

Supraclassical equivalence amounts, in effect, to the equivalence of Δ and Γ viewed as sets of ordinary rules of a (supraclassical) consequence relation. Consequently, Δ and Γ are supraclassically equivalent if and only if they have the same theories.

Lemma 4.18 *If causal theories* Γ *and* Δ *are supraclassically equivalent, and* $\Gamma \subseteq \Delta$*, then any exact theory of* Γ *is an exact theory of* Δ.

The above result clarifies the reasons why the semantics of exact theories is nonmonotonic. Indeed, if the causal rules added to a causal theory do not change the set of theories of the latter, then they extend, in general, the set of its exact theories, and therefore not all the conclusions made earlier will be preserved.

As a preparation for what follows, we introduce the following special kind of production inference relations:

Definition 4.10 *A production inference relation will be called* dense *if* $A \Rightarrow B$ *holds only if* $A \Rightarrow C$ *and* $C \Rightarrow B$*, for some proposition C.*

The above property is trivial, of course, for ordinary consequence relations (due to reflexivity), but it is nontrivial for production relations. It corresponds to the following characteristic property of the production operator:

$$\mathcal{C}(u) \subseteq \mathcal{C}(\mathcal{C}(u)).$$

The above property says, in effect, that any set of propositions of the form $\mathcal{C}(u)$ is explanatory closed. But any such set is also a theory of a regular production relation. Consequently, the set of exact theories of a dense regular production relation has an especially simple structure:

Lemma 4.19 *The nonmonotonic semantics of a dense regular production relation coincides with the set of all sets of propositions of the form* $\mathcal{C}(u)$.

Dense production relations will play an important role in our description of abduction in chapter 7.

4.3 Determinate and Literal Causal Theories

The overwhelming majority of applications of causal reasoning in AI and beyond make use of only a restricted form of causal rules, often called determinate rules, and there are deep reasons for this restriction. In fact, such causal theories will serve as the main representation tool for almost all the purposes of this book.

Definition 4.11 *A causal rule will be called* determinate *if it has the form* $A \Rightarrow l$*, where l is either a literal or a falsity constant* **f**.
A causal theory will be called determinate *if it contains only determinate rules.*

The following special kind of classical deductive theories will play an important role in describing determinate causal theories.

Definition 4.12 *A classical propositional theory will be called a* literal *one if it is a classical logical closure of a set of literals.*

Let $\mathcal{L}$ denote the set of all literals of the language. It can be easily verified that a propositional theory is literal if and only if it is a logical closure of the set of literals that belong to it:

$$u = \mathrm{Th}(\mathcal{L} \cap u).$$

This establishes a one-to-one correspondence between literal propositional theories and sets of classical literals; we will use this correspondence in what follows.

If Δ is a determinate causal theory, then all derivations that use the rules from Δ proceed via derivation of literals as intermediate steps. This justifies the following simple observation:

Lemma 4.20 *If Δ is a determinate causal theory, then $C^r_\Delta(u)$ is a literal propositional theory, for any set u of propositions.*

As an immediate consequence of the above lemma, we obtain:

Corollary 4.21 *Any exact theory of a determinate causal theory is a literal propositional theory.*

Finally, as a consequence of lemma 4.14, we obtain:

Corollary 4.22 *If Δ is a determinate causal theory, then L is the set of literals of an exact theory of Δ if and only if it satisfies the equality*

$$L = \Delta(\mathrm{Th}(L)).$$

The above characterization reduces, in effect, description of the nonmonotonic semantics for determinate causal theories to sets of literals.

Literal causal theories. A further natural restriction on the form of causal rules is provided by the following definition.

Definition 4.13 *A determinate causal rule will be called a* literal *rule if it has the form $L \Rightarrow l$, where L is a set of literals. A causal theory will be called a* literal *theory if it contains only literal rules.*

It should be noted that, even in the framework of regular inference, arbitrary nonliteral causal rules are not reducible, in general, to literal ones. In fact, reasoning in literal causal theories has some special features that do not hold for determinate theories in general. The most important such feature is that regular inference in such causal theories becomes much similar to classical logical inference on positive Horn propositional clauses, the only difference being that it admits both kinds of literals, atoms and their negations, indiscriminately.[2] This similarity is made explicit in the following theorem.

---

2. I owe this idea to Vladimir Lifschitz.

Theorem 4.23 *If Δ is a literal causal theory, then, for any literal l and any set L of literals, $L \Rightarrow^r_\Delta l$ holds if and only if $L \Rightarrow l$ is derivable from Δ using the following inference rules for literal causal rules:*

(Literal Monotonicity)   *If $L \Rightarrow l$, then $L, L' \Rightarrow l$;*
(Literal Cut)   *If $L \Rightarrow l$ and $L, l \Rightarrow l'$, then $L \Rightarrow l'$;*
(Literal Contradiction)   *$l, \neg l \Rightarrow l'$.*

Proof. Let $\mathcal{C}^l_\Delta$ denote a "literal" causal operator corresponding to the above inference rules, that is, for any set L of literals, $\mathcal{C}^l_\Delta(L)$ is the set of all literals that are derivable from L using the rules of Δ and the above inference rules. Assume now that $l \notin \mathcal{C}^l_\Delta(L)$. Due to Literal Monotonicity, L can be extended to a maximal set of literals L' such that $l \notin \mathcal{C}^l_\Delta(L')$. Note that, due to Literal Contradiction, L' is a consistent set of literals. We will show that $\Delta(L') \subseteq L'$. Indeed, otherwise there exists a literal l' such that $L' \Rightarrow l'$, but $l' \notin L'$. Since L' is a maximal set of literals that does not cause l, we have $L', l' \Rightarrow l$, but then $L' \Rightarrow l$ by Literal Cut, contrary to our assumption. Thus, $\Delta(L') \subseteq L'$.

Let u denote $\text{Th}(L')$. Clearly, u is a deductively closed theory that includes L. Moreover, $\Delta(u) = \Delta(L')$ since otherwise u would include literals that do not belong to L'. Since $\Delta(L') \subseteq L'$, this immediately implies that u is closed with respect to the rules of Δ, that is, $\Delta(u) \subseteq u$. Consequently, u is a theory of Δ. Finally, since $l \notin \mathcal{C}^l_\Delta(L')$, we have $l \notin \Delta(L')$, and hence $l \notin \Delta(u)$. Therefore, $L \not\Rightarrow^r_\Delta l$ by proposition 4.7.

Since all the literal inference rules above are derived rules of regular inference, the reverse inclusion is immediate. This completes the proof. □

Finally, we are going to show that literal causal rules are sufficient for determining exact nonmonotonic models that are literal theories.

Definition 4.14 *A literal nonmonotonic semantics of a causal theory is a set of exact theories that are also literal propositional theories.*

Our results above have shown that the general nonmonotonic semantics of determinate causal theories coincides with its literal nonmonotonic semantics. Moreover, for literal causal theories, description of the nonmonotonic semantics could be simplified further to the following:

Corollary 4.24 *If Δ is a literal causal theory, then L is the set of literals of an exact theory of Δ if and only if it satisfies the fixpoint equality:*

$$L = \Delta(L).$$

Now, for an arbitrary causal theory Δ, let $\Delta_{\mathcal{L}}$ denote the following set of literal causal rules:

$$\Delta_{\mathcal{L}} = \{L \Rightarrow l \mid l \in \text{Th}(\Delta(\text{Th}(L)))\}.$$

By proposition 4.1, $\Delta_{\mathcal{C}}$ is precisely the set of literal causal rules that are derivable from Δ by the rules of general production inference. Moreover, we have:

Theorem 4.25 *Any causal theory Δ has the same literal nonmonotonic semantics as $\Delta_{\mathcal{C}}$.*

Proof. Recall that u is an exact theory of Δ if and only if $u = \text{Th}(\Delta(u))$. Now, if u is also a literal theory, that is, $u = \text{Th}(L)$, for some set L of literals, then this equation reduces to the following condition:

$$l \in L \quad \text{iff} \quad l \in \text{Th}(\Delta(\text{Th}(L))),$$

for any literal l. However, $l \in \text{Th}(\Delta(\text{Th}(L)))$ means that the rule $L \Rightarrow l$ belongs to $\Delta_{\mathcal{C}}$, and consequently this condition reduces to $L = \Delta_{\mathcal{C}}(L)$. Due to the preceding corollary, this implies the result.   □

It can be easily verified that a literal causal rule $L \Rightarrow l$ belongs to $\Delta_{\mathcal{C}}$ if and only if Δ contains a set of rules $\{A_i \Rightarrow B_i\}$ such that $L \vDash \bigwedge A_i$ and $\bigwedge B_i \vDash l$. This gives us a convenient recipe for transforming an arbitrary causal theory to a literal one that will preserve the associated literal nonmonotonic semantics. Unfortunately, this algorithm is neither modular nor polynomial. Moreover, the complexity considerations confirm that this is as it should be, already due to the difference in complexity between arbitrary and determinate causal theories, established in Giunchiglia et al. (2004).

4.4 Basic and Causal Inference

Following Makinson and van der Torre (2000), a production inference relation will be called *basic* if it satisfies:

(Or)   If $A \Rightarrow C$ and $B \Rightarrow C$, then $A \vee B \Rightarrow C$.

The above inference rule is nothing other than the classical rule of disjunction in the antecedent. It sanctions reasoning by cases, and hence basic production relations can already be seen as systems of reasoning about complete worlds.

The Or postulate corresponds to the following characteristic condition on derivability; for any deductively closed sets u, v:

$$\mathcal{C}(u \cap v) = \mathcal{C}(u) \cap \mathcal{C}(v).$$

As a consequence of this condition, the set of propositions caused by a theory u coincides with the set of propositions that are caused by every world α containing u:

$$\mathcal{C}(u) = \bigcap \{\mathcal{C}(\alpha) \mid u \subseteq \alpha\}.$$

Let us say that a causal rule is *clausal* if it has the form $\bigwedge l_i \Rightarrow \bigvee l_j$, where l_i, l_j are classical literals. A causal theory will be called *clausal* if it contains only clausal rules.

Note that, for any production inference relation, a rule $A \Rightarrow B \wedge C$ is equivalent to the pair of rules $A \Rightarrow B$ and $A \Rightarrow C$ (due to And and Weakening). Now, for basic production inference, a rule $A \vee B \Rightarrow C$ is equivalent to a pair of rules $A \Rightarrow C$ and $B \Rightarrow C$ (by Or and Strengthening). Consequently, we obtain the following important fact about basic production inference:

Lemma 4.26 *For a basic inference, any causal theory is reducible to a clausal causal theory.*

Moreover, as a special case, we have the following:

Corollary 4.27 *For a basic inference, any determinate causal theory is reducible to a literal causal theory.*

The least basic production relation containing a causal theory Δ will be denoted by $\Rightarrow_\Delta^b$. The following characterization of this relation has been obtained in Makinson and van der Torre (2000):

Proposition 4.28 *For any set u of propositions and any proposition A:*

$$u \Rightarrow_\Delta^b A \text{ if and only if } A \in \text{Th}(\Delta(\alpha)), \text{ for every world } \alpha \supseteq u.$$

4.4.1 Possible Worlds Semantics

The semantic characterization of basic production relations can be obtained from the general classical binary semantics by restricting the set of bimodels to world-based ones: namely, to bimodels of the form (α, β), where α, β are worlds. Generalizing this setting a little, the corresponding binary semantics can be defined as a usual *relational possible worlds model* $\mathbb{W} = (W, \mathcal{B}, V)$, where W is a set of possible worlds, $\mathcal{B}$ is a binary accessibility relation on W, and V is a valuation function assigning each possible world a propositional interpretation. The following definition provides the corresponding notion of validity for causal rules:

Definition 4.15 *A rule $A \Rightarrow B$ will be said to be* valid *in a possible worlds model $(W, \mathcal{B}, V)$ if, for any pair of worlds $\alpha, \beta \in W$ such that $\alpha \mathcal{B} \beta$, if A holds in α, then B holds in β.*

We will denote by $\Rightarrow_\mathbb{W}$ the set of all causal rules that are valid in a possible worlds model $\mathbb{W}$. It can be easily verified that this set is already closed with respect to the postulate Or, so it forms a basic production relation. To prove completeness, for any basic production relation $\Rightarrow$ we can construct its *canonical possible worlds model* $\mathbb{W}_\Rightarrow$ as a triple $\mathbb{W}_\Rightarrow = (W, \mathcal{B}, V)$ such that W is a set of all worlds (maximal classically consistent sets), V is a function assigning each world its corresponding classical interpretation while $\mathcal{B}$ is a relation on W defined as follows:

$$\alpha \mathcal{B} \beta \equiv \mathcal{C}(\alpha) \subseteq \beta.$$

Then the following result shows that this semantics is strongly complete for the source production relation.

Lemma 4.29 *If* $\mathbb{W}_{\Rightarrow}$ *is the canonical semantics of a basic production relation* $\Rightarrow$*, then, for any set u of propositions and any A,*

$$u \Rightarrow A \text{ iff } A \in \beta, \text{ for any } \alpha, \beta \in W \text{ such that } \alpha \mathcal{B} \beta \text{ and } u \subseteq \alpha.$$

As a general conclusion, we obtain:

Corollary 4.30 $\Rightarrow$ *is a basic production relation if and only if it has a possible worlds semantics.*

The above semantics immediately suggests a modal translation of causal rules described in Makinson and van der Torre (2000) (see also H. Turner 1999). Namely, let $\square$ be the usual modal operator definable in a possible worlds model: $\square A$ holds in α iff A holds in all β such that $\alpha R \beta$. Then the validity of $A \Rightarrow B$ in a possible worlds model is equivalent to validity of the formula:

$$A \to \square B.$$

Consequently, for basic inference, causal rules correspond to modal formulas of the latter form.

4.4.2 Causal Inference

Now we will consider inference relations that are both basic and regular.

Definition 4.16 *A production relation will be called* causal *if it satisfies Or and Cut.*

Causal inference relations enjoy the properties of both basic and regular production relations. An important fact about causal inference relations is the following decomposition of causal rules:

Lemma 4.31 *Any causal rule* $A \Rightarrow B$ *is equivalent to a pair of rules*

$$A \wedge \neg B \Rightarrow \mathbf{f} \text{ and } A \wedge B \Rightarrow B.$$

A constraint $A \wedge \neg B \Rightarrow \mathbf{f}$ says that the classical implication $A \to B$ should hold in any causally consistent world. This constraint encodes a purely logical, or factual, information (a 'regularity') that is embodied in the causal rule $A \Rightarrow B$. In contrast, causal rules of the form $A \wedge B \Rightarrow B$ are logically trivial, but they play an important explanatory role in causal reasoning. Namely, they are purely *explanatory rules* saying that, in any world where A holds, B does not need further reason from the causal point of view.

Remark. Explanatory causal rules $A \wedge B \Rightarrow B$ could also be viewed as *weak causal claims*, claims that allow us to *explain* effect B by citing its cause A, though they do not allow us to *predict* B on the basis of A. It is relatively well known, for instance, that previous syphilis may explain subsequent paresis, even though only quite a small percentage of syphilitic patients develop this dreadful illness. An explanatory rule

$$Syphilis, Paresis \Rightarrow Paresis$$

could be used to provide a description of such cases.

The above lemma says that any causal rule can be decomposed into a constraint and an explanatory rule. This decomposition neatly delineates two kinds of information conveyed by causal rules. One is a factual (logical) information that constraints the set of admissible models while the other is an explanatory information describing what propositions are caused (explainable) in such models.[3]

Remark. The above decomposition clearly resembles the structure of Aristotelian *demonstrative syllogisms* (see section 1.1 in chapter 1). Moreover, there exists a remarkable analogy between constraints of the form $A \Rightarrow \mathbf{f}$ and *reductio ad impossibile* proofs in Aristotle's *Analytics*. Thus, in chapter I.26 of the *Posterior Analytics*, Aristotle has argued that direct demonstrations are better than those that proceed by *reductio* because the latter cannot produce scientific knowledge (*episteme*) but only knowledge of the fact. Note in this respect that, in the causal calculus, addition of factual constraints to causal theories cannot produce new causal assertions, though it can reduce the set of admissible causal models (see the next section).

Now we will turn to a semantic description of causal inference. It has been shown earlier that possible worlds models with arbitrary accessibility relations provide a semantics for basic production relations. It turns out that causal inference relations are determined by possible worlds models in which the accessibility relation is *quasi-reflexive*, that is, satisfies the following condition for any two worlds:

(Quasi-Reflexivity)   If $\alpha R \beta$, then $\alpha R \alpha$.

The following theorem shows that causal relations are actually complete for such models.

Theorem 4.32 *A production relation is causal iff it is determined by a quasi-reflexive possible worlds model.*

Remark. Incidentally, the above semantic interpretation gives clear reasons why Transitivity is a weaker property that Cut, even for basic inference relations. Indeed, Transitivity holds for all production relations determined by possible world models with a *dense* accessibility relation, that is, a relation satisfying the condition that if $\alpha R \beta$, then there exists γ such that $\alpha R \gamma$ and $\gamma R \beta$. Clearly, quasi-reflexivity is a stronger property than density.

---

3. A similar decomposition of causal rules can be found in Thielscher (1997).

In what follows, $\Rightarrow^c_\Delta$ will denote the least causal inference relation containing a causal theory Δ. In addition, $\overrightarrow{\Delta}$ will denote the set of material implications corresponding to the causal rules from Δ, namely,

$$\overrightarrow{\Delta} = \{A \to B \,|\, A \Rightarrow B \in \Delta\}.$$

As for all the kinds of productions considered earlier, Makinson and van der Torre (2000) contains a constructive description of the least causal inference relation containing a given causal theory. Such a description can be obtained from the corresponding description for regular production relations, given in proposition 4.7, simply by restricting the set of theories of Δ to worlds. Note, however, that a world α is a theory of Δ if and only if $\overrightarrow{\Delta} \subseteq \alpha$. Consequently, we obtain:

Proposition 4.33 *If $\Rightarrow^c_\Delta$ is the least causal relation containing a causal theory Δ, then*

$$u \Rightarrow^c_\Delta A \text{ iff } A \in \mathrm{Th}(\Delta(\alpha)), \text{ for any world } \alpha \supseteq u \cup \overrightarrow{\Delta}.$$

Thus, derivability in causal inference relations is reducible, in effect, to derivability in basic production relations with an additional set of assumptions $\overrightarrow{\Delta}$. This fact implies that the material implications corresponding to the causal rules can be used as auxiliary assumptions in making derivations. In other words, causal inference relations make valid the following rule:

If $A \Rightarrow B$ and $C \wedge (A \to B) \Rightarrow D$, then $C \Rightarrow D$.

4.5   Causal Nonmonotonic Semantics

If we will focus on objective understanding of causal rules as rules that describe complete worlds, it will be only natural to consider also the corresponding restriction of the nonmonotonic semantics to exact theories that are worlds. Such worlds will be both closed with respect to the causal rules and satisfy the law of universal causality according to which any fact that holds in such a world has a cause in this world.

Definition 4.17 *A causal nonmonotonic semantics of a causal theory is the set of all its exact worlds.*

Since the causal nonmonotonic semantics forms a subset of the general nonmonotonic semantics, it produces, in general, a larger set of nonmonotonic consequences. Moreover, being just a set of worlds, its logical content is exhausted by the classical propositional theory that is uniquely associated with this set of worlds (a direct description of this theory, called a propositional completion of a causal theory, will be given below). Note, however, that, unlike the general nonmonotonic semantics, the causal nonmonotonic semantics is not guaranteed to exist for every causal theory.

The interplay of the factual and explanatory content of causal rules determines, in effect, the nonmonotonic semantics, and it is responsible, in particular, for the nonmonotonic properties of the latter. Namely, the nonmonotonicity arises from the fact that the two kinds of content have opposite impacts on the set of exact worlds. Thus, addition of factual constraints leads, as expected, to a reduction of the set of admissible worlds and hence to an increase of factual information. However, addition of explanatory rules leads, in general, to an *increase* of exact worlds (among the logically possible ones) and hence to a decrease of nonmonotonically derived information (cf. corollary 4.37 below). In any case, the growth in informational content leads to a monotonic *reduction of the set of nonexact (unexplained) worlds*.

Any world is determined by its literals, so it is a literal theory (see definition 4.13). Consequently, the causal nonmonotonic semantics is also a special case of literal nonmonotonic semantics. This implies, in particular, that the causal nonmonotonic semantics of a causal inference relation is determined by the literal causal rules that belong to it.

By lemma 4.14, the causal nonmonotonic semantics of a causal theory Δ is the set of worlds α such that $\alpha = \text{Th}(\Delta(\alpha))$. Such worlds coincide with *causally explained interpretations* of McCain and Turner (1997), which determine the nonmonotonic semantics of their causal theories. Consequently the causal nonmonotonic semantics provides an adequate representation for this nonmonotonic system.

A description of exact worlds for causal theories is given below.

Corollary 4.34 *A world α is an exact world of a causal theory Δ if and only if, for any propositional atom p,*

$$p \in \alpha \;\; \textit{iff} \;\; \Delta(\alpha) \vDash p$$
$$p \notin \alpha \;\; \textit{iff} \;\; \Delta(\alpha) \vDash \neg p.$$

Now we will show that causal inference relations provide an adequate framework of reasoning with respect to the causal nonmonotonic semantics. As before, we will introduce first the following definitions.

Definition 4.18 *Causal theories Γ and Δ will be called:*

- objectively equivalent *if they have the same causal nonmonotonic semantics*
- strongly objectively equivalent *if, for any set Φ of causal rules, $\Delta \cup \Phi$ is objectively equivalent to $\Gamma \cup \Phi$*
- causally equivalent *if $\Rightarrow^c_\Delta \; = \; \Rightarrow^c_\Gamma$*

Two causal theories are causally equivalent if each theory can be obtained from the other using the inference postulates of causal inference relations.

Lemma 4.35 *Any causal theory Δ is objectively equivalent to $\Rightarrow^c_\Delta$.*

The above lemma says that the postulates of causal inference relations are adequate for reasoning with exact worlds since they preserve the latter. Moreover, the next theorem shows that causal inference relations constitute a maximal logic suitable for the causal nonmonotonic semantics.

Theorem 4.36 *Two causal theories are strongly objectively equivalent if and only if they are causally equivalent.*

Finally, just as for the general nonmonotonic semantics, the causal nonmonotonic semantics sometimes grows monotonically with the growth of these causal theories.

Definition 4.19 *Causal theories* Γ *and* Δ *will be called* classically equivalent, *if* $\overrightarrow{\Delta}$ *is logically equivalent to* $\overrightarrow{\Gamma}$.

Causal theories are classically equivalent if they are logically equivalent when viewed as sets of classical material implications. In this case, as a consequence of lemma 4.18, we obtain the corresponding restricted monotonicity property:

Corollary 4.37 *If causal theories* Γ *and* Δ *are classically equivalent, and* $\Gamma \subseteq \Delta$, *then any exact world of* Γ *is an exact world of* Δ.

4.5.1   Defaults and Exogenous Propositions

As we mentioned in chapter 1 (see section 1.2.1), causal reasoning inevitably leads to the *Agrippan trilemma*, namely, the problem that if we do not want to allow infinite regress of causal explanations, we have to accept either non-caused or self-caused facts.

Note first that an infinite chain of causal explanations is a logically sound notion, and the definition of an exact model does not exclude this possibility. Moreover, in a temporal world that does not have a first moment (of a Big Bang kind), it is even a natural possibility. Still, in normal cases, causal reasoning is used for describing finite, local situations, and consequently the corresponding causal theories are even formulated in a finite language. In such cases, any infinite chain of causal explanations is bound to contain a loop, which appears to be an even less attractive possibility.

In describing local situations, or systems, a commonly accepted way of dealing with this problem amounts to singling out a distinguished set of *exogenous* variables that are used for describing boundary conditions of a system in question and thereby become exempted from the need of further explanation. In the causal calculus, we can formally represent such an exemption by accepting an extremely simple explanatory rule

$$A \Rightarrow A.$$

The above rule makes A a self-caused (or self-explanatory) proposition. In what follows, we will call propositions that satisfy such a rule *pure defaults* or *abducibles* (especially in chapter 7). It is important to note that if a chain of causal rules of some causal theory forms a loop, then, due to transitivity of causal inference, any proposition that belongs to this loop will be a default proposition in this sense.

Remark. In chapter 9, we will introduce an alternative, relative notion of an exact model that will exploit the second horn of the Agrippan trilemma, namely, the possibility of accepting brute facts without making them defaults.

The definition of an exact world requires that any literal that holds in such a world should have a cause. This implies, in particular, that if a causal theory does not contain rules with a literal l in heads, l cannot hold in an exact model for Δ. In other words, the absence of causal rules for l implicitly restricts exact models to those where $\neg l$ holds (see the next section).

To override such implicit constraints, we will occasionally use the following stipulation:

Definition 4.20 *A* pure default completion *of a causal theory Δ is a causal theory $Dc(\Delta)$ obtained from Δ by adding rules $l \Rightarrow l$ for any literal of the language that does not appear in the heads of the rules from Δ.*

The effect of this syntactic completion construction will consist in extending the causal nonmonotonic semantics with exact worlds that admit previously "uncaused" literals, insofar as this addition is consistent with the rest of the causal rules.

Default propositions can be used as initial points of causal derivations as well as endpoints of causal explanations. Note, however, that default propositions can also be heads of other causal rules. Thus, we may have, for instance, both $A \Rightarrow A$ and $B \Rightarrow A$ in a causal theory. Though such rules are no longer needed for explanation of A, they still can provide logical (factual) constraints for the corresponding exact models. Consequently, we should still distinguish between default and exogenous propositions. This justifies the following official definition of an exogenous literal that will be used in what follows.

Definition 4.21 *A literal l will be called* exogenous *in a causal theory Δ if it is a default literal and Δ does not contain other rules with the head l.*

4.5.2 Completion

A causal theory will be called *definite* if it is determinate and any propositional atom appears in heads of no more than a finite number of its causal rules.

Clearly, any finite determinate theory will be definite, though not vice versa.

As has been established already in McCain and Turner (1997), the causal nonmonotonic semantics of definite theories coincides with the classical logical semantics of their propositional completions.

Given a definite causal theory Δ, we will define its (propositional) *completion, comp(Δ)*, as the set of all classical formulas of the forms

$$p \leftrightarrow \bigvee \{A \mid A \Rightarrow p \in \Delta\}$$

$$\neg p \leftrightarrow \bigvee \{A \mid A \Rightarrow \neg p \in \Delta\},$$

for any propositional atom p, plus the set $\{\neg A \mid A \Rightarrow \mathbf{f} \in \Delta\}$.

Note that if Δ does not have rules with the head p, then $comp(\Delta)$ contains the formula $p \leftrightarrow \bigvee \emptyset$, which amounts to $\neg p$. Similarly, if there are no causal rules with $\neg p$ in heads, the completion will include p. This implies, in particular, that if neither an atom nor its negation appear in the head of at least one causal rule from Δ, the completion of Δ will be inconsistent (since neither the atom, nor its negation would have causes). The notion of a default completion (see the preceding section) would prevent this undesirable effect.

The following result shows that the classical models of $comp(\Delta)$ precisely correspond to exact worlds of Δ. The proof follows readily from corollary 4.34.

Proposition 4.38 *The causal nonmonotonic semantics of a definite causal theory coincides with the classical semantics of its completion.*

Remark. The completion construction, described above, can actually be generalized to arbitrary, nondeterminate causal theories using an algorithm from Lee (2004). The description of the corresponding generalized completion can also be found in Bochman (2005). Then, as before, the causal nonmonotonic semantics of an arbitrary finitary causal theory will coincide with the classical semantics of its generalized completion.

The causal basis of Reiter's simple solution.   As an illustration, we will describe a causal representation that will provide "causal foundations" for the well-known Reiter's solution to the frame problem in reasoning about action and change in AI (see Reiter 1991, 2001). Despite its simplicity, this representation will contain the main ingredients of causal reasoning in dynamic domains (see chapter 12).

Let us consider a small, local causal theory for a single propositional fluent F. The temporal behavior of F is described in this theory using two propositional atoms F_0 and F_1 saying, respectively, that F holds now and F holds at the next moment.

$$C^+ \Rightarrow F_1 \qquad C^- \Rightarrow \neg F_1$$
$$F_0 \wedge F_1 \Rightarrow F_1 \qquad \neg F_0 \wedge \neg F_1 \Rightarrow \neg F_1$$
$$F_0 \Rightarrow F_0 \qquad \neg F_0 \Rightarrow \neg F_0.$$

The first pair of causal rules describes the factors (actions or natural causes) that can cause F and, respectively, $\neg F$ (C^+ and C^- can be arbitrary formulas, but they normally describe the present situation). Such rules correspond to Reiter's *effect axioms for fluent* F, though in our description they are not relativized to particular actions. Second, instead of Reiter's explanatory closure axioms, we have a pair of *inertia axioms*. The latter are instances of explanatory rules stating that if F holds (does not hold) now, then it is self-explanatory that it will hold (respectively, not hold) in the next moment. The last pair of *initial axioms* states that the truth-value of F_0 is an exogenous parameter (or boundary condition).

The above causal theory is clearly definite, and its completion is as follows:

$$F_1 \leftrightarrow C^+ \vee (F_0 \wedge F_1)$$

$$\neg F_1 \leftrightarrow C^- \vee (\neg F_0 \wedge \neg F_1).$$

Now, it can be easily verified that the latter formulas are logically equivalent to the following two:

$$\neg(C^+ \wedge C^-)$$

$$F_1 \leftrightarrow C^+ \vee (F_0 \wedge \neg C^-).$$

The above formulas provide an abstract description of Reiter's simple solution: the first formula corresponds to his *consistency condition*, while the second one—to the *successor state axiom* for F.

4.5.3 On Nondeterminate Causal Theories and Scientific Laws

It has been shown earlier that determinate, and even literal, causal rules are sufficient for describing both the literal and causal nonmonotonic semantics of an arbitrary causal theory. Moreover, we will show in chapter 6 that even indeterminate causality that apparently requires causal rules with indeterminate effects can eventually be reduced to determinate causal rules. These facts confirm that the idea of determination, and determinism in general, are germane to our concept of causation and, moreover, that they can be safely placed at the basis of our causal formalism.[4]

It should be kept in mind, however, that nondeterminate causal theories are not *logically* reducible to determinate ones, and they actually have representation capabilities that significantly extend the capabilities of the latter. Though we will leave this important topic to future studies, we will sketch below a couple of promising directions of research for a further development of a logical theory of causality.

It has already been shown above that any causal theory can be transformed into a literal causal theory while preserving the associated literal nonmonotonic semantics. It should be kept in mind, however, that we do not have strong equivalence here. The following example illustrates this.[5]

Example 4.2 (Two Gears) *Two gears are powered by separate motors that can turn them in opposite (i.e., compatible) directions. In addition, when the gears are connected, each can also turn the other.*

---

4. Just as this has been done in Pearl (2000), where structural equations provide a deterministic description of a causal model, whereas the uncertainty is encoded in probabilistic distributions that are imposed on such models.

5. This is a simplified version of the example due to Marc Denecker that has been described in McCain and Turner (1997).

This domain can be given the following causal representation Δ:

$$Connected \Rightarrow Turning1 \leftrightarrow Turning2$$

$$Motor1 \Rightarrow Turning1 \qquad Motor2 \Rightarrow Turning2,$$

plus the assumptions that *Motor1*, *Motor2*, *Connected* and their negations are exogenous (self-explainable) propositions.

Due to the first rule, the causal theory Δ is not determinate, but it can be transformed to a literal theory Δ_L obtained from Δ by replacing the first rule with the following two rules:

$$Connected \wedge Motor1 \Rightarrow Turning2 \quad Connected \wedge Motor2 \Rightarrow Turning1.$$

The causal theories Δ and Δ_L have the same causal nonmonotonic semantics (see theorem 4.25), namely, the worlds where both gears are turning and either both motors are working or else one of them is working and the gears are connected. Note that this causal theory does not have causal models in which one of the gears is not turning; such worlds cannot be causally explained by the theory.

Let us add, however, a causal rule

$$\neg Turning2 \Rightarrow \neg Turning2$$

to each of these causal theories. Since the added rule is purely explanatory, it can only extend the set of exact worlds. And indeed, when added to Δ, the resulting theory admits a new exact world α in which the gears are connected, the two motors are not working, and both gears are not turning. This is because $\neg Turning2$ is now self-explanatory while $\neg Turning1$ is explained by the rule $Connected \wedge \neg Turning2 \Rightarrow \neg Turning1$ that is derivable from the first rule of Δ and $\neg Turning2 \Rightarrow \neg Turning2$. However, if the latter rule is added to Δ_L, the world α will not be an exact world of the resulting theory, since it cannot explain why the first gear is not turning. Accordingly, Δ and Δ_L are not strongly equivalent, and hence also not causally equivalent (see theorem 4.36).

The above example suggests a promising way of representing *reversible causation* where a physical setup can determine opposite directions of causal influence among two or more variables, depending on which of them is actually affected by external factors. Note that such setups produce, at best, only *cyclic* causal graphs in the structural account (see chapter 5). In general, however, they even cannot be represented by a single causal graph (Hausman, Stern, and Weinberger 2014; Dash and Druzdzel 2001).

Such systems are abundant in physical models described by usual mathematical equations that determine a functional relation among a number of physical factors or variables. It should be clear that such mathematical equations does not describe *directly* causal relations among the variables involved. Still, they are not completely noncausal since they can be used to determine such causal relations in every actual instantiation of the model. This use of "non-directional" mathematical equations could be captured in the causal calculus by representing them via rules of the form[6]

$$\mathbf{t} \Rightarrow A,$$

where A provides a formal description of the equation in an appropriate (presumably first-order) language,[7] and the whole procedure of using such rules in causal reasoning turns out to be much similar to the methods of generating a *causal order* in a system described by a set of mathematical equations (see, e.g., Simon 1953; Nayak 1994). We will illustrate the use of such laws in causal descriptions on the following example.

Example 4.3 (The ideal gas law) *The ideal gas law,*

$$PV = kT,$$

determines a functional relationship among the variables of pressure (P), volume (V), and temperature (T) of a fixed amount of gas. Being restricted to propositional language, the inferential role of this law can be described as follows: the law singles out triples of propositional atoms $\{P = p, V = v, T = t\}$ corresponding to assignments of values to these three variables that conform to the equation. Since the equation determines functional relations, the law asserts that any two propositions in such a triple imply the third. This information can be encoded using the following causal rules:

$$\mathbf{t} \Rightarrow ((P = p \land V = v) \rightarrow T = t)$$

$$\mathbf{t} \Rightarrow ((V = v \land T = t) \rightarrow P = p)$$

$$\mathbf{t} \Rightarrow ((P = p \land T = t) \rightarrow V = v).$$

Taken by themselves, the above causal rules still do not determine direct causal relations among our three parameters. This means, in particular, that they cannot determine by themselves a valid causal (exact) model.

Suppose, however, that we have an experimental setup that comprises a gas trapped in a chamber with a movable piston while the chamber itself can be heated or cooled from outside, so we can control both the volume and temperature of the gas. In our "toy" language, we can represent this setup by adding causal rules of the form

$$Push \Rightarrow V=v \qquad Heat \Rightarrow T=t.$$

Combined with the law (more precisely, the second causal rule above), these rules produce now a further causal rule

$$Push, Heat \Rightarrow P=p.$$

---

6. For a more liberal understanding of the underlying logic, this causal rule is a consequence of the axiom $\vdash A$ of the background supraclassical consequence relation.

7. See chapter 5 on how the causal calculus can be generalized to the classical first-order language.

Moreover, if we consider *Push* and *Heat* to be exogenous propositions, the resulting causal theory will already have an exact causal model $\{Push, Heat, V=v, T=t, P=p\}$.

Suppose now that we lock the piston in place. This means that we "cancel" the rule $Push \Rightarrow V=v$ and replace it with the rule

$$\mathbf{t} \Rightarrow V=v$$

that makes the volume a fixed parameter, a part of the setup. In this case, the ideal gas law reduces to (nondeterminate) rules of the form

$$\mathbf{t} \Rightarrow (P = p \leftrightarrow T = t)$$

for appropriate values of pressure and temperature. Combined with the "heating" rule, we can now derive

$$Heat \Rightarrow P=p.$$

As before, the world $\{Push, Heat, V=v, T=t, P=p\}$ will be an exact model of this causal theory (though *Push* no longer influences the volume).

4.6   On Causal Interpretation of Logic Programs

In this section we will show that the causal calculus provides a natural formal basis for logic programming. On the causal interpretation, described below, any general logic program can be seen as a causal theory satisfying the principle of negation as default (alias the closed world assumption). Moreover, given this principle, the correspondence between logic programs and causal theories will turn out to be bidirectional in the sense that, for an appropriate causal logic, any causal theory will be reducible to some logic program. In addition, the correspondence will turn out to be adequate for a broad range of logic programming semantics.[8]

Speaking generally, the causal interpretation of logic programs is based on a recurrent idea that logic program rules express causal, or explanatory, relationships among propositional atoms (see, e.g., Dix, Gottlob, and Marek 1994; Schlipf 1994). The declarative meaning of a logic program contains, however, an additional component: namely, an asymmetric treatment of positive and negative information, which is reflected in viewing **not** as *negation as failure*. It turns out that such an understanding can be uniformly captured in the causal calculus by accepting the Default Negation postulate below that gives a formal expression to the closed word assumption.

Definition 4.22 *A causal inference relation will be called* negatively closed, *if it satisfies*

(Default Negation) $\neg p \Rightarrow \neg p$, *for any propositional atom p.*

---

8. A more detailed description as well as proofs of all the results in this section can be found in Bochman (2005).

The Default Negation postulate stipulates, in effect, that negations of atomic propositions are defaults. As a result, the principle of sufficient reason is reduced in such systems to the necessity of explaining only positive facts. The postulate can be seen as giving a formal expression to Reiter's *closed world assumption* (Reiter 1978) and reflects the main distinctive feature of reasoning behind logic programs and databases.

For a world α, we will denote by $At(\alpha)$ the set of propositional atoms that belong to α. The next definition provides a semantics for negatively closed causal relations.

Definition 4.23 *A possible worlds semantics* $\mathbb{W} = (W, R, V)$ *will be called* inclusive *if, for any* $\alpha, \beta \in W$, $\alpha R \beta$ *holds only if* $At(\beta) \subseteq At(\alpha)$.

Theorem 4.39 *A basic production relation is negatively closed if and only if it has an inclusive possible worlds semantics.*

As we will see later, negatively closed causal relations constitute a causal logic corresponding to the stable semantics of logic programs.

It turns out that the supported (completion) semantics of logic programs presupposes a somewhat different, stronger causal logic that is described in the next definition.

Definition 4.24 *A causal inference relation will be called* semi-classical *if it satisfies*

(Semi-Classicality) $l \wedge m \Rightarrow l \vee m$, *for distinct literals l and m,*

and tight *if it is semi-classical and negatively closed.*

Semi-classical causal relations are "highly explanatory" since they automatically explain disjunctions of any two distinct literals that hold in a given world. As a result, only single literals have to be explained in such causal systems. As with the postulate of Default Negation, however, Semi-Classicality is not a structural postulate since it does not hold for arbitrary propositions.

The following definition gives a semantic description of semi-classical causal relations.

Definition 4.25 *A possible world semantics will be called* pointwise *if, for any worlds* $\alpha, \beta \in W$, $\alpha R \beta$ *holds only if the symmetric difference* $At(\beta) \triangle At(\alpha)$ *contains no more than one propositional atom; and it will be called* tight *if it is pointwise and inclusive.*

Theorem 4.40 *A causal relation is semi-classical [tight] if and only if it has a pointwise [tight] quasi-reflexive possible worlds semantics.*

Causal rules of semi-classical relations satisfy the following reduction property: if c is a consistent set of literals, then a rule $A \Rightarrow \bigvee c$ of a semi-classical causal relation is equivalent to the following set of determinate rules:

$$\{A, \neg(c \backslash \{l\}) \Rightarrow l \mid l \in c\}.$$

As a consequence, any causal theory is *logically* reducible to a determinate theory.

For tight causal relations, an even stronger reduction holds: if a, c are disjoint sets of propositional atoms, then a rule $A \Rightarrow \bigvee(\neg a \cup c)$ is equivalent to the following set of rules:

$$\{A, a, \neg(c \backslash \{p\}) \Rightarrow p \mid p \in c\}.$$

A determinate causal theory will be called *positive* if it contains only rules of the form $A \Rightarrow \tilde{p}$, where $\tilde{p}$ is either an atom or the falsity constant $\mathbf{f}$. Then we obtain:

Corollary 4.41 *For tight causal inference, any causal theory is equivalent to some positive determinate theory.*

In what follows, we will denote by *DN* the set of causal rules

$$\neg p \Rightarrow \neg p,$$

for all propositional atoms p of the underlying language.

The stable interpretation.   The *stable (causal) interpretation* of logic programs amounts to interpreting a program rule $\mathbf{not}\, d, c \leftarrow a, \mathbf{not}\, b$ as the following causal rule:

$$d, \neg b \Rightarrow \bigwedge a \rightarrow \bigvee c.$$

The interpretation assigns a purely classical understanding to **not**. Consequently, the nonclassicality of logic programs reduces under this representation solely to the nonclassicality of the causal connective $\Rightarrow$.

For a program Π, we will denote by $st(\Pi)$ the set of causal rules corresponding by this interpretation to the rules of Π while $ST(\Pi)$ will denote the union of $st(\Pi)$ and *DN*. Then the following correspondence can be established:

Theorem 4.42 *The causal nonmonotonic semantics of $ST(\Pi)$ coincides with the stable semantics of Π.*

The above result shows that the stable causal interpretation of program rules provides an adequate description of logic programs under the stable semantics. Moreover, negatively closed causal inference relations constitute a precise causal logic behind stable logic programming.

Recall now that any causal rule is reducible to a set of clausal causal rules $\bigwedge a \Rightarrow \bigvee b$, where a and b are sets of classical literals. Moreover, under the stable interpretation of logic programs, each such causal rule can be identified with some program rule. Accordingly, under this interpretation, any causal theory is reducible to a logic program, and vice versa. Note, however, that even simple normal program rules correspond to nondeterminate causal rules under this interpretation. Consequently, this interpretation gives us indirectly yet another perspective on a potential use of nondeterminate causal rules.

The supported interpretation.   Since stable and supported semantics of logic programs are based on a different understanding of program rules, the latter should have a different causal interpretation under the supported semantics. The corresponding causal interpretation of logic programs can be obtained by interpreting a program rule **not** $d, c \leftarrow a,$ **not** b as the following causal rule:

$$a, d, \neg b \Rightarrow \bigvee c.$$

We will call this causal interpretation a *supported interpretation* of program rules. The only difference with the stable interpretation, described earlier, amounts to treating positive premises in a as additional causal explanations rather than as part of what is explained.

Let us denote by $su(\Pi)$ the set of causal rules corresponding by this interpretation to the rules of a program Π. As we already mentioned, however, a full meaning of a logic program presupposes also the principle of Default Negation. Accordingly, let us denote by $SU(\Pi)$ a causal theory which is the union of $su(\Pi)$ and the set DN of default negation rules. Then we obtain

Theorem 4.43 *The causal nonmonotonic semantics of $SU(\Pi)$ coincides with the set of supported models of Π.*

The above result shows that the supported causal interpretation of program rules provides an adequate description of logic programs under the supported semantics.

Note now that the supported causal interpretation produces only head-positive causal rules. As has been shown earlier, the reasoning with such rules admits a stronger causal logic of tight causal relations. In fact, tight causal relations constitute a maximal causal logic for the supported semantics. Moreover, for such causal relations, any causal theory is reducible to a positive determinate theory, whereas under the supported interpretation any such rule corresponds to either a normal program rule or a constraint. Consequently every causal theory is reducible to a normal logic program under this interpretation.

4.7 Negative Causation and Negative Completion

From nothing, from a mere negation, no consequences can proceed.
—Mill (1872, bk. III, chap. 5, sec. 3)

The same thing is the cause of contrary results. For that which by its presence brings about one result is sometimes blamed for bringing about the contrary by its absence. Thus we ascribe the wreck of a ship to the absence of the pilot whose presence was the cause of its safety.
—Aristotle, *Physics*, II.3, 333

The above two quotations express, respectively, two apparently conflicting views on the role of negation in causal reasoning. Mill's thesis can be seen as a "logical" formulation of the principle *ex nihilo nihil fit* (nothing comes from nothing), which has been viewed by many as just a negative formulation of the principle of sufficient reason. Incidentally, the first known formulation of the principle *ex nihilo nihil fit* can also be found in Aristotle's *Physics* (where it was attributed to Parmenides).

The problem of representing negation has emerged as one of the main problems of non-monotonic reasoning in general, and it has immediate implications for causal reasoning. Accordingly, a proper treatment of negation in causal contexts should be viewed as an important part of an adequate analysis of causality in general.

So far, in this chapter we have largely ignored the distinction between positive and negative literals and have treated both indiscriminately (except for the preceding representation of logic programs). This uniformity can be viewed as a significant theoretical advantage, and the adherents of both regularity and counterfactual approaches to causality have justly celebrated it.

A large group of philosophers has rejected, however, this generalization of causal relation to absences and negation, or at least its uniformity, though for varying reasons. Beside some general metaphysical and conceptual objections (that are largely outside the scope of this book), the corresponding studies have pointed out some important differences between these two kinds of causal assertions, as well as specific difficulties that arise in interpreting and justifying negative causal claims. Thus, Hall (2004b) has even suggested distinguishing two kinds of causation, *production* and *dependence*, that is roughly correlated with the distinction between positive and negative causation. In a similar vein (though on different metaphysical grounds), Moore (2009) has argued that if we do not consider omissions and preventions as "true" causes, we would be relieved, for instance, from causal perplexities about the omission overdetermination cases where we lack intuitions about whether these are cases of symmetric overdetermination or preemption (see chapter 8). According to Moore, we lack such intuitions precisely because these are not causal issues at all, and the desert-determiner for such cases is just counterfactual dependence.

The shift from the standard possible-world semantics for classical logic to the local, situation-based semantics that we have suggested in chapter 2 of this book creates necessary logical space for an alternative understanding of negation, including the concept of default negation. As we already mentioned earlier, the latter is based on the idea that a negative proposition can be accepted in a situation when we do not have *reasons* for accepting the corresponding positive proposition. In this respect, the nonmonotonic semantics of exact models, described earlier in this chapter, complements this idea in that it embodies a particular, *causal* closed world assumption, according to which the current causal theory provides an *exhaustive* description of all the causal factors that could be used as a reason for acceptance of propositions in a model. Accordingly, the task of describing negative causation amounts to the problem how to combine this causal closed world assumption with the principle of default negation.

The concept of negation as default (which is formalized in logic programming) covers a significant portion of our understanding and use of negation in local situations. Still, it does not fully reflect the behavior of negation in the context of causal reasoning. The difference can be roughly described as follows. When negation is viewed as default, negative propositions are exempted from the need of causal explanation; in other words, they do not need causes for their acceptance. The only thing we should care about is consistency of such negative propositions with other, positive assertions and known causal rules. Explanatory rules $\neg p \Rightarrow \neg p$ provide precisely this functionality. Commonsense causal reasoning, however, tends to preserve the symmetry between positive and negative assertions and treat also the latter as something that can be caused and be causes themselves. This symmetric vision can even be found in some legal positions. Thus, US Supreme Court Justice Antonin Scalia has urged that there is no conceptual distinction (and a fortiori no moral distinction) between "positive" harms caused and "negative" benefits prevented (Moore 2009). This understanding is also clearly visible in our initial quote from Aristotle's *Physics*, which confirms also that this is not a pure theoretical invention of modern approaches to causality. In fact, the fully symmetric treatment of positive and negative assertions is implicit in Pearl's approach to causality (see chapter 5). The causal calculus also does not impose any a priori distinctions between the use of positive and negative propositions in causal rules, though, as we have seen, it can express negation as default.

Now, the commonsense reality displays a certain "mix" of these two views of negation. On the one hand, negative causal claims are commonly used in our reasoning (a lot of examples of their use will be discussed in chapters 8 and 9). On the other hand, the assertions of positive and negative causation *are* different, and this creates an obvious problem for any attempt of a uniform description that purports to cover both kinds. As Lewis (2000) has put it, this is everyone's problem.

From now on, we will restrict our attention to determinate causal theories. For this case, a more sophisticated mix of the two views of negation suggests itself. We will call it the principle of negative causation.

Definition 4.26 (Negative Causation Principle) *A rule $B \Rightarrow \neg p$ is* acceptable *with respect to a determinate causal theory* Δ *if any causal rule of the form $A \Rightarrow p$ that belongs to Δ is such that A is (classically) incompatible with B.*

According to this principle, B causes $\neg p$ when it undermines all potential causes of p in Δ. This principle should be applied, of course, only to endogenous propositions that are viewed as fully determined by their causes (see also chapter 5).

In some sense, the above principle could also be viewed as a "positive" reformulation of the ancient principle *ex nihilo nihil fit*, namely,

> *Negation (absence) of effects follows from negation (absence) of causes.*

Note that, like the concept of default negation itself, this principle is also nonmonotonic: an acceptable negative causal claim can become unacceptable with an addition of new positive causal rules to the causal theory.

Remark. Once more we can find Aristotle's footsteps around the above principle. To begin with, he considered affirmative propositions as prior to and better known than the corresponding negative propositions (since affirmation explains denial just as being is prior to not-being—see *Metaphysics*, 996b14–16). Accordingly, he also has considered negative demonstration as inferior in general to positive demonstration. Moreover, in *Posterior Analytics* Aristotle provides the following reasons for rejecting an alleged demonstration that walls do not breathe because they are not animals:

> If this were the cause of not breathing, being an animal would have to be the cause of breathing, i.e., if the negation is the cause of something's not being attributed, then the affirmation is the cause of its being attributed . . . , and, similarly, if the affirmation is the cause of something's being attributed, then the negation is the cause of its not being attributed. (*APo* I.13, 78b13–28)

Some earlier commentators of Aristotle have even suspected him in committing a logical fallacy here (see Angioni 2018). And indeed, this principle is obviously invalid for classical deductive inference. But it is a sound and useful principle of causal reasoning.

As we will see in due course, the above principle of negative causation is actually directly encoded in Pearl's structural approach to causality (when applied to two-valued Boolean endogenous variables). It is also compatible, however, with philosophical approaches to causality according to which positive causation (or causation between real events) is the only "genuine" causation, whereas negative causation is, at best, a derivative notion (see, e.g., Armstrong 1997; Dowe 2000).

A somewhat simpler description of the principle of negative causation can be obtained for definite causal theories. The proofs are immediate.

Lemma 4.44 *Let A_p denote the disjunction of all bodies of the rules from a causal theory Δ that have p as its head, that is*

$$A_p = \bigvee \{C \mid C \Rightarrow p \in \Delta\}.$$

Then $B \Rightarrow \neg p$ is an acceptable rule if and only if B is incompatible with A_p.

Corollary 4.45 *The set of acceptable rules $B \Rightarrow \neg p$ for a particular propositional atom p coincides with the set of causal rules that are derivable from the rule $\neg A_p \Rightarrow \neg p$ by general production inference.*

In view of the above results, all acceptable negative causal rules are subsumed by rules of the form $\neg A_p \Rightarrow \neg p$ for each atom p. This sanctions the following notion of *negative completion*:

Definition 4.27 *A negative causal completion of a definite causal theory Δ is a causal theory $Nc(\Delta)$ obtained from Δ by adding rules of the form*

$$\neg A_p \Rightarrow \neg p,$$

for all atoms p that appear in the heads of causal rules from Δ.

The above notion of negative completion will be extensively used in what follows. It will allow us, in effect, to write only a positive part of a causal theory and refer to negative completion as a way of obtaining a complete theory.

In chapters 8 and 9 we will restrict our language to literal causal rules. This restriction will require a slight modification of the above definition of negative completion, though the main idea will remain the same.

In the following definition, for a set L of literals, we will use $\neg L$ to denote the set $\{\neg l \mid l \in L\}$.

Definition 4.28 *A literal negative completion of a literal causal theory* Δ *is a literal causal theory* $Lc(\Delta)$ *obtained from* Δ *by adding all rules of the form*

$$\neg L \Rightarrow \neg p,$$

where L is a set of literals forming a prime implicate[9] *of* A_p.

On a mathematical formulation, the relevant sets L in the above definition are *minimal hitting sets* (see Reiter 1987a) of the set of sets $\{L_i \mid L_i \Rightarrow p \in \Delta\}$.

It can be verified that, for any atom p, the above modification of the notion of negative completion agrees with our earlier definition in the sense that it produces a set of literal rules that is equivalent to $\neg A_p \Rightarrow \neg p$ with respect to basic inference.

The above notion of a negative completion will be used in examples of chapters 8 and 9 for completing positive literal causal theories that do not contain negative literals in the heads of their rules.

4.8 Defeasible Causality and Deep Representation

As we have argued in chapter 1, causal rules of the causal calculus should be viewed as *default assumptions*, claims that we accept unless there are reasons for their rejection. An important part of this understanding is that causal rules are *defeasible*, that is, they can be canceled by other causal rules or facts.

This understanding still has not received a syntactic expression in our formal representation. Moreover, our formal description of various causal situations in chapters 8 and 9 will show that, in many cases of interest, we even will not need it. The reason is that, in most such cases, causal rules will either remain unchallenged, or they will be defeated just by refuting some of their antecedent conditions. It will be required, however, for describing situations where there is a complex interaction between accepted causal rules and other existing factors that can defeat them *without refuting their antecedents*. For an

---

9. The prime implicates of a classical formula are its logically strongest clausal consequences—see, for example, Darwiche and Marquis (2002).

adequate description of such cases, we will need a representation that will make causal rules explicitly defeasible.

Guided by solutions suggested for the problem of defeasible entailment in nonmonotonic reasoning (see section 3.3), we will represent defeasible causal rules as rules of the form

$$C, n \Rightarrow E,$$

where n denotes the underlying causal mechanism or process that, given an "input" C, produces an "output" E. This more detailed representation will be called a *deep representation* of causal rules, whereas a representation that does not explicitly mention the underlying mechanisms will be called a *surface representation*. This terminology will be justified by the fact that, in most cases of interest, the names of mechanisms can be eliminated (by "forgetting"—see section 8.3.1 in chapter 8), and thereby a deep representation can be transformed into a surface representation.

Remark. To prevent possible misunderstandings, it should be stressed already at this stage that we are considering surface and deep representations simply as different *levels of approximation* we use in describing (causal) situations, much in the same sense as the classical Newton's mechanics and Einstein's special relativity theory can be used for representing the same physical situations. Thus, questions like "What is the right, or ultimate, representation?" are wrong questions to ask in this respect; all that matters is whether the representation chosen is sufficiently adequate for the relevant situation and the questions asked. In particular, the defeasible nature of causal rules should be made explicit only in representing situations where such rules (more precisely, their underlying mechanisms) could be undermined by other rules and mechanisms *present* in these local situations.

Propositional atoms $m, n \ldots$ will be reserved in what follows for "naming" of mechanisms. All such atoms will be tacitly assumed to be *default assumptions*, which will mean that they can be accepted whenever they are not explicitly canceled (see below). Note, however, that (in contrast to exogenous propositions) $\neg n$ already will not be viewed as a default proposition, so it will require a cause for its acceptance. Still, it is interesting to note that if we would assume both n and $\neg n$ to be default propositions, the defesible causal rule $C, n \Rightarrow E$ would become "functionally equivalent" to a purely explanatory rule $C, E \Rightarrow E$ or, in other words, to a weak causal assertion (see section 4.4.2).

In some sense, this use of names for mechanisms in the antecedents of causal rules will play the same role as Mill's omnibus negative conditions (see section 1.2.1). Due to their status as (defeasible) assumptions, however, our representation will be immune to the objections that has been raised against Mill's proposal. Moreover, it will allow us to "hide" not only negative preventing conditions but also usual, or normal, conditions that are necessary for the obtaining of a causal relation, such as the presence of oxygen for a fire (see chapter 9). This will make our causal rules much similar to the corresponding commonsense causal assertions.

Remark. The idea of omitting negative conditions in the description of causes can also be found in Mill's *System of Logic* where he contemplated a "convenient modification of the meaning of the word cause, which confines it to the assemblage of positive conditions." Similar to Aristotle, Mill justified such a modification by the fact that causes invariably prevent the effects of other causes by virtue of the same laws according to which they produce their own, which enables us to dispense with the consideration of negative conditions entirely. According to Mill, on this modification, instead of unconditional regularities, the definition of causes must refer only to regularities that are "subject to no other than negative conditions."

Defeasible causal rules $C, n \Rightarrow E$ can be actually defeated by causal rules that undermine either the cause C or the mechanism n. In the framework of the causal calculus, these are the rules that cause either $\neg C$ or $\neg n$, respectively. This will naturally lead us to the notion of prevention that we will discuss in chapter 8.

Finally, to secure a proper logical behavior for such default propositions, including a possibility of their elimination, we will generalize the principle of negative causation (see section 4.7) to all default assumptions. This time, however, the corresponding default completion will add only *positive* causal rules of the form $B \Rightarrow n$.

Definition 4.29 (Default Causation Principle) *If n is a default assumption, then a rule $B \Rightarrow n$ is* acceptable *with respect to a determinate causal theory Δ if and only if any causal rule $A \Rightarrow \neg n$ that belongs to Δ is such that A is (classically) incompatible with B.*

Thus, a causal rule $B \Rightarrow n$ is acceptable if B refutes the antecedents of all cancellation rules for n. As a technical stipulation, we will assume also that if a causal theory Δ does not contain cancellation rules for n, the latter will be viewed as a pure default $n \Rightarrow n$.

As before, let A_n denote disjunction of all bodies of the rules from Δ that cancel n, that is,

$$A_n = \bigvee \{C \mid C \Rightarrow \neg n \in \Delta\}.$$

Then $B \Rightarrow n$ is an acceptable rule if and only if B is incompatible with A_n.

Definition 4.30 *A default causal completion of a definite causal theory Δ is a causal theory obtained from Δ by adding rules of the form*

$$\neg A_n \Rightarrow n$$

for all default assumptions n.

This default causal completion will also be used in describing causal situations in chapters 8 and 9. It will allow us, in particular, to eliminate the names of mechanisms in deep causal representations and thereby transform them into corresponding surface representations.

4.9 Causal Reasoning as Argumentation

In this last section we will situate the causal calculus in the landscape of the nonmonotonic formalisms described in chapter 3. More precisely, we will show that the causal calculus can be identified with a particular kind of argumentation. Proofs of all the results in this section can be found in Bochman (2005).

Recall the notion of a *propositional attack relation* that has been defined in chapter 3 as an attack relation on classical propositions that respects classical entailment (see definition 3.7). This argumentation system has been shown to encompass default logic. Now, the causal calculus turns out to correspond to a special case of propositional argumentation that satisfies further natural properties:

(Left Or)   If $A \hookrightarrow C$ and $B \hookrightarrow C$, then $A \vee B \hookrightarrow C$;
(Right Or)   If $A \hookrightarrow B$ and $A \hookrightarrow C$, then $A \hookrightarrow B \vee C$;
(Self-Defeat)   If $A \hookrightarrow A$, then $\mathbf{t} \hookrightarrow A$.

Definition 4.31 *A propositional attack relation will be called* probative *if it satisfies Left Or,* basic, *if it also satisfies Right Or, and* causal, *if it is basic and satisfies Self-Defeat.*

Probative argumentation allows for argumentation by cases. Its semantic interpretation can be obtained by restricting classical bimodels to pairs (v, α), where α is a world. The corresponding binary semantics will also be called *probative*. Similarly, the semantics for basic argumentation is obtained by restricting bimodels to world pairs (α, β); such a binary semantics will be called *basic*. It can already be generalized to possible worlds models of Section 4.4.1. Finally, the *causal* binary semantics is obtained from the basic semantics by requiring further that (α, β) is a bimodel only if (α, α) is also a bimodel.

The following result shows that these three systems of argumentation correspond precisely to their associated binary semantics.

Proposition 4.46 *A propositional attack relation is probative [basic, causal] iff it is determined by a probative [resp. basic, causal] binary semantics.*

Basic propositional argumentation can already be given a purely four-valued semantic interpretation, in which the classical negation $\neg$ has the following semantic description:

$\neg A$ is accepted iff A is not accepted

$\neg A$ is rejected iff A is not rejected.

A syntactic characterization of this connective in collective argumentation can be obtained by imposing the rules

$$A, \neg A \hookrightarrow \qquad \hookrightarrow A, \neg A$$

$$\text{If } a, A \hookrightarrow b \text{ and } a, \neg A \hookrightarrow b \text{ then } a \hookrightarrow b \qquad (A_\neg)$$

$$\text{If } a \hookrightarrow b, A \text{ and } a \hookrightarrow b, \neg A \text{ then } a \hookrightarrow b.$$

Then a basic propositional attack relation can be alternatively described as a collective attack relation satisfying the rules $(A_\wedge)$ and $(A_\neg)$ from chapter 3. Moreover, the global negation $\sim$ can be added to this system just by adding the corresponding postulates (AN). It turns out, however, that the global negation is *eliminable* in this setting via to the following reductions:

$$a, \sim A \hookrightarrow b \equiv a \hookrightarrow b, \neg A \qquad a \hookrightarrow \sim A, b \equiv a, \neg A \hookrightarrow b$$

$$a, \neg \sim A \hookrightarrow b \equiv a \hookrightarrow b, A \qquad a \hookrightarrow \neg \sim A, b \equiv a, A \hookrightarrow b. \qquad (R_\sim)$$

As a result, the basic attack relation can be safely restricted to an attack relation in a classical language.

Finally, the rule Self-Defeat of causal argumentation gives a formal representation for an often expressed desideratum that self-conflicting arguments should not participate in defeating other arguments (see, e.g., Bondarenko et al. 1997). This aim is achieved in our setting by requiring that such arguments are attacked even by tautologies and hence by any argument whatsoever.

Probative attack relations turn out to be equivalent to general production inference relations of the causal calculus. The correspondence can be established directly on the syntactic level using the following definitions:

$$A \Rightarrow B \equiv \neg B \hookrightarrow A \qquad \text{(PA)}$$

$$A \hookrightarrow B \equiv B \Rightarrow \neg A. \qquad \text{(AP)}$$

Under these correspondences, the rules of a probative attack relation correspond precisely to the postulates for production relations. Moreover, the correspondence extends also to a correspondence between basic and causal argumentation, on the one hand, and basic and causal production inference, on the other. Hence the following result is straightforward.

Lemma 4.47 *If $\hookrightarrow$ is a probative [basic, causal] attack relation, then (PA) determines a [basic, causal] production inference relation, and vice versa, if $\Rightarrow$ is a [basic, causal] production inference relation, then (AP) determines a probative [basic, causal] attack relation.*

Remark. A seemingly more natural correspondence between propositional argumentation and production inference can be obtained using the following definitions:

$$A \Rightarrow B \equiv A \hookrightarrow \neg B \qquad A \hookrightarrow B \equiv A \Rightarrow \neg B.$$

By these definitions, *A* explains *B* if it attacks ¬*B*, and vice versa: *A* attacks *B* if it explains ¬*B*. Unfortunately, this correspondence, though plausible by itself, does not take into account the intended understanding of *arguments as default assumptions*. As a result, it cannot be extended directly to the correspondence between the associated nonmonotonic semantics, described below.

Turning to semantic descriptions, it can be immediately seen that the logical semantics of basic and causal attack relations coincide, respectively, with the binary semantics of basic and causal inference relations. Moreover, this semantic correspondence extends actually to the correspondence between the causal nonmonotonic semantics of the causal calculus and stable semantics of argumentation. The precise correspondence between exact worlds and stable sets of assumptions is established in the next theorem.

Theorem 4.48 *If Δ is a causal theory, and Δ_a its corresponding argument theory given by (AP), then a world α is an exact world of Δ iff $\tilde{\alpha}$ is a stable set of assumptions in Δ_a.*

The above result shows, in effect, that propositional argumentation subsumes the causal calculus as a special case. Moreover, it can be shown that causal attack relations constitute a strongest argumentation system suitable for the causal nonmonotonic semantics.

5 Structural Equation Models

In this chapter we are going to provide a logical representation of structural causal models from Pearl (2000) in the causal calculus.

A common assumption behind both the causal calculus and the structural approach to causation is that reasoning in the relevant domains cannot be expressed in the plain language of (classical) logic but requires the explicit use of causal concepts and a general language of causation. In both cases, the corresponding causal formalisms have provided working concepts of reasoning that have turned out to be essential for an adequate representation of the respective areas and for broad correspondence with commonsense descriptions. Nevertheless, these studies so far have used apparently different formalisms and pursued quite different objectives.

In this chapter we are going to show that, despite the differences, these two causal formalisms are based on essentially the same understanding of causation. We will do this by demonstrating that the central notion of Pearl's theory, the notion of a structural causal model (which is based, in turn, on the notion of a structural equation), can be naturally represented in the causal calculus, and especially in the first-order generalization of the latter, introduced in Lifschitz (1997). Moreover, this representation creates a powerful generalization of Pearl's formalism that relaxes many of its more specific restrictions (such as acyclicity and uniqueness of solutions or existence of a single mechanism behind every endogenous variable). As we will show in part III in discussing actual causality, this additional generality will turn out to be crucial for an adequate formal representation of causal assertions. In addition, our logical representation will allow us to clarify some issues and confusions related to the use of causal concepts in Pearl's theory, such as the distinction between plain mathematical and causal understanding of structural equations. In sum, we believe that our representation provides all the necessary conditions for returning Pearl's theory of causal reasoning to logic (broadly understood), with all the expected benefits such a logical representation could provide for its analysis, generalization, and further development.

5.1 Structural Equations and Causal Models

According to Pearl (2000, chapter 7), a causal model is a triple $M = \langle U, V, F \rangle$, where

- U is a set of *background* (or *exogenous*) variables,
- V is a set $\{V_1, V_2, \ldots, V_n\}$ of *endogenous* variables that are determined by other variables in $U \cup V$, and
- F is a set of functions $\{f_1, f_2, \ldots, f_n\}$ such that each f_i is a mapping from $U \cup (V \backslash V_i)$ to V_i, and the entire set, F, forms a mapping from U to V.

Symbolically, F can be represented as a set of *structural* equations

$$V_i = f_i(PA_i, U_i) \quad i = 1, \ldots, n,$$

where PA_i is the unique minimal set of variables in $V \backslash \{V_i\}$ (parents) sufficient for representing f_i, and similarly for $U_i \subseteq U$. Each such equation stands for a set of "structural" equalities

$$v_i = f_i(pa_i, u_i) \quad i = 1, \ldots, n,$$

where v_i, pa_i and u_i are, respectively, particular realizations of V_i, PA_i and U_i. Such an equality assigns a specific value v_i to a variable V_i depending on the values of its parents and relevant exogenous variables.

In Pearl's account, every instantiation $U = u$ of the exogenous variables determines a particular "causal world" of the causal model. Such worlds stand in one-to-one correspondence with the solutions to the above equations in the ordinary mathematical sense. However, structural equations also encode causal information in their very syntax by treating the variable on the left-hand side of the = as the effect and treating those on the right as causes. Accordingly, the equality signs in structural equations convey the asymmetrical relation of "is determined by." This causal reading does not affect the set of solutions of a causal model, but it plays a crucial role in determining the effect of external interventions and evaluation of counterfactual assertions with respect to such a model (see Pearl 2012). This ability of structural equation models (SEMs) to predict the effect of manipulation has been known in econometrics since the inception of the concept of SEMs. The seminal paper of Simon (1953) has explicated this property and gave it a causal interpretation.

Each structural equation in a causal model is intended to represent a stable and autonomous physical mechanism, which means that it is conceivable to modify (or cancel) one such equation without changing the others. In Pearl's theory, this modularity plays an instrumental role in determining the answers to three types of queries that can be asked with respect to a causal model:

- *predictions* (e.g., would the pavement be slippery if we find the sprinkler off?),
- *interventions* (e.g., would the pavement be slippery if we make sure that the sprinkler is off?), and

- *counterfactuals* (e.g., would the pavement be slippery had the sprinkler been off, given that the pavement is in fact not slippery and the sprinkler is on?).

The answers to prediction queries can be obtained using plain deductive inferences from a logical description of the causal worlds. However, in order to obtain answers to the intervention (action) and counterfactual queries, we have to consider what was termed by Pearl submodels of a given causal model. Given a particular instantiation x of a set of variables X from V, a *submodel* M_x of M is described as the causal model that is obtained from M by replacing its set of functions F by the following set:

$$F_x = \{f_i \mid V_i \notin X\} \cup \{X = x\}.$$

In other words, F_x is formed by deleting from F all functions f_i corresponding to members of the set X and replacing them with the set of constant functions $X = x$. A submodel M_x can be viewed as a result of performing an action $do(X = x)$ on M that produces a minimal change required to make $X = x$ hold true under any u. This submodel is used in Pearl's theory for evaluating counterfactuals of the form, "Had X been x, whether $Y = y$ would hold?"

The above brief description does not exhaust, of course, the content of the entire Pearl's theory of causal reasoning. Nevertheless, it will be sufficient for providing a logical reformulation of its key concepts.

We will now recast Pearl's ideas in a logical form convenient for our analysis, starting with the propositional case.

Propositional case. As before, propositional formulas are formed from propositional atoms and the logical constants **f, t** using the classical connectives $\wedge$, $\vee$, $\neg$, and $\rightarrow$.

Definition 5.1 *Assume that the set of propositional atoms is partitioned into a set of* background *(or* exogenous*) atoms and a finite set of* explainable *(or* endogenous*) atoms.*

- *A Boolean structural equation is an expression of the form $p = F$, where p is an endogenous atom, and F is a propositional formula in which p does not appear.*
- *A Boolean causal model is a set of Boolean structural equations $p = F$, one for each endogenous atom p.*

Definition 5.2 *A* solution *(or a* causal world*) of a Boolean causal model M is any propositional interpretation satisfying the equivalences $p \leftrightarrow F$ for all equations $p = F$ in M.*

Example 5.1 (Firing squad) *(Pearl 2000, chapter 7) Let U, C, A, B, D stand, respectively, for the following propositions: "Court orders the execution," "Captain gives a signal," "Rifleman A shoots," "Rifleman B shoots," and "Prisoner dies." The story is formalized* *using the following causal model M, in which U is the only exogenous atom:*

$$\{C = U, \ A = C, \ B = C, \ D = A \vee B\}.$$

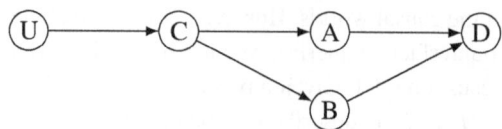

It has two solutions: in one of them all atoms are true, in the other all atoms are false. This causal model allows us to answer "factual" queries concerning the domain. For instance, M implies $\neg A \rightarrow \neg D$ in the sense that this implication is satisfied by every causal world of M. That gives us a *prediction*:

S1. If rifleman A did not shoot, the prisoner is alive.

It implies also the implication $\neg D \rightarrow \neg C$, which amounts to *abduction*:

S2. If the prisoner is alive, the captain did not signal.

And it implies $A \rightarrow B$, which amounts to *transduction*:

S3. If rifleman A shot, then B shot as well.

Definition 5.3 *Given a Boolean causal model M, a subset X of the set of endogenous atoms, and a truth-valued function I on X, the* submodel M_X^I *of M is the causal model obtained from M by replacing every equation $p = F$, where $p \in X$, with $p = I(p)$.*

Example 5.1 (Firing squad, continued) Consider the action sentence:

S4. If the captain gave no signal and rifleman A decides to shoot, the prisoner will die and B will not shoot.

To evaluate it, we have to consider the submodel $M_{\{A,C\}}^I$ with $I(A) = \mathbf{t}$ and $I(C) = \mathbf{f}$:

$$\{C = \mathbf{f}, \ A = \mathbf{t}, \ B = C, \ D = A \vee B\}.$$

This submodel[1] has a single causal world, and in this world both D and $\neg B$ hold, so S4 is justified.

First-order case. *Terms* are formed from object constants and function symbols as usual in first-order logic.

Definition 5.4 *Assume that the set of object constants is partitioned into a set of* rigid *symbols, a set of of* background *(or exogenous) symbols, and a finite set of* explainable *(or endogenous) symbols.*

---

1. We have omitted U from the submodel since it already does not influence other variables.

- *A structural equation is an expression of the form c = t, where c is an endogenous symbol and t is a ground term in which c does not appear.*
- *A causal model consists of an interpretation of the set of rigid symbols and function symbols (in the sense of first-order logic) and a set of structural equations c = t, one for each endogenous symbol c.*

Definition 5.5 *A solution (or a* causal world*) of a causal model M is an extension of the interpretation of rigid symbols and function symbols in M to the exogenous and endogenous symbols that satisfies all equalities c = t in M.*

Definition 5.6 *Given a causal model M, a subset X of the set of endogenous symbols, and a function I from X to the set of rigid constants, the submodel M_X^I of M is the causal model obtained from M by replacing every equation c = t, where c $\in$ X, with c = I(c).*

Example 5.2 (Special gas laws) *Let us consider a closed gas container with variable volume that can be heated. P, V, T will denote, respectively, the pressure, volume, and temperature of the gas. In this particular setup, it is natural to treat P and V as endogenous while T as an exogenous symbol.*

The corresponding causal model will involve two structural equations that are actually two directional instances of the ideal gas law:

$$P = k \cdot \frac{T}{V} \qquad V = k \cdot \frac{T}{P} ,$$

where k is a (rigid) constant. The language may include also names of real numbers; we classify them as rigid. A causal model is constructed by combining the above equations with the interpretation that has the set of positive real numbers as its universe, and interprets the function symbols as corresponding mathematical operations (i.e., multiplication and division).

As can be seen, the above causal model is cyclic with respect to its endogenous parameters P and V. However, if we fix the volume V, we obtain a submodel

$$P = k \cdot \frac{T}{V} \qquad V = v.$$

This submodel provides a description of *Gay-Lussac's law*, according to which pressure is proportional to temperature. Note, however, that this submodel is still "structural" in that the temperature of the gas is not *determined* by its pressure.

Similarly, by fixing pressure, we obtain another submodel,

$$P = p \qquad V = k \cdot \frac{T}{P} ,$$

that represents *Charles's law* by which the volume is proportional to temperature (though not the other way round).

Remark. The above example provides a somewhat "toy" description of the ideal gas law (or, better, of the special gas laws that have preceded it); it should be viewed only as an illustration of our formal definitions. As we have mentioned in chapter 4, a proper representation of scientific laws in the causal calculus requires the use of more complex, nondeterminate causal rules (see section 4.5.3).

5.2 Representing Structural Equations by Causal Rules

We will describe now a formal representation of Pearl's causal models as causal theories.[2] The representation itself is fully modular, and the nonmonotonic semantics of the resulting causal theory will correspond to solutions (causal worlds) of the causal model.

5.2.1 Propositional Case

For the propositional language, the corresponding translation can be defined as follows:

Definition 5.7 *For any Boolean causal model M, Δ_M is the propositional causal theory consisting of the rules,*

$$F \Rightarrow p \text{ and } \neg F \Rightarrow \neg p,$$

for all equations $p = F$ in M, and the rules,

$$p \Rightarrow p \text{ and } \neg p \Rightarrow \neg p,$$

for all exogenous atoms p of M.

Theorem 5.1 *The causal worlds of a Boolean causal model M are identical to the exact models of Δ_M.*

Proof. It can be verified that the propositional completion of Δ_M, $comp(\Delta_M)$, amounts to the set of all formulas of the form $p \leftrightarrow F$. Hence, the result follows from proposition 4.38 in the preceding chapter. $\square$

The above result could also be obtained as a special case of the general first-order correspondence result, described below.

Due to the above equivalence, the terms "exact worlds" and "causal worlds" will be used indiscriminately in this book.

Remark. The above representation has been chosen from a number of alternative (logically non-equivalent) translations producing the same nonmonotonic semantics. The choice reflected the symmetric character of (Boolean) structural equations, according to which both truth and falsity assignments to an endogenous atom should be determined by the

---

2. The connection between structural equation models and the causal theory by McCain and Turner has been pointed out already in Geffner (1997).

corresponding function. It has turned out to be adequate also for establishing a logical correspondence between the two formalisms.

Example 5.1 (Firing squad, continued) If M is the causal model from the firing squad example, then Δ_M consists of the causal rules

$$U \Rightarrow C, \ \neg U \Rightarrow \neg C, \ C \Rightarrow A, \ \neg C \Rightarrow \neg A,$$
$$C \Rightarrow B, \ \neg C \Rightarrow \neg B, \ A \vee B \Rightarrow D, \ \neg(A \vee B) \Rightarrow \neg D,$$
$$U \Rightarrow U, \ \neg U \Rightarrow \neg U.$$

This causal theory has two exact models, identical to the solutions (causal worlds) of M.

Definition 5.8 *Given a determinate causal theory Δ, a set X of atoms, and a truth-valued function I on X, the subtheory Δ_X^I of Δ is the determinate causal theory obtained from Δ by (i) removing all rules $A \Rightarrow B$ and $A \Rightarrow \neg B$ with $B \in X$, (ii) adding the rule $\mathbf{t} \Rightarrow B$ for each $B \in X$ such that $I(B) = \mathbf{t}$, and (iii) adding the rule $\mathbf{t} \Rightarrow \neg B$ for each $B \in X$ such that $I(B) = \mathbf{f}$.*

Subtheories of propositional causal theories correspond to submodels of Boolean causal models: the causal theory $\Delta_{M_X^I}$ is essentially identical to the subtheory $(\Delta_M)_X^I$ of Δ_M. The only difference is that the former contains additional trivial rules with the body $\mathbf{f}$.

Example 5.1 (Firing squad, continued) The submodel $M_{\{A,C\}}^I$ with $I(A) = \mathbf{t}$ and $I(C) = \mathbf{f}$ that was used for evaluating the action sentence S4 corresponds to the subtheory $\Delta_{\{A,C\}}^I$:

$$\mathbf{t} \Rightarrow \neg C, \ \mathbf{t} \Rightarrow A,$$
$$C \Rightarrow B, \ \neg C \Rightarrow \neg B, \ A \vee B \Rightarrow D, \ \neg(A \vee B) \Rightarrow \neg D.$$

As before, this causal theory has a single exact model $\{\neg C, A, \neg B, D\}$, where both D and $\neg B$ hold, so S4 is justified.

Pearl causal theories. Our representation of Pearl's causal models has produced quite specific causal theories. More precisely, in the propositional case it presupposes, in effect, that for any explainable atom p there exists a propositional formula F in which p does not appear such that $F \Rightarrow p$ and $\neg F \Rightarrow \neg p$ are the only causal rules of the corresponding causal theory that involve p in heads. This means, in particular, that such causal theories satisfy the principle of negative causation (see section 4.7) in the sense that any such theory will already be closed with respect to negative causal completion, that is, $\Delta = Nc(\Delta)$.

A slight generalization of these restrictions will lead us to the following special kind of causal theories:

Definition 5.9 *A propositional causal theory will be called a* causal Pearl theory *if it is determinate and satisfies the following conditions:*

- *no endogenous atom can appear both in the head and the body of a causal rule and*
- *two rules $A \Rightarrow p$ and $B \Rightarrow \neg p$ belong to a causal theory only if $A \wedge B$ is classically inconsistent.*

Using our terminology from chapter 4, the first condition above excludes purely explana-tory causal rules of the form $A \wedge p \Rightarrow p$ (see section 4.4.2) while the main effect of the second condition amounts to exclusion of nontrivial derived factual constraints in such causal theories. Note in this respect that if both $A \Rightarrow p$ and $B \Rightarrow \neg p$ belong to a causal the-ory, then we can derive $A \wedge B \Rightarrow \mathbf{f}$, which imposes a factual constraint to the effect that $\neg(A \wedge B)$ should hold in any causal model of this theory.

The above class of causal theories will be sufficient for important correspondence results between structural equation models and counterfactuals that will be described later in this chapter. It should also be noted that the second condition above can be traced back at least to Darwiche and Pearl (1994) where it played a key role in constructing symbolic causal networks satisfying the causal Markov property.

5.2.2 First-Order Case

We will generalize now the above representation to first-order languages. It will be based on a first-order generalization of the causal calculus, presented in Lifschitz (1997).

According to Lifschitz (1997), a *first-order causal rule* is an expression of the form $G \Rightarrow F$, where F and G are first-order formulas. A *first-order causal theory* Δ is a finite set of first-order causal rules coupled with a list $\mathbf{c}$ of object, function, and/or predicate constants, called the *explainable* symbols of Δ.

The nonmonotonic semantics of first-order causal theories has been defined in Lifschitz (1997) by a syntactic transformation that turns the causal theory Δ into a second-order sentence D_Δ. That sentence provided a precise formal description of the requirement that the explainable symbols should be explained, or *determined*, by Δ.

This transformation is defined as follows. Let $v\mathbf{c}$ denote a list of new variables similar to $\mathbf{c}$,[3] and let $\Delta(v\mathbf{c})$ denote the conjunction of the formulas

$$\forall \mathbf{x}(G \to F^{\mathbf{c}}_{v\mathbf{c}}),$$

for all rules $G \Rightarrow F$ of Δ, where $\mathbf{x}$ is the list of all free variables of F and G and $F^{\mathbf{c}}_{v\mathbf{c}}$ denotes the result of substituting the variables $v\mathbf{c}$ for the corresponding constants $\mathbf{c}$ in F. Then D_Δ is the second-order sentence

$$\forall v\mathbf{c}(\Delta(v\mathbf{c}) \leftrightarrow (v\mathbf{c} = \mathbf{c})).$$

The sentence D_Δ (and its classical interpretations) is viewed then as describing the non-monotonic semantics of the causal theory Δ. Informally speaking, these are the models of Δ in which the interpretation of the explainable symbols $\mathbf{c}$ is the only interpretation of these symbols that is determined, or "causally explained," by the rules of Δ.

---

3. That is to say, the lists $\mathbf{c}$ and $v\mathbf{c}$ have the same length; object constants in the former correspond to object variables in the latter, function symbols correspond to function variables, and predicate symbols to predicate variables.

The process of literal completion, defined initially for propositional causal theories (see section 4.5.2), has been extended to two classes of first-order causal theories in Lifschitz (1997) and Lifschitz and Yang (2013). We consider here the special case of the definition from the second paper when every explainable symbol of Δ is an object constant and Δ consists of rules of the form

$$G(x) \Rightarrow c = x,$$

one for each explainable symbol c, where $G(x)$ is a formula without any free variables other than x. The *(functional) completion* of Δ is defined in this case as the conjunction of the sentences

$$\forall x (c = x \leftrightarrow G(x))$$

for all rules of Δ.

Now we turn to the translation of Pearl's causal models into the above first-order causal calculus.

Definition 5.10 *For any first-order causal model M, Δ_M is the first-order causal theory whose explainable constants are the endogenous symbols of M and whose rules are*

$$x = t \Rightarrow x = c,$$

for every structural equation $c = t$ from M (where x is a variable).

Then the following theorem establishes a key correspondence result.

Theorem 5.2 *An extension of the interpretation of rigid and function symbols in M to the exogenous and endogenous symbols on a universe of cardinality > 1 is a solution of M iff it is a nonmonotonic model of Δ_M.*

The proof of the above result follows directly from the results on functional completion, described in Lifschitz and Yang (2013).

The first-order causal calculus is closer to the causal models of Pearl than the propositional causal calculus, not only because it is based on a richer language but also because it relaxes the requirement of total explainability of the latter and restricts it to explainable symbols only. It has been noted in Lifschitz (1997), however, that we can easily turn, for example, an unexplainable (exogenous) predicate $Q(x)$ into an explainable predicate by adding the following two causal rules:

$$Q(x) \Rightarrow Q(x) \qquad \neg Q(x) \Rightarrow \neg Q(x).$$

This will not change the nonmonotonic semantics. Still, it will allow us to reduce partial explainability to the universal explainability of the propositional causal calculus.

Remark. The first-order generalization of the causal calculus will not play any significant role in this book, which is confined to the propositional language. It should also be noted that the first-order causal calculus is still severely underdeveloped. For instance, describing the monotonic (logical) basis of the first-order causal calculus remains at this point an open problem. Nevertheless, the very existence of the above translation of Pearl's causal models in their full generality into the causal calculus implies that the restriction to the propositional language is not an essential limitation of our approach to causality. It also suggests a promising direction of research for future studies.

5.3   Causal Counterfactuals

Causal models have been used by Pearl to provide a novel semantics for counterfactual conditionals. Unlike Lewisian semantics for counterfactuals that we have discussed in chapter 2, this semantics is not based on the possible-worlds semantics of comparative similarity; in fact, it does not even rely on a similarity heuristic. Instead, it "rests directly on the mechanisms (or 'laws,' to be fancy)" (Pearl 2000, p. 239) and thereby relies primarily on causal inference.[4]

There is a growing literature on this new semantics of counterfactuals, including important applications in linguistic semantics. This topic certainly deserves a separate study, but it has been largely left beyond the primary scope of this book. Still, in what follows we will introduce a formal definition of counterfactuals in the causal calculus that will correspond to the original definition in the structural approach (under our translation of the latter). This will give us an opportunity to study their representation capabilities and, more generally, their role in causal reasoning.

5.3.1   Interventions, Contractions, and Revisions

In the framework of Pearl's causal models, counterfactuals are defined using the notions of intervention and submodel. In the translation of Bochman and Lifschitz (2015) that has been described earlier, the latter corresponds to subtheories of a causal theory. We will provide below a somewhat different definition of these notions that will exploit a striking similarity between interventions and belief revision operations of the so-called AGM theory of belief change (Alchourrón, Gärdenfors, and Makinson 1985). However, due to some well-known difficulties in defining interventions with respect to arbitrary logical formulas, these "causal revisions," as well as the corresponding counterfactuals will be restricted to literals in this study.

As before, for a set L of literals, we will denote by $\neg L$ the set of classical literals corresponding to $\{\neg l \mid l \in L\}$.

Definition 5.11 *Given a determinate causal theory Δ and a set L of literals,*

- *the* contraction *$\Delta{-}L$ of Δ with respect to L is the determinate causal theory obtained from Δ by removing all rules $A \Rightarrow l$ for $l \in L$, and*

---

4. Cf. the Central Claim in Schulz (2011).

- *the* revision $\Delta{*}L$ *of* Δ *is the determinate causal theory obtained from the contraction* $\Delta{-}\neg L$ *by adding the rule* $\mathbf{t}\Rightarrow l$ *for each* $l\in L$.

Contraction can be viewed as an operation that transforms some endogenous (explainable) literals into exogenous ones, namely, literals that are not determined by rules of the causal theory.

It can be immediately verified that revisions of causal theories exactly correspond to submodels of Boolean causal models.

Example 5.1 (Firing squad, continued) In the firing squad example, let us consider the following action sentence (in the terminology of Pearl 2000):

S4. If the captain gave no signal and rifleman A decides to shoot, the prisoner will die and B will not shoot.

To evaluate it, we have to consider the revision $\Delta{*}\{\neg C, A\}$ of the original causal theory:

$$\mathbf{t}\Rightarrow\neg C,\ \mathbf{t}\Rightarrow A,$$
$$C\Rightarrow B,\ \neg C\Rightarrow\neg B,\ A\vee B\Rightarrow D,\ \neg(A\vee B)\Rightarrow\neg D,$$
$$U\Rightarrow U,\ \neg U\Rightarrow\neg U.$$

Both D and $\neg B$ hold in all causal worlds of the above theory, so S4 is justified.

In chapter 9, the above idea of revision of causal theories will be generalized to an alternative semantic framework for representing actual causality.

5.3.2 The Formalization and Some Comparisons

By a *counterfactual* we will mean an expression of the form $L{>}A$, where L is a finite set of literals and A a proposition. Traditionally, counterfactuals are defined semantically with respect to worlds. The interventionist definition suggests, however, a powerful and useful generalization of validity for counterfactuals with respect to causal theories.

Definition 5.12 *Counterfactual* $L{>}A$ *will be said to* hold in a causal theory Δ *(notation* $L{>}_\Delta A$*), if A holds in all causal worlds of the revision* $\Delta * L$.

As in the structural account, acyclic causal theories always determine a unique causal world for any interpretation of the exogenous variables. Accordingly, for a causal world α of a causal theory Δ, we will call the set of exogenous literals that hold in α the *basis* of this world (cf. Veltman 2005; Schulz 2011). This notion of a basis will allow us to extend the above definition of causal counterfactuals to worlds.

Let Δ^α denote the causal theory obtained from Δ by adding rules $\mathbf{t}\Rightarrow l$ for each literal from the basis of α.

Definition 5.13 (World-based counterfactuals) *Counterfactual* $L{>}B$ *will be said to* hold in a causal world α *of a causal theory* Δ *if it holds in* Δ^α.

It can be easily verified that the above definition corresponds to the definition of causal counterfactuals in structural equation models (see, e.g., Halpern 2016a).

Example 5.1 (Firing squad, continued) Following Pearl (2000), given the actual world $\alpha = \{U, C, A, B, D\}$ of the firing squad example in which the prisoner is dead, let us evaluate the following counterfactual $\neg A > D$:

The prisoner would be dead even if rifleman A had not shot.

By our definition, this world-based counterfactual should be evaluated with respect to Δ^{α}, which can be safely reduced to the following causal theory:

$$\mathbf{t} \Rightarrow U,\ U \Rightarrow C,\ C \Rightarrow A,\ C \Rightarrow B,\ A \vee B \Rightarrow D.$$

The revision of this causal theory with $\neg A$, that is, $\Delta^{\alpha} * \neg A$, amounts to

$$\mathbf{t} \Rightarrow U,\ U \Rightarrow C,\ \mathbf{t} \Rightarrow \neg A,\ C \Rightarrow B,\ A \vee B \Rightarrow D.$$

The latter causal theory has a unique causal world $\{U, C, \neg A, B, D\}$, and since D holds in this world, $\neg A > D$ holds in α.

As can be seen, the above definitions provide feasible tools for evaluating counterfactuals both with respect to specific worlds and causal theories in general.

It turns out that the completion construction for determinate causal theories (see section 4.5.2) can be adapted to the case of counterfactuals.

Recall that a pure default completion of a causal theory Δ is a causal theory $Dc(\Delta)$ obtained from Δ by adding rules $l \Rightarrow l$ for any literal of the language that does not appear in the heads of the rules from Δ (see definition 4.20). Then we have:

Theorem 5.3 *A counterfactual $L > B$ holds in a determinate causal theory Δ iff*

$$comp(Dc(\Delta - \neg L)) \vDash \wedge L \to B.$$

The above result reduces, in effect, checking counterfactual assertions in the causal setting to classical entailment. The proof follows immediately from the definition of the propositional completion $comp(\Delta)$.

Some comparisons. A formal comparison of the causal semantics and Lewisian comparative similarity semantics for counterfactuals can be found in Halpern (2000). Halpern has shown, in particular, that the two formalizations are largely incomparable. Thus, though counterfactual theories based on acyclic causal structures are representable in (a very special case of) the comparative similarity semantics, once we step beyond acyclic theories, causal counterfactuals become not representable, in general, in Lewis's semantics. The following example from Halpern (2000), translated into our logical framework, illustrates this.

Example 5.3 *Let us consider the following causal theory:*

$$r, \neg q \Rightarrow p \quad p, \neg r \Rightarrow q \quad q, \neg p \Rightarrow r$$

plus its negative causal completion. The only exact world of this theory is $\{\neg p, \neg q, \neg r\}$, *and it justifies the following counterfactuals:*

$$r \wedge \neg q > p, \quad p \wedge \neg r > q, \quad q \wedge \neg p > r.$$

However, the latter counterfactual theory does not have a comparative similarity model.

In fact, the two semantics for counterfactuals are incomparable already on the level of representations they provide (or cannot provide) for particular situations. Moreover, it turns out that the causal calculus creates in this respect enhanced representation capabilities that cannot be obtained even using structural equations models.

To begin with, the causal interpretation of counterfactuals immediately allows us to resolve an old and famous puzzle:

Example 5.4 (Striking a match) *(Goodman 1947) Suppose that a match lights whenever it is struck, unless it is wet. Suppose also that in the actual world the match is not struck, it is not wet, and is not lit. Would it light if it were struck?*

The commonsense answer "yes" is obvious, but the comparative similarity account has a peculiar problem with it. In order to obtain this answer, it should keep the fact that the match is not wet in the counterfactual world(s) where the match is struck while at the same time does not keep the fact that the match is not lit. In other words, it should somehow make the world in which the match is struck, is not wet, and is lit more similar to the actual world than the world where it is struck, wet, and not lit. This asymmetry cannot be obtained, for instance, from the classical logical description of the relevant "match" law. On the causal interpretation, however, such an asymmetry can be immediately derived from the following causal law:

$$Struck, \neg Wet \Rightarrow Lights.$$

$\neg Struck$ and $\neg Wet$ form the basis of the actual world $\alpha = \{\neg Struck, \neg Wet, \neg Lights\}$, and therefore the revision $\Delta^{\alpha} * Struck$ has a single causal world $\{Struck, \neg Wet, Lights\}$. Consequently, $Struck > Lights$ holds in the actual world.

Yet another well-known and problematic example in the literature on counterfactuals will allow us to compare all three semantics for counterfactuals, mentioned above.

Example 5.5 (Jones' Hat) *(Tichy 1976) Jones is possessed of the following dispositions regarding wearing his hat. Bad weather invariably induces him to wear a hat. Fine weather, on the other hand, affects him neither way: on fine days he puts his hat on or leaves it on the peg, completely at random. Suppose moreover that actually the weather is bad, so Jones* is *wearing his hat. The question is, "If the weather had been fine, would Jones have been wearing his hat?"*

A natural answer to the above question would be "no" (or, better, "don't know"). The similarity account, however, and even the Ramsey test for conditionals both require that the counterfactual situation should be maximally similar to the actual one, which immediately implies that the answer to the above counterfactual question is "yes" since Jones is wearing his hat in the actual world (see Veltman 2005).

The following causal theory Δ provides a proper *causal* description of this situation:

$$BadWeather \Rightarrow WearingHat$$

$$\neg BadWeather, WearingHat \Rightarrow WearingHat$$

$$\neg BadWeather, \neg WearingHat \Rightarrow \neg WearingHat.$$

If the actual world is $\alpha = \{BadWeather, WearingHat\}$, then its revision with respect to $\neg BadWeather$, that is, $\Delta^\alpha * \neg BadWeather$ has two exact worlds:

$$\{\neg BadWeather, WearingHat\} \text{ and } \{\neg BadWeather, \neg WearingHat\}.$$

Consequently, $\neg BadWeather > WearingHat$ does not hold.

Note that the structural account also cannot provide an adequate representation for this example, already because the above theory is not a Pearl causal theory (see definition 5.9). More specifically, this theory describes a situation in which *WearingHat* could be viewed as an exogenous variable *when the weather is good*.

Lewis's own response to puzzles of this kind (see Lewis 1979) has been that similarities with respect to particular facts have little or no importance for establishing similarity between worlds. However, Veltman (2005) has suggested the following variant of Tichy's puzzle implying that this proposal should be qualified.

> Suppose that Jones always flips a coin before he opens the curtains to see what the weather is like. Heads means he is going to wear his hat in case the weather is fine, whereas tails means he is not going to wear his hat in that case. Like above, bad weather invariably makes him wear his hat. Now suppose that today heads came up when he flipped the coin, and that it is raining. So, again, Jones is wearing his hat.

This time the relevant causal theory Δ is

$$BadWeather \Rightarrow WearingHat$$

$$\neg BadWeather, CoinHeads \Rightarrow WearingHat$$

$$\neg BadWeather, \neg CoinHeads \Rightarrow \neg WearingHat,$$

and the actual world is $\beta = \{BadWeather, CoinHeads, WearingHat\}$.

Now $\Delta^\beta * \neg BadWeather$ has a single exact world,

$$\{\neg BadWeather, CoinHeads, WearingHat\},$$

and consequently ¬*BadWeather*>*WearingHat* will hold. The fact *WearingHat* has been preserved this time because it has a *reason* why it should still hold, namely, *CoinHeads*.

5.4 Intervention-Equivalence and Basic Inference

In this section we will explore expressive capabilities of the counterfactual language in describing causation. To simplify the exposition, we will restrict our discussion to finite causal theories.

The difference between causal and purely mathematical understanding of structural equations can be revealed by performing interventions on a causal model. Thus, two different causal models may "accidentally" determine the same causal worlds, but their difference can be revealed by performing the same "surgery" on both of them, as in the following example from Woodward (2003).

Example 5.6 *Let us consider two arithmetical structural models:*

$$\{A = B;\ B = D\} \ and \ \{A = (B + D)/2;\ B = D\}.$$

On a reasonable understanding, these systems of structural equations are "objectively" equivalent since they determine the same solutions or causal worlds. Nevertheless, if we will fix $B = 0$, we will obtain, respectively, the following structural models:

$$\{A = B = 0\} \ and \ \{A = D/2;\ B = 0\},$$

which are not equivalent.

According to Pearl (2000), every causal model stands not for just one but for a whole set of its submodels that embody interventional contingencies. These submodels determine, in a sense, the "causal content" of a given causal model in Pearl's approach. In accordance with this, the following definition has been introduced (though in a different terminology) in Bochman and Lifschitz (2015):

Definition 5.14 *Determinate causal theories Γ and Δ are* intervention-equivalent *(for short, i-equivalent) if, for every set L of literals, the revision $\Gamma * L$ has the same nonmonotonic semantics as the revision $\Delta * L$.*

Now, at least in the finite case, it is easy to show that intervention-equivalence of two causal theories amounts to coincidence of their associated counterfactuals:

Theorem 5.4 *Determinate causal theories Δ and Γ are i-equivalent iff they determine the same counterfactuals: for any set L of literals, and any A,*

$$L >_\Delta A \ \textit{iff} \ L >_\Gamma A.$$

Thus, intervention-equivalence provides a useful tool for the study of causal counterfactuals and their expressive capabilities.

In Bochman and Lifschitz (2015), an attempt has been made to connect intervention equivalence with causal equivalence in the causal calculus. The results, however, were not entirely satisfactory. Thus, though it has been shown that intervention equivalence implies equivalence with respect to causal inference, the reverse implication has been shown to hold only for a very narrow class of *modular* causal theories. In contrast, we are going to show now that intervention equivalence is intimately related to a stronger equivalence with respect to basic causal inference.

To begin with, our next result will show that interventions cannot distinguish causal theories that are basically equivalent:

Theorem 5.5 *Basically equivalent determinate causal theories are intervention equivalent.*

Proof. Assume that Δ and Γ are basically equivalent determinate causal theories, but for some set L of literals, the revision $\Gamma * L$ does not have the same nonmonotonic semantics as the revision $\Delta * L$. This means that the completions of $\Gamma * L$ and $\Delta * L$ are not logically equivalent. By the definition of propositional completion, this is possible only if there exists a literal l (not in L) such that

$$\bigvee \{A_i \mid A_i \Rightarrow l \in \Delta\}$$

is not logically equivalent to

$$\bigvee \{B_i \mid B_i \Rightarrow l \in \Gamma\}.$$

However, due to proposition 4.28 from chapter 4, this will imply that Δ and Γ do not determine the same basic production relation $\Rightarrow^b$, which means that they are not basically equivalent, contrary to our assumption. $\qquad \square$

Being combined with our preceding result, the above theorem implies, in effect, that the language of counterfactuals does not allow us to distinguish basically equivalent sets of causal rules.

It has been shown in Bochman and Lifschitz (2015) (using a suitable counterexample) that causal equivalence of causal theories does not imply intervention-equivalence. In other words, there are causal theories that are causally equivalent, but their revisions with the same literal determine different causal worlds (and counterfactuals). Strengthening this result, the following example shows that even regular equivalence does not imply intervention equivalence.

Example 5.7 *The causal theories*

$$\Delta = \{p \Rightarrow q,\ p \wedge q \Rightarrow r\}$$

and

$$\Gamma = \{p \Rightarrow q,\ p \Rightarrow r\}$$

are regularly equivalent. However, their respective revisions $\Delta * \neg q = \{\mathbf{t} \Rightarrow \neg q,\ p \wedge q \Rightarrow r\}$ *and* $\Gamma * \neg q = \{\mathbf{t} \Rightarrow \neg q,\ p \Rightarrow r\}$ *are already not regularly equivalent. Moreover, if p is exogenous, they have different causal worlds:* $\{p, \neg q, r\}$ *is a causal world of* $\Gamma * \neg q$ *but not of* $\Delta * \neg q$.

Theorem 5.5 above implies that the set of counterfactuals that hold with respect to a causal theory cannot uniquely determine the source causal theory already because it cannot distinguish basically equivalent theories. Still, the following lemma will show that, up to the basic equivalence, there is indeed a one-to-one correspondence between Pearl causal theories and their associated counterfactuals.

Lemma 5.6 *If* Δ *is a Pearl causal theory, then, for any literal l and any set L of literals such that* $l \notin L$,

$$L \Rightarrow_\Delta^b l \quad \text{if and only if} \quad L' >_\Delta l,$$

for any consistent set of literals $L' \supseteq L$ *such that* $\neg l \notin L'$.

Proof. If $L \Rightarrow_\Delta^b l$, then, by the preceding theorem, the rule $L \Rightarrow l$ can be safely added to Δ without changing the associated counterfactuals. But if this rule belongs to a causal theory, then any revision of this theory with a set of literals L' that does not include $\neg l$ will not remove this rule, which gives us the direction from left to right. In the other direction, if $L \not\Rightarrow_\Delta^b l$, then by proposition 4.28 there exists a world α that includes L and such that $\Delta(\alpha)$ does not imply l. Since Δ is a determinate theory, this means that α makes false the antecedents of all causal rules in Δ with a head l. Let L' denote the set of all literals in α, except l. Then the revision $\Delta * L'$ will falsify l, and consequently $L' >_\Delta l$ will not hold. $\square$

The above result implies that the set of counterfactuals that hold with respect to a causal theory uniquely determines the basic closure ($\Rightarrow_\Delta^b$) of the latter. As an immediate consequence, we obtain the following.

Corollary 5.7 *Pearl causal theories are intervention equivalent if and only if they are basically equivalent.*

It should be noted, however, that the above correspondence cannot be extended to arbitrary causal theories. In fact, the following counterexample, given in Bochman and Lifschitz (2015), can also be used in the present context:

Example 5.8 *Causal theories* $\{p \Rightarrow p\}$ *and* $\{\mathbf{t} \Rightarrow p\}$ *are not equivalent even for causal inference relations. Still, it is easy to verify that they are intervention equivalent, since all their possible revisions have the same causal worlds.*

Basic equivalence in the structural account. It has often been argued that structural equation models and causal counterfactuals are essentially equivalent formalisms—see, e.g., Galles and Pearl (1998), Hitchcock (2007), and Woodward (2016). The above results can now be used to formally justify this claim.

Our last result above makes basic inference an "internal" causal logic of interventions and counterfactuals in Pearl causal theories. It should be noted, however, that basic equivalence (i.e., the validity of the Or rule of inference) is in some sense a "built-in" feature of Pearl's structural account of causation itself due to the underlying assumption that any endogenous variable is determined by a *single* causal mechanism (formulated as a structural equation). Indeed, according to this assumption, all the alternative causal factors that determine a given (Boolean) endogenous variable should be conjoined by disjunction into a single formula by the very definition of the structural equation. Consequently, basically equivalent causal descriptions are indistinguishable in the language of structural equations. But then the above results will imply, in effect, that there is indeed a one-to-one correspondence between Pearl's causal models and their associated sets of counterfactuals.

As we are going to show in chapter 8, however, the assumption of a single mechanism behind every endogenous variable and the ensuing "collapse" of basically equivalent causal descriptions could be viewed as the main source of the problem of structural equivalents in the structural accounts of actual causation.

5.5   Causal Models and Dialectical Argumentation

General connections between the causal calculus and argumentation theory have been described in chapter 4. As a further illustration of these connections, we will describe in this section a direct correspondence between Pearl's causal models and abstract dialectical frameworks of Brewka and Woltran (2010) and Brewka et al. (2013).

Abstract dialectical frameworks (ADFs) have been developed as an abstract argumentation formalism purported to capture more general forms of argument interaction than just attacks among arguments, which form the basis of the original Dung's argumentation frameworks (see chapter 3). To achieve this, each argument in an ADF has been associated with an *acceptance condition*, which is some propositional function determined by arguments that are linked to it. Using such acceptance conditions, ADFs allow us to express that arguments may jointly support another argument, or that two arguments may jointly attack a third one, and so on. Dung's argumentation frameworks are recovered in this setting by acceptance condition saying that an argument is accepted if none of its parents is.

The authors of ADFs have repeatedly stressed that they primarily see their formalism not as a knowledge representation tool but rather as a convenient and conceptually neutral abstraction tool ("argumentation middleware") that is intended to encompass a broad range of more specific argumentation and other nonmonotonic formalisms. On the other hand, Strass (2013) has considered ADFs as a particular knowledge representation formalism. In this section also, we will view ADFs as a specific knowledge representation formalism and show its close conceptual connections with the formalism of causal reasoning. This will help to single out some basic principles behind the construction of ADFs and their semantics and further situate the causal calculus in the range of closely related formalisms of knowledge representation in AI.

5.5.1 Abstract Dialectical Frameworks

An ADF is a directed graph whose nodes represent statements or positions which can be accepted or not. The links represent dependencies: the status of a node s only depends on the status of its parents [denoted $par(s)$], that is, the nodes with a direct link to s. In addition, each node s has an associated acceptance condition C_s specifying the exact conditions under which s is accepted. C_s is a function assigning to each subset of $par(s)$ one of the truth values $\mathbf{t}, \mathbf{f}$. Intuitively, if for some $R \subseteq par(s)$ we have $C_s(R) = \mathbf{t}$, then s will be accepted provided the nodes in R are accepted and those in $par(s) \setminus R$ are not accepted.

Definition 5.15 *An abstract dialectical framework is a tuple $D = (S, L, C)$, where*

- *S is a set of statements (positions, nodes),*
- *$L \subseteq S \times S$ is a set of links; and*
- *$C = \{C_s\}_{s \in S}$ is a set of total functions $C_s : 2^{par(s)} \to \{\mathbf{t}, \mathbf{f}\}$, one for each statement s. C_s is called acceptance condition of s.*

A more "logical" representation of ADFs can be obtained simply by assigning each node s a *classical* propositional formula corresponding to its acceptance condition C_s (see Ellmauthaler 2012). In this case we can tacitly assume that the acceptance formulas implicitly specify the parents a node depends on. It is then not necessary to give the links L, so an ADF D amounts to a tuple (S, C) where S is a set of statements and C is a set of propositional formulas, one for each statement from S. The notation $s[C_s]$ has been used by the authors to denote the fact that C_s is the acceptance condition of s.

A two-valued interpretation v is a (two-valued) *model* of an ADF (S, C) whenever for all statements $s \in S$ we have $v(s) = v(C_s)$, that is, v maps exactly those statements to true whose acceptance conditions are satisfied under v. This notion of a model provides a natural semantics for ADFs. In addition to this semantics, however, the authors have defined appropriate generalizations for all the major semantics of Dung's argumentation frameworks. In Brewka et al. (2013), all these semantics were defined by generalizing the two-valued interpretations to three-valued ones. All of them were formulated using the basic operator Γ_D over three-valued interpretations that was introduced, in effect, already in Brewka and Woltran (2010). In the formulation of Brewka et al. (2013), for an ADF D and a three-valued interpretation v, the interpretation $\Gamma_D(v)$ is given by the mapping

$$s \mapsto \prod \{w(C_s) \mid w \in [v]_2\},$$

where $\prod$ is the product operator on interpretations while $[v]_2$ is the set of all two-valued interpretations that extend v.

For each statement s, the operator Γ_D returns the consensus truth value for its acceptance formula C_s, where the consensus takes into account all possible two-valued interpretations w that extend the input valuation v. If v is two valued, we get $\Gamma_D(v)(s) = v(C_s)$, so v is a two-valued model for D iff $\Gamma_D(v) = v$. In other words, two-valued models of D are precisely those classical interpretations that are fixed points of Γ_D.

The *grounded model* of an ADF D can now be defined as the least fixpoint of Γ_D. This fixpoint is in general three valued, and it always exists since the operator Γ_D is monotone in the information ordering $\leq_i$, as shown in Brewka and Woltran (2010). This grounded semantics has been viewed by the authors as the greatest possible consensus between all acceptable ways of interpreting the ADF at hand.[5]

The operator Γ_D also provides a proper basis for defining admissible, complete, and preferred semantics for arbitrary ADFs.

Definition 5.16 *A three-valued interpretation v for an ADF D is*

- *admissible iff $v \leq_i \Gamma_D(v)$;*
- *complete iff $\Gamma_D(v) = v$; and*
- *preferred iff it is $\leq_i$-maximal admissible.*

As can be shown, the above definitions provide proper generalizations of the corresponding semantics for Dung's argumentation frameworks and, moreover, preserve much of the properties and relations of the latter. Thus, the grounded semantics is always a complete model, and each complete model is admissible. In addition, as it is the case for Dung's argumentation frameworks, all preferred models are complete, the grounded model is the $\leq_i$-least complete model, and the set of all complete models forms a complete meet-semilattice with respect to the information ordering $\leq_i$.

In Brewka and Woltran (2010), the standard Dung semantics of stable extensions was generalized only to a restricted type of ADFs called bipolar, but Brewka et al. (2013) has suggested a new definition that avoids unintended features of the original definition and covers arbitrary ADFs, not just bipolar ones (see also Strass 2013). This new definition is based on the notion of a *reduct* of an ADF, similar to the Gelfond-Lifschitz transformation of logic programs. We will discuss the representation of the stable semantics in ADFs later in this section.

5.5.2   The Causal Representation of ADFs

Now we are going to provide a uniform and modular translation of ADFs into the causal calculus. An essential precondition of this causal representation, however, will consist in transforming the underlying semantic interpretations of ADFs in terms of three-valued models, used in Brewka et al. (2013), into ordinary classical logical descriptions. This latter transformation will also allow us to clarify to what extent the various semantics suggested for ADFs admit a classical logical reading. In fact, the very possibility of such a classical reformulation stems from the crucial fact that the basic operator Γ of an

---

5. We will qualify this claim in what follows.

ADF, described earlier, is defined, ultimately, in terms of ordinary classical interpretations extending a given three-valued one. Nevertheless, our reformulation will also reveal a significant discrepancy between these semantics and their immediate causal counterparts.

To begin with, any three-valued interpretation v on the set of statements S can be encoded using an associated set of literals $[v] = S_0 \cup \neg S_1$ such that $S_0 = \{p \in S \mid v(p) = \mathbf{t}\}$ and $S_1 = \{p \in S \mid v(p) = \mathbf{f}\}$. Moreover, this set of literals generates a unique deductively closed theory $\mathrm{Th}([v])$ that corresponds in this sense to the source three-valued interpretation v. Conversely, any *literal* deductively closed theory u (see section 4.3 in chapter 4) will correspond to a unique three-valued interpretation v such that $u = \mathrm{Th}([v])$. These simple facts establish a precise bidirectional correspondence between three-valued interpretations and classical literal theories. Moreover, we will see in what follows that the main operator Γ of ADFs will correspond under this reformulation to a "literal" restriction of the causal operator $\mathcal{C}$ of basic production inference.

As our starting point, we note a striking similarity between the official definition of an ADF and the notion of a causal model from Pearl (2000). More precisely, our definition of Boolean causal models, given earlier, is much similar to the logical reformulation of ADFs, with equations $p = F$ playing essentially the same role as the acceptance conditions $p[F]$. The differences are that only endogenous atoms are determined by their associated conditions in causal models, but on the other hand, there are no restrictions on appearances of atoms on both sides in ADF's acceptance conditions. Moreover, plain (two-valued) models of ADFs correspond precisely to causal worlds of the causal model.

Now, a modular representation of Boolean causal models as causal theories of the causal calculus, described earlier in this chapter, can now be seamlessly transformed into the following causal representation of ADFs:

Definition 5.17 (Causal representation of an ADF) *For any ADF D, Δ_D is the causal theory consisting of the rules*

$$F \Rightarrow p \quad and \quad \neg F \Rightarrow \neg p$$

for all acceptance conditions $p[F]$ in D.

The above representation is fully modular, and it will be taken as a uniform basis for the correspondences described below.

To begin with, we immediately establish the following theorem.

Theorem 5.8 *The two-valued semantics of an ADF D corresponds precisely to the causal nonmonotonic semantics of Δ_D.*

As a consequence, the full system of causal inference provides a precise logical basis for this nonmonotonic semantics.

5.5.3   General Correspondences

Now we are going to show that the above causal representation also survives the transition to three-valued models of ADFs. To this end, however, we will have to retreat from the system of causal inference to a weaker system of basic production inference.

A broader correspondence between various semantics of ADFs and general nonmonotonic semantics of the causal calculus arises from the fact that the operator Γ of an ADF naturally corresponds to a particular causal operator of the associated causal theory.

Let $\mathcal{L}$ denote the set of classical literals of the underlying language. We will denote by $\mathcal{C}^{\mathcal{L}}$ the restriction of a causal operator $\mathcal{C}$ to literals, that is, $\mathcal{C}^{\mathcal{L}}(u) = \mathcal{C}(u) \cap \mathcal{L}$. As we are going to show, the operator Γ of ADFs corresponds precisely to this "literal restriction" of the causal operator associated with a basic production inference. As before, $[v]$ will denote the set of literals corresponding to a three-valued interpretation v.

Lemma 5.9 *For any three-valued interpretation v,*

$$[\Gamma_D(v)] = \mathcal{C}_D^{\mathcal{L}}([v]),$$

where $\mathcal{C}_D$ is a basic causal operator corresponding to Δ_D.

The above equation has immediate consequences for the broad correspondence between the semantics of ADFs that are defined in terms of the operator Γ_D and natural sets of propositions definable wrt associated causal theory. Thus, we have the following.

Theorem 5.10 *Complete models of an ADF D correspond precisely to the fixed points of $\mathcal{C}_D^{\mathcal{L}}$:*

$$v = \Gamma_D(v) \quad \textit{iff} \quad [v] = \mathcal{C}_D^{\mathcal{L}}([v]).$$

As a result, we immediately conclude that preferred models of an ADF correspond to maximal fixpoints of $\mathcal{C}_D^{\mathcal{L}}$ (with respect to set inclusion) while the grounded model corresponds to the least fixpoint of $\mathcal{C}_D^{\mathcal{L}}$.

It turns out, however, that when viewed in a classical logical setting, the restriction of the causal operator to literals inadvertently leads to an information loss. More precisely, though disjunctive formulas can appear in acceptance conditions used by Γ in an ADF, the operator itself records, in effect, only literals that are produced, and thereby disregards all other information that can be obtained from its output. The following example illustrates this.

Example 5.9 *Let us consider the following ADF D:*

$$q[p] \qquad r[\neg p] \qquad s[q \vee r].$$

The grounded model of this ADF is empty (all atoms are unknown). However, the associated causal theory Δ_D comprises the following rules:

$$p \Rightarrow q \qquad \neg p \Rightarrow r \qquad q \vee r \Rightarrow s$$

$$\neg p \Rightarrow \neg q \qquad p \Rightarrow \neg r \qquad \neg q \wedge \neg r \Rightarrow \neg s.$$

In view of lemma 4.13, the least exact theory of $\mathcal{C}_D$ is precisely the set of propositions that are provable from the above theory using the postulates of causal *inference (since it is both basic and regular). Now, the first two rules imply $\mathbf{t} \Rightarrow q \vee r$ (by Or), and hence $\mathbf{t} \Rightarrow s$ by Cut. Similarly, the fourth and fifth rule imply $\mathbf{t} \Rightarrow \neg q \vee \neg r$. As result, the least exact theory of $\mathcal{C}_D$ is much more informative, namely $\mathrm{Th}(\{q \leftrightarrow \neg r, s\})$.*

It can also be seen from the above example that the restriction of exact theories to literals does not necessary produce fixed points of the corresponding literal operator $\mathcal{C}^{\mathcal{L}}$. Still, it can be shown that for any fixed point of the latter (that is, for any complete model an ADF) there exists a least exact theory that contains it. The latter theory may contain, however, more information than its literal source.

The above considerations and results suggest a natural generalization of an ADF to acceptance conditions of the form $A[B]$, where both A and B are classical formulas. This would supply the abstract argumentation frameworks with further representation capabilities and thereby even contribute to its original purpose of providing a powerful and widely applicable abstraction tool for argumentation.

5.6 Conclusions

It has been noted by Pearl (2000) in his epilogue, "The Art and Science of Cause and Effect," that many scientific discoveries have been delayed for the lack of a mathematical language that can amplify ideas and let scientists communicate results. In this respect, we hope that our representation of Pearl's causal models in the causal calculus has provided a missing *logical* language for Pearl's causal reasoning, a language that could return this reasoning back to logic, albeit a nonmonotonic one.

The fact that Pearl's causal models are interpreted in this study as *nonmonotonic* causal theories allows us to clarify a large part of confusions surrounding a causal reading of structural equations. To begin with, the causal calculus has its own *causal logic*, a fully logical, though nonclassical, formalism of causal inference relations that has an appropriate (possible worlds) *logical semantics*. This causal logic and its semantics provide a formal interpretation for the causal rules. In addition to that, however, the causal calculus includes a nonmonotonic "overhead" that is determined by the corresponding *nonmonotonic semantics*. Moreover, in our particular case, this nonmonotonic semantics can be captured by purely classical logical means, namely by forming the *completion* of the source causal theory, which is an ordinary classical deductive theory that provides a complete description for the nonmonotonic semantics. In the first-order case it directly describes the corresponding equations in the usual mathematical sense. Still, this completion construction is global and nonmodular with respect to the source causal theory, and it changes nonmonotonically with changes of the latter. That is why the nonmonotonic causal reasoning is not reducible to a standard, deductive logical reasoning. Causal reasoning is more than deductive reasoning; it includes that latter as its essential part but is not reducible to it.

Despite the established connection between the causal calculus and Pearl's structural equation formalism, there are obvious differences in the respective objectives of these theories as well as in required means of expression. Thus, Pearl causal models correspond to a restricted kind of causal theories, and as we shall see later, these restrictions create problems in application of such theories for describing actual causality and dynamic domains in reasoning about action and change (see part IV). In this respect, the causal calculus constitutes a highly expressive logical replacement of structural equations also for many of the purposes envisaged by Pearl.

6 Indeterminate Causation

In this chapter we will explore the idea of indeterminate causation and, more specifically, the use of *disjunctive* (multiple-conclusion) causal rules for its formal representation.

We provide first a logical formalization of such rules in the form of a disjunctive causal inference relation and describe its logical semantics. Then we consider a nonmonotonic semantics for such rules, suggested in Turner's logic of universal causation (see H. Turner 1999). It will be shown, however, that, under this semantics, disjunctive causal rules admit a stronger logic in which these rules are reducible to ordinary, singular causal rules. This semantics also tends to give an exclusive interpretation of disjunctive causal effects and so excludes some reasonable models in particular cases. To overcome these shortcomings, we will introduce an alternative nonmonotonic semantics for disjunctive causal rules called a covering semantics that permits an inclusive interpretation of indeterminate causal information. Still, it will be shown that even in this case there exists a systematic procedure that we will call normalization that allows us to capture precisely the covering semantics using only singular causal rules. This normalization procedure can also be viewed as a kind of nonmonotonic completion, and it generalizes established ways of representing indeterminate effects in current theories of actions and change in artificial intelligence.

Proofs of all the results in this chapter that are not given in the text can be found in the appendix to the chapter.

6.1 Introduction

The ability to represent actions with indeterminate effects constitutes an important objective of any general theory of actions and change in AI. And indeed, most of these theories provide specific tools and methodologies for representing such actions in their formalisms.

The basic difficulty in representing indeterminate actions boils down to the fact that a plain classical logical description of the overall "disjunctive" effect of such an action is patently insufficient for describing (and predicting) the actual effects of this action in particular action domains; what remains to be specified is what are the particular fluents that can (or cannot) be affected by the action.

This difficulty is quite general, so it transcends the boundaries of specific action theories. Accordingly, most such theories have adopted a twofold approach to representing indeterminate actions which stipulates both the disjunctive effect of an action, and the particular fluents that can be influenced by (or depend on) this action (see, e.g., Lang, Lin, and Marquis 2003; Castilho, Herzig, and Varzinczak 2002).

It has been shown in McCain and Turner (1997) and Giunchiglia et al. (2004) that the causal rules of the causal calculus are capable of describing such indeterminate causal effects, but also in this case, the corresponding causal description involved two kinds of rules, a rule (more precisely, a constraint) that describes the disjunctive effect of an action and a number of explanatory rules of the form $A \wedge F \Rightarrow F$ saying that a literal F is causally explainable if it holds after action A.

Fangzhen Lin has suggested in Lin (1996) an alternative approach, according to which an action with indeterminate effects could be fully described using a single disjunctive causal rule of the form:

$$A \Rightarrow B_1, \ldots, B_n,$$

saying roughly that whenever A holds, one of B_i is caused. This idea has been taken up in Bochman (2003c) where the formalism of the causal calculus has been generalized to such disjunctive causal rules.

Taken as independent concept, a disjunctive causal rule can be naturally viewed as a qualitative counterpart of probabilistic causation, the only difference being that the latter also assigns probabilities to each particular effect. Accordingly, such rules have much in common with probabilistic causal rules of CP-logic of Vennekens, Denecker, and Bruynooghe (2009) (see also Bogaerts et al. 2014).

The appropriateness of causal rules of this kind for representing indeterminate causation can be justified as follows. Suppose that we have two ordinary causal rules $A \wedge C \Rightarrow B$ and $A \wedge \neg C \Rightarrow \neg B$.[1] The rules can be interpreted as saying that A causes either B or $\neg B$ depending on whether the additional condition C holds. Suppose now that C is absent from our vocabulary (it is a "hidden parameter"). Still, the above description conveys a nontrivial information about the situation in question, namely that when A holds, either B is caused or $\neg B$ is caused, that is, $A \Rightarrow B, \neg B$. Note that this information is completely lost in a singular causal rule $A \Rightarrow B \vee \neg B$, which always holds for causal inference relations.

A further interesting aspect of the above example is that the use of disjunctive causal rules is still not essential for the above situation: due to the fact that B and $\neg B$ are logically incompatible, the disjunctive rule $A \Rightarrow B, \neg B$ turns out to be logically equivalent to a pair of explanatory causal rules $A \wedge B \Rightarrow B$ and $A \wedge \neg B \Rightarrow \neg B$ (see lemma 6.30 below). Nevertheless, such a reduction is impossible for general disjunctive rules.

---

1. Say that A is "The switch is flipped," B is "The switch is down," and C is "The switch is initially up."

From a purely technical point of view, disjunctive causal rules form a natural, though largely unused, fragment of Turner's logic of universal causation (see H. Turner 1999), and there are deep reasons for this lack of use. Namely, it has been shown in Bochman (2003c), that the nonmonotonic semantics of UCL sanctions a straightforward reduction of such rules to ordinary singular causal rules (see theorem 6.18 below). Further, it has a shortcoming in that it tends to give an exclusive interpretation to disjunctive heads of causal rules. That is why we will introduce below a new *covering semantics* for such causal theories that will provide an inclusive interpretation of disjunctive causal information.

It will be shown, however, that in many regular cases a disjunctive causal theory can be "normalized," that is, transformed into a nondisjunctive causal theory in such a way that the standard nonmonotonic semantics of the latter will coincide with the covering semantics of the source disjunctive theory. Moreover, this transformation will be shown to be closely related to canonical ways of representing indeterminism in causal theories, employed in McCain and Turner (1997) and Giunchiglia et al. (2004). Accordingly, disjunctive causal theories coupled with the new semantics will suggest themselves as a systematic framework for representing indeterminate causal information.

6.2 Disjunctive Causal Relations

To begin with, we will generalize the causal calculus to indeterminate causal rules that involve multiple heads.

As before $a, b, \ldots$ will denote finite sets of propositions while $u, v, \ldots$ will denote arbitrary such sets. For a finite set of propositions a, we will denote by $\bigwedge a$ the conjunction of all propositions from a; as a special case, $\bigwedge \emptyset$ will denote $\mathbf{t}$. For any set u of propositions, we will denote by $\bar{u}$ the complement of u in the set of all propositions and by $\neg u$ the set $\{\neg A \mid A \in u\}$.

We will consider disjunctive causal rules as rules holding primarily between finite sets of propositions: $a \Rightarrow b$ will be taken to mean that if all the propositions in a hold, then at least one of the propositions in b is caused. We will use also an ordinary notation for premise and conclusion sets in causal rules. Thus, $a, b \Rightarrow c$ will stand for $a \cup b \Rightarrow c$, and $a, A \Rightarrow$ will mean $a \cup \{A\} \Rightarrow \emptyset$, and so on.

Definition 6.1 *A disjunctive causal relation is a binary relation $\Rightarrow$ on finite sets of classical propositions satisfying the following conditions:*

(Left Monotonicity)   *If $a \Rightarrow b$, then $A, a \Rightarrow b$;*
(Right Monotonicity)   *If $a \Rightarrow b$, then $a \Rightarrow b, A$;*
(Cut)   *If $a \Rightarrow b, A$ and $A, a \Rightarrow b$, then $a \Rightarrow b$;*
(Strengthening)   *If $A \vDash B$ and $a, B \Rightarrow b$, then $a, A \Rightarrow b$;*
(Weakening)   *If $A \vDash B$ and $a \Rightarrow b, A$, then $a \Rightarrow b, B$;*
(Left And)   *If $a, A \wedge B \Rightarrow b$, then $a, A, B \Rightarrow b$;*
(And)   *If $a \Rightarrow b, A$ and $a \Rightarrow b, B$, then $a \Rightarrow b, A \wedge B$;*

(Or)   *If $A, a \Rightarrow b$ and $B, a \Rightarrow b$, then $A \lor B, a \Rightarrow b$;*

(Falsity) $\mathbf{f} \Rightarrow$*;*

(Truth) $\Rightarrow \mathbf{t}$*.*

Similarly to ordinary consequence relations, a disjunctive causal relation can be extended to arbitrary sets of propositions by requiring *compactness*:

(Comp) $u \Rightarrow v$ iff $a \Rightarrow b$, for some finite $a \subseteq u, b \subseteq v$.

As can be seen, a disjunctive causal relation forms a subsystem of the classical sequent calculus. However, the former is only an applied logical formalism that is not purported to give meaning to the logical connectives. In fact, the *classical* meaning of these connectives is secured by the use of the classical entailment in the above postulates. Note, in particular, that Strengthening and Weakening imply that classically equivalent propositions are interchangeable both in bodies and heads of the causal rules. Some further "classical" properties are determined by the other postulates. Thus, it is straightforward to show that the postulates imply the following two general rules:

(Logical Strengthening) If $c \vDash A$ and $A, a \Rightarrow b$, then $c, a \Rightarrow b$.

(Logical Weakening) If $c \vDash A$ and $a \Rightarrow b, C$, for every $C \in c$, then $a \Rightarrow b, A$.

The first rule implies that a finite set of premises in a causal rule can be replaced by their conjunction:

$$a \Rightarrow b \text{ iff } \bigwedge a \Rightarrow b.$$

However, the heads of disjunctive causal rules cannot be replaced with their classical disjunctions; we have only that $a \Rightarrow b$ always implies $a \Rightarrow \bigvee b$, though not vice versa. Also, only the following structural rule for negation holds for disjunctive causal relations:

(Reduction) If $a \Rightarrow b, A$, then $a, \neg A \Rightarrow b$.

Actually, it turns out that, given the rest of the postulates, Reduction can replace the Cut postulate:

Lemma 6.1 *For disjunctive causal relations, Cut is equivalent to Reduction.*

Proof. If $a \Rightarrow b, A$, then $a, \neg A \Rightarrow b, A$ by Left Monotonicity. In addition, $A, \neg A \Rightarrow$ by Falsity and Logical Strengthening, and hence $A, \neg A, a \Rightarrow b$. Therefore $a, \neg A \Rightarrow b$ by Cut. Thus, Reduction holds.

Suppose now that Reduction holds, and we have $a \Rightarrow b, A$ and $A, a \Rightarrow b$. Then $a, \neg A \Rightarrow b$ by Reduction, and therefore $a \Rightarrow b$ by Or and Logical Strengthening. Hence Cut holds. $\square$

As before, an arbitrary set of causal rules will be called a *causal theory*. Any causal theory Δ determines a unique least disjunctive causal relation that includes Δ; it will be denoted by $\Rightarrow_\Delta$. The latter consists of all the causal rules that are derivable from Δ by the postulates for disjunctive causal relations.

Bitheories. Bitheories, introduced below, will play the same role they played for regular and causal inference relations in chapter 4.

Definition 6.2 *A* bitheory *is a pair* (α, u), *where u is a deductively closed set and α a world (maximal consistent set) such that $u \subseteq \alpha$. A* bitheory of a disjunctive causal relation $\Rightarrow$ *is any bitheory* (α, u) *such that* $\alpha \not\Rightarrow \overline{u}$.

Bitheories of $\Rightarrow_\Delta$ will also be called *bitheories of a causal theory* Δ. Bitheories of a causal relation can be seen as bitheories that are closed with respect to all its causal rules; this understanding is justified by the following simple result:

Lemma 6.2 *A bitheory* (α, u) *is a bitheory of a causal relation* $\Rightarrow$ *if and only if $b \cap u \neq \emptyset$, for any causal rule $a \Rightarrow b$ from $\Rightarrow$ such that $a \subseteq \alpha$.*

Similarly, bitheories of a causal theory Δ are precisely bitheories that are closed with respect to the rules from Δ. A proof can be found in the appendix to this chapter.

Lemma 6.3 *If $\Rightarrow_\Delta$ is a least disjunctive causal relation containing Δ, then a bitheory (α, u) is a bitheory of $\Rightarrow_\Delta$ if and only if $b \cap u \neq \emptyset$, for any causal rule $a \Rightarrow b$ from Δ such that $a \subseteq \alpha$.*

The next definition describes the worlds that are consistent with respect to a causal relation.

Definition 6.3 *A world α will be said to be* causally consistent *with respect to a disjunctive causal relation $\Rightarrow$ if $\alpha \not\Rightarrow$.*

The following lemma gives some useful equivalent conditions of causal consistency.

Lemma 6.4 *A world α is causally consistent with respect to $\Rightarrow$ iff one of the following conditions holds:*

- (α, α) *is a bitheory of* $\Rightarrow$;
- (α, u) *is a bitheory of* $\Rightarrow$, *for some u.*

Proof. The two conditions above are trivially equivalent. Moreover, each of them implies causal consistency by Right Monotonicity. So, it is sufficient to show that α is causally consistent only if (α, α) is a bitheory of $\Rightarrow$.

Assume that (α, α) is not a bitheory of $\Rightarrow$, that is, $\alpha \Rightarrow \overline{\alpha}$. Then by compactness $a \Rightarrow b$, for some finite sets a, b such that $a \subseteq \alpha$ and $b \cap \alpha = \emptyset$. By Reduction, we have $a, \neg b \Rightarrow$. Moreover, since α is a world, $\neg b \subseteq \alpha$. Consequently, $a \cup \neg b \subseteq \alpha$, and therefore $\alpha \Rightarrow$. Hence α is not causally consistent. $\qquad\square$

As follows directly from the definition, if (α, u) is a bitheory of $\Rightarrow$ and v an arbitrary deductively closed set such that $u \subseteq v \subseteq \alpha$, then (α, v) will also be a bitheory of $\Rightarrow$. This indicates that the set of bitheories is determined, in effect, by its inclusion minimal elements; we will call them *minimal* bitheories in what follows.

A simplified description of minimal bitheories is given in the next lemma.

Lemma 6.5 *If α is a world, then a pair (α, u) is a minimal bitheory of $\Rightarrow$ if and only if $\alpha \not\Rightarrow \overline{u}$, and $\alpha \Rightarrow v$, for any $v \supset \overline{u}$.*

The following consequence of the above result gives a convenient alternative description of minimal bitheories.

Lemma 6.6 *If α is a causally consistent world, then a pair (α, u) is a minimal bitheory of $\Rightarrow$ if and only if, for any proposition A,*

$$A \in u \ \ iff \ \ \alpha \Rightarrow \overline{u}, A.$$

The importance of minimal bitheories stems from the following fact:

Lemma 6.7 *If $v \not\Rightarrow w$, then there exists a minimal bitheory (α, u) of $\Rightarrow$ such that $v \subseteq \alpha$ and $w \cap u = \emptyset$.*

The above result will be used in the next section in the proof of the completeness theorem.

To end this section, we will give a characterization of propositions that are "minimally produced" by worlds.

Lemma 6.8 *If α is a causally consistent world, then $A \in u$, for some minimal bitheory (α, u) if and only if there exists a finite set b such that $\alpha \Rightarrow b, A$ and $\alpha \not\Rightarrow b$.*

Logical semantics.   We will describe now a logical (monotonic) semantics of disjunctive causal relations. It will be just a slight modification of the semantics for causal theories, given in H. Turner (1999). The latter semantics has been described in terms of possible worlds models of the form (i, W), where i is a propositional interpretation while W is a set of propositional interpretations that contains i. However, any pair (i, W) determines a unique bitheory (α, u), where α is the world corresponding to the interpretation i while u is a theory containing the propositions that hold in all interpretations from W. Accordingly, by a *causal semantics* we will mean a set of bitheories. The validity of disjunctive causal rules with respect to this semantics is defined as follows.

Definition 6.4 *A causal rule $a \Rightarrow b$ will be said to be* valid *in a causal semantics $\mathcal{B}$ if, for any bitheory (α, u) from $\mathcal{B}$, $a \subseteq \alpha$ implies $b \cap u \neq \emptyset$.*

We will denote by $\Rightarrow_{\mathcal{B}}$ the set of all causal rules that are valid in a causal semantics $\mathcal{B}$. It can be easily verified that this set is closed with respect to the postulates for disjunctive causal relations, and hence $\Rightarrow_{\mathcal{B}}$ forms a disjunctive causal relation. To prove completeness, for any disjunctive causal relation $\Rightarrow$, we can consider its *canonical causal semantics* defined as the set of all minimal bitheories of $\Rightarrow$. Then the following theorem shows that this semantics is *strongly complete* for the source disjunctive causal relation.

Theorem 6.9 *If $\mathcal{B}_{\Rightarrow}$ is the canonical semantics for a disjunctive causal relation $\Rightarrow$, then, for any sets of propositions v, w,*

$$v \Rightarrow w \text{ iff } w \cap u \neq \emptyset, \text{ for any } (\alpha, u) \in \mathcal{B}_{\Rightarrow} \text{ such that } v \subseteq \alpha.$$

Proof. If $v \Rightarrow w$ and $v \subseteq \alpha$, for some world α, then clearly $\alpha \Rightarrow w$. Suppose that (α, u) is a minimal bitheory such that $w \cap u = \emptyset$. Then $w \subseteq \overline{u}$, and consequently $\alpha \Rightarrow \overline{u}$, contrary to our assumptions. Hence $w \cap u \neq \emptyset$, as required.

In the other direction, if $v \nRightarrow w$, then by lemma 6.7 there is a minimal bitheory (α, u) such that $v \subseteq \alpha$ and $w \cap u = \emptyset$. This completes the proof. $\qquad\square$

Combining the above results, we finally conclude with the following.

Corollary 6.10 *A relation on sets of propositions is a disjunctive causal relation if and only if it is determined by a causal semantics.*

Due to the direct correspondence between the causal semantics and the possible worlds semantics from H. Turner (1999), disjunctive causal relations correspond to a subsystem of Turner's UCL. Namely, a causal rule $a \Rightarrow B_1, \ldots, B_n$ corresponds to the modal UCL formula $\bigwedge a \to \mathbf{C}B_1 \vee \cdots \vee \mathbf{C}B_n$. Thus, disjunctive causal relations cover Turner's causal theories that involve modal formulas with only positive occurrences of the causal operator $\mathbf{C}$ (see also Lin 1996).

6.3 Singular Causal Inference

Disjunctive causal rules having a single proposition in their heads will be called *singular* in what follows. Such causal rules have the same meaning as the causal rules of the causal calculus. In this section we will give a precise characterization of disjunctive causal relations that are generated by singular causal rules.

To begin with, the following fact can be easily verified:

Lemma 6.11 *The set of singular rules belonging to a disjunctive causal relation forms a causal inference relation.*

In what follows, the above causal inference relation will be called the *normal subrelation* of a disjunctive causal relation (by analogy with normal logic programs). Actually, there is a quite simple and modular recipe how such a subrelation could be obtained from a given set of disjunctive causal rules.

For any set of disjunctive causal rules Δ, let us consider the following set of singular rules:

$$N(\Delta) = \{a, \neg b \Rightarrow \bigvee c \mid a \Rightarrow b, c \in \Delta\}.$$

Let $\Rightarrow^c_{N(\Delta)}$ denote the least causal inference relation that includes $N(\Delta)$. Then the following result shows that the set $N(\Delta)$ captures the "singular content" of Δ.

Theorem 6.12 $\Rightarrow^c_{N(\Delta)}$ *is the normal subrelation of* $\Rightarrow_\Delta$.

Disjunctive causal relations that are generated by singular causal rules constitute a disjunctive counterpart of causal inference relations from chapter 4.

Definition 6.5 *A disjunctive causal relation will be called* singular *if it is a least disjunctive causal relation containing some set of singular causal rules.*

In what follows, we are going to give more instructive descriptions of such disjunctive causal relations.

A causal semantics $\mathcal{B}$ will be called *functional* if for any world α there is no more than one theory u such that $(\alpha, u) \in \mathcal{B}$. Then we have the following theorem.

Theorem 6.13 *A disjunctive causal relation is singular if and only if its canonical causal semantics is functional.*

Proof. As has been shown earlier, the canonical semantics of a causal relation determined by a set of singular causal rules Δ is the set of all bitheories of the form $(\alpha, \mathrm{Th}(\Delta(\alpha)))$. Clearly, this semantics is functional. In the other direction, if the canonical semantics is functional, and, for some minimal bitheory (α, u), $\alpha \not\Rightarrow b$ and $\alpha \not\Rightarrow c$, then $b, c \subseteq \overline{u}$, and consequently $\alpha \not\Rightarrow b, c$. Hence the causal relation is singular by the preceding corollary. $\square$

Unfortunately, the above claim cannot be extended to arbitrary causal semantics since functional causal semantics do not always produce singular causal relations.

The next result states that, for singular causal relations, worlds produce only determinate effects.

Lemma 6.14 *A disjunctive causal relation is singular iff, for any world* α, $\alpha \Rightarrow b, c$ *only if either* $\alpha \Rightarrow b$ *or* $\alpha \Rightarrow c$.

In order to give yet another important description of singular causal relations, let us reconsider first the following example.

Example 6.1 *Two singular rules* $A \wedge C \Rightarrow B$ *and* $A \wedge \neg C \Rightarrow \neg B$ *imply* $A \Rightarrow B, \neg B$ *(by Right Monotonicity and Or), though they imply neither* $A \Rightarrow B$, *nor* $A \Rightarrow \neg B$.

The above example shows that singular causal rules can generate nontrivial disjunctive rules that cannot be decomposed directly into singular rules. As we are going to show, however, this example describes, in a sense, the only way of producing nontrivial disjunctive rules from singular ones. Thus, the following theorem shows that, for singular causal relations, disjunctive effects are always "separable" by adding some further assumptions to the premises.

Theorem 6.15 *A disjunctive causal relation is singular if and only if it satisfies the following condition:*

$$\text{If } a \Rightarrow b, c, \text{ then } a, A \Rightarrow b \text{ and } a, \neg A \Rightarrow c, \text{ for some proposition } A.$$

The above theorem basically says that indeterminate effects arise in singular disjunctive causal relations only due to "forgetting" of some relevant parameters. This characteristic property is even more vivid in the following corollary (which has actually been used in the proof of the theorem; see the appendix to this chapter):

Corollary 6.16 *A disjunctive causal relation is singular if and only if it satisfies the following condition:*

$$a \Rightarrow B_1, \ldots, B_n \text{ iff there are pairwise incompatible propositions } A_1, \ldots, A_n \text{ such that their disjunction is a tautology, and } a, A_i \Rightarrow B_i, \text{ for any } i \leq n.$$

As a last result of this section, we will show that any disjunctive causal relation can be viewed as a language restriction of some singular causal relation. We provide a constructive proof of this claim by transforming any set Δ of disjunctive causal rules in a language $\mathcal{L}$ into a set of singular rules Δ_s in some extended language as follows.

For any causal rule $r = a \Rightarrow B_1, \ldots, B_n$ from Δ, we will introduce n new propositional atoms $r_1, \ldots, r_n$, and consider the following set of singular rules:

$$a, \neg r_1, \ldots, \neg r_n \Rightarrow \mathbf{f}$$

$$a, r_i \Rightarrow B_i, \text{ for all } i \leq n.$$

Δ_s will denote the set of all singular rules obtained in this way from Δ. Then it can be shown that the generated causal relation $\Rightarrow_\Delta$ is exactly the restriction of $\Rightarrow_{\Delta_s}$ to $\mathcal{L}$. As a result, we obtain the following theorem.

Theorem 6.17 *For any disjunctive causal relation $\Rightarrow$ in a language $\mathcal{L}$, there exists a singular causal relation $\Rightarrow_s$ in an extended language $\mathcal{L}_s \supseteq \mathcal{L}$ such that $\Rightarrow$ coincides with the restriction of $\Rightarrow_s$ to the language $\mathcal{L}$.*

The above theorem shows that any disjunctive causal relation could be seen as a singular causal relation in which some relevant factors are "hidden."

It is important to note also that a number of further causal rules involving the new propositional atoms can be added to the translation without affecting the proofs. For example, we may require further that that the new atoms are mutually incompatible by adding the following causal constraints

$$r_i, r_j \Rightarrow \mathbf{f}$$

for any $i \neq j$. Moreover, each new atom could be made exogenous by adding the rules:

$$r_i \Rightarrow r_i \qquad \neg r_i \Rightarrow \neg r_i.$$

The above result lends, in effect, a formal logical support to the claim that deterministic causation (and determinism in general) provides a comprehensive basis for a general theory of causation—see, e.g., Pearl (2000) for a similar position.

6.4   Stable Semantics

Now we will turn to the main subject of interest in causal theories, namely, their nonmonotonic semantics.

The nonmonotonic semantics of causal theories arises from a general requirement that adequate models of such theories should satisfy two conditions. First, they should be closed with respect to the causal rules. Second, they should be *causally explainable*: propositions that hold in such models should have justifications in the sense of being produced (caused) by the causal rules that are "active" in the model.

The nonmonotonic semantics for singular causal theories has been generalized to arbitrary theories of universal causal logic in H. Turner (1999). Since disjunctive causal relations form a subsystem of UCL, this semantics can be immediately translated into our framework.

Definition 6.6 *A world α will be said to be* exact *with respect to a disjunctive causal relation $\Rightarrow$ if (α, α) is a minimal bitheory of $\Rightarrow$. The set of all exact worlds will be said to form a* stable nonmonotonic semantics *of $\Rightarrow$.*

The above definition is equivalent to the definition of causally explained interpretations, given in H. Turner (1999).

The above stable semantics appears as a natural, and even almost inevitable, extension of the nonmonotonic semantics for singular causal theories to the disjunctive case. This impression could also be supported by the fact, established in H. Turner (1999), that in the general correspondence between UCL and disjunctive default logic of Gelfond et al. (1991), causally explained interpretations correspond to extensions of disjunctive default theories. Moreover, by the same correspondence, simple disjunctive theories can be translated into disjunctive logic programs, and then causally explained interpretations will exactly correspond to answer sets of such programs. Summing up, there is a tight correspondence between the above semantics and respectable semantics of nonmonotonic formalisms and disjunctive logic programming.

Despite the above impressive support, it has been noticed in the literature that semantics of this kind are problematic in their treatment of indeterminate information. Namely, due to minimization that is involved in their definitions, these semantics tend to give an exclusive interpretation to disjunctions. It turns out that the same shortcoming is preserved by the above stable semantics for disjunctive causal theories. The following example has been suggested by Ray Reiter:

Example 6.2 (Dropped Pin) *Suppose that we drop a pin on a board painted black and white. As a result, the pin lands on the board in such a way that it touches either black or white area, or both. Moreover, if the pin is large (or black and white areas are small), then a most probable effect is that the pin touches both black and white areas.*

A natural representation of this situation could be given using the main disjunctive rule

$$Drop \Rightarrow Black, White$$

and a couple of auxiliary causal assertions that need not bother us for now. Unfortunately, the stable semantics of the resulting causal theory does not explain the world in which the pin touches both black and white areas, though it readily justifies worlds in which the pin touches only one of them.

Turner has mentioned a similar problem.

Example 6.3 (MaybeFlip) *(H. Turner 1999) An action MaybeFlipUp is a nondeterministic action that moves one of the two latches of a suitcase up. However, if we describe this action using something like a disjunctive rule*

$$MaybeFlipUp \Rightarrow LeftLatchUp, RightLatchUp,$$

then according to the stable semantics (coupled with natural persistence assumptions), the action would never cause the right latch to go up if the left latch was already up.

In order to overcome the above problem, it is important to determine what causes it. In this respect, the present causal formalism has two main components. First of all, we have a logical relation of disjunctive causal inference that determines not only the internal logic of causal rules but also, and most importantly, a natural knowledge representation framework for describing causal information. If we were challenged this part of the formalism, we would have to claim that either our causal logic need to be corrected or else that the above disjunctive causal rules does not give an adequate representation of the source non-determinate information. For many good reasons, we will not adopt this approach. Instead, we will question the second component of the system, namely, the associated stable nonmonotonic semantics.

As a general theoretical background for a discussion, we will show now that the stable semantics imposes quite a radical view on the role of disjunctive causal rules. Namely, such rules can be considered as an inessential "syntactic sugar" that does not change our representation capabilities. This is an immediate consequence of the following quite surprising result:

Theorem 6.18 *The stable semantics of a disjunctive causal relation coincides with the nonmonotonic semantics of its normal subrelation.*

In accordance with the above theorem, in order to compute a stable semantics of a disjunctive causal theory Δ, we can transform Δ into a set of singular causal rules $N(\Delta)$, as described earlier, and then compute the nonmonotonic semantics for the latter singular causal theory. In this sense, any disjunctive causal rule $a \Rightarrow b$ can be directly replaced by a set of singular causal rules.

Example 6.4 (Dropped Pin, continued) *Returning to Reiter's* Dropped Pin *example, the rule Drop $\Rightarrow$ Black, White implies the following three singular rules:*

$$Drop, \neg Black \Rightarrow White \qquad Drop, \neg White \Rightarrow Black$$

$$Drop \Rightarrow Black \vee White.$$

Since the last rule is not determinate, only the first two rules determine the causal nonmonotonic semantics (see section 4.5 in chapter 4). Note that these two rules cannot be applied simultaneously, due to the constraint Drop $\wedge \neg Black \wedge \neg White \Rightarrow \mathbf{f}$ that is implied by each of them. As a result, such rules cannot give an explanation for the possible fact Black $\wedge$ White.

Extending the above results, we are going to show now that they are actually a by-product of a deeper, *logical* reduction of disjunctive causal rules in an appropriate causal logic that is sanctioned by the stable nonmonotonic semantics.

Stable disjunctive causal relations.   It turns out that, unlike the causal calculus, disjunctive causal relations do not form a maximal causal logic that is adequate for the stable semantics. Instead, the latter corresponds to a rather peculiar stronger logic described in the next definition.

Definition 6.7 *A disjunctive causal relation will be called* stable *if it satisfies the following postulate:*

(Saturation) $A, B \Rightarrow A, A \rightarrow B$

To begin with, the following lemma gives a number of useful equivalent conditions of Saturation.

Lemma 6.19 *A disjunctive causal relation is stable if and only if it satisfies one of the following conditions:*

1. $A \Rightarrow B \vee A, \neg B \vee A$;
2. *If* $a \Rightarrow b, C \vee D$, *then* $a, C, D \Rightarrow b, C, D$;
3. *If* $a \Rightarrow \bigvee b, c$, *then* $a, b \Rightarrow b, c$;
4. *If* $a, \neg b_1 \Rightarrow \bigvee b_2$, *for any partition* (b_1, b_2) *of* b, *then* $a \Rightarrow b$.

The last condition above is especially interesting since it shows, in effect, that any disjunctive rule is equivalent to the set of its singular consequences with respect to a stable causal relation. Indeed, we have seen earlier that the set of singular causal rules derived from a given disjunctive causal rule $a \Rightarrow b$ amounts to $\{a, \neg b_1 \Rightarrow \bigvee b_2\}$ while the condition (4) says that the latter are sufficient for deriving $a \Rightarrow b$ in stable causal relations. As a result, we obtain the following lemma.

Lemma 6.20 *For stable causal relations, any disjunctive causal theory Δ is equivalent to its singular reduction $N(\Delta)$.*

The above result shows that stable disjunctive causal relations are essentially equivalent to singular causal production relations. More exactly, for stable disjunctive causal relations, any disjunctive rule $a \Rightarrow b$ can be seen as a shortcut for the set of singular rules $\{a, \neg b_1 \Rightarrow \bigvee b_2\}$.

Now we are going to give a semantic interpretation for stable causal relations.

Definition 6.8 *A bitheory (α, u) will be be called* saturated *if $u = \alpha \cap \beta$ for some world β. A causal semantics will be called* saturated *if it contains only saturated bitheories.*

A bitheory (α, u) is saturated if u either coincides with α or is a maximal subtheory of α. The following result shows that stable causal relations are sound and complete for saturated causal semantics.

Theorem 6.21 *A disjunctive causal relation is stable if and only if it is generated by a saturated causal semantics.*

Our next result will show that stable causal relations do not introduce new derived singular rules.

For a disjunctive causal relation $\Rightarrow$, we will denote by $\Rightarrow^{st}$ the least stable causal relation containing $\Rightarrow$. Similarly, $\Rightarrow^{st}_\Delta$ will denote the least stable causal relation containing a set of causal rules Δ.

Theorem 6.22 *The normal subrelation of a disjunctive causal relation $\Rightarrow$ coincides with the normal subrelation of $\Rightarrow^{st}$.*

Now we are going to show that stable causal relations constitute a strongest causal logic that is adequate for the stable semantics.

As has been defined in chapter 4, two singular causal theories are *causally equivalent* if they determine the same causal inference relation. Now we will define the corresponding notions for (stable) disjunctive causal relations.

Definition 6.9 *Causal theories Δ and Γ will be called*

- stably equivalent *if $\Rightarrow^{st}_\Delta$ coincides with $\Rightarrow^{st}_\Gamma$;*
- nonmonotonically equivalent *if they determine the same stable semantics; and*

- strongly equivalent *if, for any set* Φ *of causal rules,* $\Delta \cup \Phi$ *is nonmonotonically equivalent to* $\Gamma \cup \Phi$.

Two disjunctive theories are stably equivalent if each can be obtained from the other using the postulates for stable disjunctive causal relations. Strongly equivalent theories are interchangeable in any larger causal theory without changing the stable nonmonotonic semantics. Consequently, strong equivalence can be seen as an equivalence with respect to the monotonic logic of causal theories. As we are going to show, these two notions of equivalence coincide. But first we will present a couple of auxiliary results.

First, as a consequence of the above result, we obtain the following corollary.

Corollary 6.23 *Causal theories* Δ *and* Γ *are stably equivalent if and only if* $N(\Delta)$ *is causally equivalent to* $N(\Gamma)$.

Proof. Lemma 6.20 shows, in effect, that any set of rules Δ is stably equivalent to $N(\Delta)$. Consequently, Δ and Γ are stably equivalent if and only if $N(\Delta)$ and $N(\Gamma)$ determine the same stable causal relation. But by theorem 6.22, this holds if and only if they are causally equivalent.   □

Second, we have the following lemma.

Lemma 6.24 Δ *is strongly equivalent to* $N(\Delta)$.

Proof. Γ is always nonmonotonically equivalent to $N(\Gamma)$, and consequently $N(\Delta) \cup \Phi$ is equivalent to $N(N(\Delta) \cup \Phi)$ while the latter coincides with $N(\Delta \cup \Phi)$. Therefore, $N(\Delta) \cup \Phi$ is nonmonotonically equivalent to $\Delta \cup \Phi$, which shows that Δ is strongly equivalent to $N(\Delta)$.   □

Now, as a consequence of the above results, we can establish the following theorem.

Theorem 6.25 *Two disjunctive causal theories are strongly equivalent if and only if they are stably equivalent.*

6.5   Covering Semantics

As we already mentioned, Reiter's example presents difficulties not only for causal reasoning but also for disjunctive logic programming, default logic, and theories of update. Briefly put, since the intended models of all these formalisms are required to be minimal, they usually enforce an exclusive interpretation of disjunctive information. However, in the history of nonmonotonic reasoning there have been a number of attempts to define nonmonotonic semantics that would preserve the usual inclusive understanding of logical disjunctions, the most prominent suggestions being Minker's weak generalized closed world assumption (GCWA) (see Lobo, Minker, and Rajasekar 1992) and Sakama's possible model semantics (Sakama 1989) in logic programming, an alternative semantics for circumscription, suggested in Eiter, Gottlob, and Gurevich (1993), and a theory of disjunctive updates proposed in Zhang and Foo (1996). The covering semantics, described below, will belong to the same class.

Basically, our desiderata will be the same as those described above for the stable nonmonotonic semantics. Namely, we want to single out models that are closed with respect to the causal rules but also such that any fact that holds in the model is explainable, ultimately, by the causal rules that hold in the model. Our point of departure, however, will be that the notion of explainability determined by disjunctive rules could be given a different interpretation.

A disjunctive rule $p \Rightarrow q, r$ says that p causes either q or r. In the deterministic setting this *normally* means that there are additional facts, say q_0 and r_0, such that $p \wedge q_0$ causes q, $p \wedge r_0$ causes r and, in addition, either q_0, or r_0 holds in any situation in which p holds. This gives us a way of explaining why q or r occurs, provided we know that p holds. Note, however, that there are no a priori constraints on the compatibility of q_0 and r_0, so they could hold simultaneously, in which case both q and r are caused (and hence explainable).

Our qualification about normality above indicates, however, that the above interpretation may sometimes lead us astray. Thus, it may well be that the disjunctive rule $p \Rightarrow q, r$ is simply a logical consequence of a valid singular causal rule $p \Rightarrow q$, in which case it should not give us an explanation for an occurrence of r. This means that the disjunctive rule should at least be *irreducible* in order to serve as an explanation for its indeterminate conclusions.

Unfortunately, even this qualification is not sufficient. A more complex, but equally plausible situation might be that, though $p \Rightarrow q, r$ is an irreducible rule, q is actually caused by another rule, say $q_0 \Rightarrow q$, in which case the presence of p cannot serve as a proper explanation for r.

The above considerations suggest that indeterminate causal rules can provide explanations for their conclusions only in situations when they are not superseded by other active causal rules that bring about more determinate effects. Only in such cases the explanation provided by the indeterminate rules will be the only explanation possible. The definition of causally covered models, given below, is an attempt to make this idea precise.

If (α, u) is a minimal bitheory of a disjunctive causal relation $\Rightarrow$, we will say that u is a *causal projection* of α with respect to $\Rightarrow$. A world is *causally consistent* if it has a causal projection in this sense, but it may have a number of causal projections. Still, there are worlds that are uniquely determined by their causal projections. Such worlds will constitute the alternative nonmonotonic semantics of disjunctive causal theories.

Definition 6.10 *A world α will be said to be* causally covered *with respect to a disjunctive causal relation $\Rightarrow$ if it is the unique world that includes all its causal projections. The set of all causally covered worlds will constitute a nonmonotonic* covering semantics *of $\Rightarrow$.*

Yet another description of causally covered worlds is given in the next lemma.

Lemma 6.26 *A world α is causally covered if and only if*

$$\alpha = \mathrm{Th}(\bigcup\{u \mid u \text{ is a causal projection of } \alpha\}).$$

Proof. Any world β includes all the models of α if and only if it contains the logical closure of the union of these models. Hence the result.   □

The above lemma gives a precise meaning to the requirement that facts holding in intended interpretations of a causal theory should be explained by the causal rules of the theory. Namely, any proposition belonging to a causally covered world should be a logical consequence of the facts that have a causal explanation in this world.

The following results will describe the relationship between the covering and stable semantics. Thus, the next lemma shows that the stable semantics is stronger than the covering semantics.

Lemma 6.27 *Any exact world is also causally covered.*

Proof. Follows from the fact that if α is a causally explained world, then α itself is the only model of α.   □

The next result shows that both semantics coincide for singular causal relations.

Theorem 6.28 *If $\Rightarrow$ is a singular causal relation, then its stable semantics coincides with its covering semantics.*

Proof. By theorem 6.13, any causally consistent world of a singular causal relation has a unique model. Consequently, a world is causally covered if and only if it coincides with its model, which means that it is causally explained.   □

Thus, the difference between the two semantics can show itself only for genuinely disjunctive causal relations.

Example 6.5 *Let us consider the following disjunctive causal theory:*

$$p \Rightarrow q, r \quad p \Rightarrow p \quad \neg p \Rightarrow \neg p \quad \neg q \Rightarrow \neg q \quad \neg r \Rightarrow \neg r.$$

The world α that is determined by the literals $\{p, q, r\}$ is not exact, so it is not included in the stable semantics. However, it has two causal projections, namely $\mathrm{Th}(p, q)$ and $\mathrm{Th}(p, r)$. Consequently, it is causally covered with respect to this theory. As a result, the stable semantics for this example validates an intuitively implausible conclusion $\neg(q \wedge r)$, though it is not valid in the covering semantics.

As a last result in this section, we will show that disjunctive causal relations constitute a maximal logic that is adequate for the covering semantics.

Definition 6.11 *Disjunctive causal theories Γ and Δ will be called*

- causally equivalent, *if they determine the same disjunctive causal relation;*
- c-equivalent, *if they determine the same covering semantics; and*
- strongly c-equivalent, *if, for any set Φ of causal rules, $\Gamma \cup \Phi$ is c-equivalent to $\Delta \cup \Phi$.*

Disjunctive theories are causally equivalent if each can be obtained from the other using the postulates for disjunctive causal relations while strongly c-equivalent theories are interchangeable in any larger causal theory without changing the covering semantics. In other words, strong c-equivalence can be seen as a maximal logical (monotonic) equivalence that preserves the covering semantics.

Theorem 6.29 *Disjunctive causal theories are strongly c-equivalent if and only if they are causally equivalent.*

The above theorem says, in effect, that, unlike the case of the stable semantics, any logical distinction between disjunctive causal theories, allowed by the formalism of disjunctive causal relations, is relevant for determining the covering semantics either for these theories themselves, or for their extensions.

6.5.1 Normalization

In this final section we are going to show that in regular circumstances, a disjunctive causal relation can be systematically extended to a singular causal inference relation in such a way that the covering semantics of the former will coincide the nonmonotonic semantics of the latter.

To begin with, there are many cases when disjunctive causal rules are *logically* reducible to singular rules even in the general logic of disjunctive causal inference. For example, the rule $A \Rightarrow B, \neg B$ is equivalent to a pair of singular causal rules $A, B \Rightarrow B$ and $A, \neg B \Rightarrow \neg B$. This is a consequence of the following general fact:

Lemma 6.30 *A rule $a \Rightarrow B_1, \ldots, B_n$ is reducible to the set of singular rules $\{a, B_i \Rightarrow B_i \mid i = 1, \ldots, n\}$ and a constraint $a, \neg B_1, \ldots, \neg B_n \Rightarrow \mathbf{f}$ if and only if $a, B_i, B_j \Rightarrow B_i$, for any $i \neq j$.*

Note that the above characteristic condition for reduction holds, in particular, when all B_i are incompatible, given a, that is, when $a, B_i, B_j \Rightarrow \mathbf{f}$. For example (cf. Lin 1996), a disjunctive rule $a \Rightarrow B \wedge C, B \wedge \neg C, \neg B \wedge C$ is reducible to the following set of singular rules:

$$a, B \Rightarrow B \quad a, \neg B \Rightarrow \neg B \quad a, C \Rightarrow C$$

$$a, \neg C \Rightarrow \neg C \quad a, \neg B, \neg C \Rightarrow \mathbf{f}.$$

The above result shows, in effect, that irreducibly disjunctive rules could arise only in cases when their heads contain mutually compatible propositions. For such rules, we need a different strategy.

Let $\mathcal{B}$ be a causal semantics. For a world α, we will denote by u_α the least theory containing all the causal projections of α in $\mathcal{B}$, that is,

$$u_\alpha = \mathrm{Th}(\bigcup \{u \mid u \text{ is a causal projection of } \alpha\}).$$

Now, let $n(\mathcal{B})$ denote the set of all bitheories of the form (α, u_α). Clearly, $n(\mathcal{B})$ is a functional causal semantics, so it will produce a singular causal relation. We will call $n(\mathcal{B})$ the *normalization* of $\mathcal{B}$.

Using the above notions, the basic idea behind the constructions below can be described quite simply. Suppose that $\mathcal{B}$ is the canonical semantics of some disjunctive causal relation $\Rightarrow$. Then it is easy to verify that a world α is causally covered in $\mathcal{B}$ if and only if it is an exact world in its normalization $n(\mathcal{B})$ (cf. lemma 6.26). Consequently, if $n(\mathcal{B})$ is a canonical semantics of some singular causal relation $\Rightarrow^n$, then the stable semantics of $\Rightarrow^n$ will coincide with the covering semantics of $\Rightarrow$.

Unfortunately, the above construction works only under severe finiteness constraints. Accordingly, to simplify our discussion at this stage, we will occasionally restrict our attention to finite causal theories and causal relations in some finite propositional language.

We will begin with the following notions.

Definition 6.12 *Let $\mathcal{B}$ be the canonical semantics of a disjunctive causal relation $\Rightarrow$ and $n(\mathcal{B})$ its normalization. The (singular) causal inference relation generated by $n(\mathcal{B})$ will be called the* normalization *of $\Rightarrow$, and it will be denoted by $\Rightarrow^n$.*

The following lemma shows that the normalization of a disjunctive causal relation $\Rightarrow$ always includes the normal subrelation of the latter.

Lemma 6.31 *If $\Rightarrow_n$ is a normal subrelation of $\Rightarrow$, then $\Rightarrow_n \subseteq \Rightarrow^n$.*

Proof. If $A \Rightarrow B$ and $A \in \alpha$, for some causally consistent world α, then B belongs to all models of α, and therefore $B \in u_\alpha$. Hence the result. $\square$

Moreover, the following result holds in the finite case:

Theorem 6.32 *If $\Rightarrow$ is a disjunctive causal relation in a finite language, then its covering semantics coincides with the stable semantics of its normalization.*

Thus, for a class of finite disjunctive causal relations, the covering semantics coincides with the stable semantics of their normalizations. It should be kept in mind, however, that this normalization involves, in effect, an extension of the source disjunctive causal relation with new singular rules that are not derivable logically from it.

Normalization as a nonmonotonic completion. The normalization can be viewed as a (presumably unusual) kind of *nonmonotonic completion* of the source disjunctive causal relation, which is obtained by restricting its logical semantics to functional bitheories. As

such, it is similar to other nonstandard forms of completion that exist in the AI literature, such as the Markov assumption (see chapter 10 in this book). Thus, as for the latter,
this semantic restriction does not lead to a change of the underlying logic of disjunctive
causal inference; this follows immediately from the fact that any disjunctive causal relation is a language restriction of some singular causal relation—see theorem 6.17. Still, the
restriction of the set of models means that more causal derivations become valid, so more
information is obtained in particular cases.

6.5.2 Causal Covers

We still need to provide a more constructive description of normalization. To this end, we
introduce the following definition.

Definition 6.13 *A proposition C is a causal cover of a set b of propositions in a disjunctive
causal relation if $C \Rightarrow b$ and $D \wedge \neg C \Rightarrow$, for any D such that $D \Rightarrow b$.*

The causal cover of a set of propositions is a weakest causally sufficient condition for
this set. Notice, in particular, that falsity $\mathbf{f}$ is a cover of the empty set $\emptyset$.

An alternative characterization of covers is given in the next lemma. The proof is
immediate.

Lemma 6.33 *C is a causal cover of b iff, for any causally consistent world α, $\alpha \Rightarrow b$ if and
only if $C \in \alpha$.*

Based on this lemma, it can be easily verified that in the finite case any set of propositions
has a causal cover.

The following consequence of the lemma connects causal covers with normalization:

Corollary 6.34 *If C is a causal cover of b, and $a \Rightarrow b, A$, then $a, \neg C \Rightarrow^n A$.*

Proof. If $a \Rightarrow b, A$, $a \subseteq \alpha$, and $\neg C \in \alpha$, for some causally consistent world α, then $\alpha \not\Rightarrow b$
and $\alpha \Rightarrow b, A$, and therefore A belongs to some model of α. Consequently, $A \in u_\alpha$, and
hence $a, \neg C \Rightarrow A$ holds in any bitheory from $n(\mathcal{B})$. $\qquad\square$

The notion of a causal cover allows us to provide a relatively simple description of
normalization.

Given a finite disjunctive causal theory Δ, we define Δ^s as the set of all singular causal
rules of the form:

$$a, \neg C_b \Rightarrow A,$$

such that $a \Rightarrow b, A \in \Delta$, and C_b is a causal cover of b. Using this set, we will construct the
following singular causal theory:

$$\Delta^n = \Delta^s \cup \{a \Rightarrow \mathbf{f} \mid a \Rightarrow \emptyset \in \Delta\}.$$

The next result shows that the above singular causal theory generates precisely the normalization of Δ.

Theorem 6.35 *If Δ is finite, then $\Rightarrow_\Delta^n$ coincides with $\Rightarrow_{\Delta^n}$.*

In accordance with the above result, the covering semantics of Δ will coincide with the standard nonmonotonic semantics of a singular causal theory Δ^n. However, this description of normalization is still not fully constructive since it depends on determining causal covers.

Construction of covers. As a final step in our description, we describe algorithms of constructing covers for finite causal theories. We consider first a simplest case of causal rules that do not involve logical connectives and then generalize the method to arbitrary causal theories.

A causal rule $a \Rightarrow b$ will be called a *literal* one if a and b are sets of literals. A causal theory will be called *literal* if it contains only literal causal rules. Finally, a disjunctive causal relation will be called *literal* if it is generated by a literal causal theory. It should be noted that most examples of indeterminate causation appearing in the literature are representable by causal theories of this kind.

By a *reduction* of a causal rule $a \Rightarrow b$ we will mean any rule $a, \neg b_0 \Rightarrow b_1$ such that $b = b_0 \cup b_1$ and $b_0 \cap b_1 = \emptyset$. $\Delta^\neg$ will denote the set of all reductions of the rules from Δ.

Note that all the rules from $\Delta^\neg$ are derivable from Δ (by Reduction). It should be clear also that if Δ is a literal theory, then $\Delta^\neg$ will also be a literal one.

The construction below is based on the following fact about literal causal relations:

Lemma 6.36 *If Δ is a literal causal theory, then for any causally consistent world α and any set b of literals, $\alpha \Rightarrow_\Delta b$ only if $d \subseteq b$, for some rule $a \Rightarrow d \in \Delta^\neg$ such that $a \subseteq \alpha$.*

Now, given a finite literal causal theory Δ, we can define a causal cover for a finite set of literals b as follows:

$$C_b = \vee\{\wedge a \mid a \Rightarrow d \in \Delta^\neg \ \& \ d \subseteq b\}.$$

Lemma 6.37 *If Δ is a literal theory, then C_b is a causal cover of b in $\Rightarrow_\Delta$.*

Recall that for a finite causal theory Δ, the normalization can be constructed using the set of singular rules $a, \neg C_b \Rightarrow A$, for every disjunctive rule $a \Rightarrow b, A$. Now, the above construction can be used for determining all causal covers required for defining the normalization of a literal causal theory. Note also that if Δ is a literal causal theory, then its normalization will be a determinate causal theory, that is, it will contain only causal rules of the form $a \Rightarrow l$, where l is a literal. As has been shown already in McCain and Turner (1997), the nonmonotonic semantics of such theories coincides with the set of models of their classical completion.

The following examples show that normalization of literal disjunctive causal theories corresponds to established ways of describing nondeterminate information by singular causal rules.

Example 6.6 *Let us return to the disjunctive causal theory:*

$$p \Rightarrow q, r; \quad p \Rightarrow p; \quad \neg p \Rightarrow \neg p; \quad \neg q \Rightarrow \neg q; \quad \neg r \Rightarrow \neg r.$$

To construct the normalization, the rule $p \Rightarrow q, r$ should be replaced with two singular rules $p, \neg C_{\{q\}} \Rightarrow r$ and $p, \neg C_{\{r\}} \Rightarrow q$. The relevant causal covers can be calculated by the above formula for C_b, and we immediately obtain $C_{\{q\}} = p \wedge \neg r$ and $C_{\{r\}} = p \wedge \neg q$. As a result, the above singular rules are reduced, respectively, to $p \wedge r \Rightarrow r$ and $p \wedge q \Rightarrow q$.

As can be seen, the latter rules correspond exactly to the way of representing indeterminate information suggested in McCain and Turner (1997) and Giunchiglia et al. (2004).

Example 6.7 *Suppose that we add a rule $s \Rightarrow q$ to the causal theory in the preceding example. Then the causal cover $C_{\{q\}}$ for $\{q\}$ changes to $(p \wedge \neg r) \vee s$. As a result, the first singular rule of the normalization will now be $p \wedge r \wedge \neg s \Rightarrow r$. In other words, the explainability of r will depend now not only on p but also on the absence of s; this is because s causes q and thereby supersedes the indeterminate rule $p \Rightarrow q, r$.*

The next example shows that the normalization procedure gives us a reasonable solution also in more complex situations than what can be handled intuitively.

Example 6.8 *Consider the following theory:*

$$p \Rightarrow q, r; \quad p \Rightarrow s, \neg r.$$

The relevant causal covers are: $C_{\{q\}} = p \wedge \neg r$, $C_{\{r\}} = p \wedge \neg q$, $C_{\{s\}} = p \wedge r$, $C_{\{\neg r\}} = p \wedge \neg s$. As a result, the normalization of the above causal theory is

$$p \wedge q \Rightarrow q; \quad p \wedge r \Rightarrow r; \quad p \wedge s \Rightarrow s; \quad p \wedge \neg r \Rightarrow \neg r.$$

Finally, we will briefly describe a construction of causal covers for arbitrary finite causal theories.

To begin with, we will introduce a special notation for dealing with "disjunctive" sets of propositions. For two sets of propositions a, b, we will denote by $a\&b$ the set $\{A_i \wedge B_j \mid A_i \in a, B_j \in b\}$. In addition, we will use a "disjunctive" entailment relation $\vDash^\vee$ on sets of propositions, defined as follows: $a \vDash^\vee b$ holds iff for any $A \in a$ there exists $B \in b$ such that $A \vDash B$.

By a *join* of two causal rules $a_1 \Rightarrow b_1$ and $a_2 \Rightarrow b_2$ we will mean a rule $a_1, a_2 \Rightarrow b_1 \& b_2$. $\Delta^\&$ will denote the set of all finite joins of the rules from Δ while $\Delta^\oplus$ will denote $(\Delta^\&)^\neg$, that is, the set of all reductions of the rules from $\Delta^\&$.

Using the above derived set of rules, the following fact can be proved for arbitrary causal theories.

Lemma 6.38 *If Δ is a causal theory, then for any causally consistent world α and any set b, $\alpha \Rightarrow_\Delta b$ only if $d \vDash^\vee b$, for some rule $a \Rightarrow d \in \Delta^\oplus$ such that $a \subseteq \alpha$.*

A causal cover C_b of a set b of propositions can now be described as follows:

$$C_b = \vee\{\wedge a \mid a \Rightarrow d \in \Delta^\oplus \ \& \ d \vDash^\vee b\}.$$

Theorem 6.39 C_b *is a causal cover of b in $\Rightarrow_\Delta$.*

As a by-product of the above construction, we immediately obtain a constructive proof of the fact that any finite disjunctive causal theory is normalizable.

6.6 Summary

The main results of this chapter are twofold. On the one hand, confirming the initial idea of Lin (1996), it has been shown that disjunctive causal rules provide a highly expressive and versatile tool for representing indeterminate causation. Moreover, viewed under a more appropriate covering nonmonotonic semantics, such rules are not reducible logically to ordinary, singular causal rules, so they contain more information. On the other hand, however, it has been shown that even in the latter case this additional information can be expressed using ordinary causal rules by way of *systematic* completion of the source causal theory with new singular causal rules that provide the required information about causes (or excuses) for particular causal effects. As we have seen in the examples, in simple cases such additional rules correspond to existing ways of describing indeterminate actions, given in the literature. However, the described normalization procedure is fully general, so it allows us to automate an important and nontrivial part of representing nondeterminate actions.

Speaking more generally, the above results suggest a certain primacy of ordinary, singular rules for representing causation. More precisely, they show that the restriction to "deterministic" causal rules can be sustained as a convenient and useful representation decision or *representation assumption*. Nevertheless, this general assumption does not deprave truly disjunctive causal rules of their potentially useful role in representing indeterminate causation. Thus, it has been shown that the corresponding "translation" to singular causal rules is in general far from being trivial. It remains to be seen whether disjunctive causal rules can fulfill this representation role in describing complex causal or action domains that go beyond toy examples.

6.A Appendix: Proofs of the Main Results

Lemma (6.3) *If $\Rightarrow_\Delta$ is a least disjunctive causal relation containing Δ, then a bitheory (α, u) is a bitheory of $\Rightarrow_\Delta$ if and only if $b \cap u \neq \emptyset$, for any causal rule $a \Rightarrow b$ from Δ such that $a \subseteq \alpha$.*

Proof. It is sufficient to show that if a bitheory (α, u) satisfies the above condition, it will be closed with respect to any causal rule $a \Rightarrow b$ from $\Rightarrow_\Delta$. The latter claim will be proved by induction on the derivation of $a \Rightarrow b$ from Δ.

If $a \Rightarrow b \in \Delta$ or it is a rule $\Rightarrow \mathbf{t}$ or $\mathbf{f} \Rightarrow$, then the claim is trivial.

The proof for most of the postulates is straightforward, so we will show the claim only for Reduction and And.

Reduction. Assume that $a, \neg B \Rightarrow b$ has been obtained by Reduction from $a \Rightarrow b, B$, and $a, \neg B \subseteq \alpha$. Then $(b \cup \{B\}) \cap u \neq \emptyset$ by the inductive assumption. But u is included in α, so $B \notin u$, and consequently $b \cap u \neq \emptyset$, as required.

And. Assume that $a \Rightarrow b, B \wedge C$ has been obtained by And from $a \Rightarrow b, B$, and $a \Rightarrow b, C$. If $a \subseteq \alpha$, then $(b \cup \{B\}) \cap u \neq \emptyset$ and $(b \cup \{C\}) \cap u \neq \emptyset$ by the inductive assumption. Since u is deductively closed, we have that either $b \cap u \neq \emptyset$, or $B \wedge C \in u$, as required. $\square$

Lemma (6.5) *If α is a world, then a pair (α, u) is a minimal bitheory of $\Rightarrow$ if and only if $\alpha \nRightarrow \overline{u}$, and $\alpha \Rightarrow v$, for any $v \supset \overline{u}$.*

Proof. Only the direction from right to left needs to be shown. To this end, we have to prove that u is a deductively closed set that is included in α.

Assume first that $u \nsubseteq \alpha$. Then there exists $A \in u$ such that $\neg A \in \alpha$. By the above condition, $\alpha \Rightarrow \overline{u}, A$, and therefore $\alpha \Rightarrow \overline{u}$ by Reduction—a contradiction. Hence u is included in α.

Assume now that $a \vDash A$, for some $a \subseteq u$. Then $\alpha \Rightarrow \overline{u}, A_i$, for any $A_i \in a$. Consequently, $\alpha \Rightarrow \overline{u}, A$ by Logical Weakening, and therefore $A \in u$. This shows that u is a deductively closed set. $\square$

Lemma (6.6) *If α is a causally consistent world, then a pair (α, u) is a minimal bitheory of $\Rightarrow$ if and only if, for any proposition A,*

$$A \in u \ \text{ iff } \ \alpha \Rightarrow \overline{u}, A.$$

Proof. Assume first that (α, u) is a minimal bitheory and $A \in u$. Then, due to minimality of u, we have $\alpha \Rightarrow \overline{u}, A$. In the other direction, if $A \notin u$, then $\overline{u} \cup \{A\} = \overline{u}$ and hence $\alpha \nRightarrow \overline{u}, A$.

Assume now that α is causally consistent and the above condition holds. Suppose first that u is the set of all propositions of the language. Since $\mathbf{f} \in u$, the condition implies $\alpha \Rightarrow \mathbf{f}$. But the latter implies $\alpha \Rightarrow \emptyset$ by Falsity and Cut, contrary to our assumptions. Thus, there exists $A \notin u$, which immediately implies $\alpha \nRightarrow \overline{u}$. Moreover, since $\alpha \Rightarrow \overline{u}, A$, for any $A \in u$, we clearly have $\alpha \Rightarrow v$, for any $v \supset \overline{u}$. This completes the proof. $\square$

Lemma (6.7) *If $v \nRightarrow w$, then there exists a minimal bitheory (α, u) of $\Rightarrow$ such that $v \subseteq \alpha$ and $w \cap u = \emptyset$.*

Proof. Assume that $v \nRightarrow w$. Due to compactness, v is included in a maximal set v_0 such that $v_0 \nRightarrow w$. By Logical Strengthening, this set must be deductively closed. Moreover, the postulate Or implies that v_0 must actually be a world. In addition, due to compactness again, w is included in some maximal set w_0 such that $\alpha \nRightarrow w_0$. Then the pair $(v_0, \overline{w}_0)$ will clearly be a required minimal bitheory.   □

Lemma (6.8) *If α is a causally consistent world, then $A \in u$, for some minimal bitheory (α, u) if and only if there exists a finite set b such that $\alpha \Rightarrow b, A$ and $\alpha \nRightarrow b$.*

Proof. If $A \in u$, for some minimal bitheory (α, u), then $\alpha \Rightarrow \overline{u}, A$. By compactness, $\alpha \Rightarrow b, A$ for some finite $b \subseteq \overline{u}$. But $\alpha \nRightarrow \overline{u}$, and hence $\alpha \nRightarrow b$.

If $\alpha \Rightarrow b, A$ and $\alpha \nRightarrow b$, then by the preceding lemma there is a minimal bitheory (α, u) such that $b \subseteq \overline{u}$. But we have also $\alpha \Rightarrow \overline{u}, A$, and hence $A \in u$ by lemma 6.6. This completes the proof.   □

Theorem (6.12) $\Rightarrow^n_{N(\Delta)}$ *is the normal subrelation of $\Rightarrow_\Delta$.*

Proof. We will use $\Rightarrow^n$ as a shortcut for $\Rightarrow^n_{N(\Delta)}$ in the proof.

All the rules from $N(\Delta)$ are derivable from Δ by Reduction and Weakening. Moreover, any postulate for causal production relations is also valid for disjunctive causal relations. Consequently, $\Rightarrow^n$ is included in $\Rightarrow_\Delta$.

If $a \nRightarrow^n A$, then there exists a world α such that $a \subseteq \alpha$ and $A \notin \mathcal{C}^n(\alpha)$, where $\mathcal{C}^n$ is the provability operator corresponding to $\Rightarrow^n$. Let u be a maximal subset of α that includes $\mathcal{C}^n(\alpha)$ and does not imply A. Clearly, u will be deductively closed. Moreover, (α, u) will be a bitheory of $\Rightarrow_\Delta$. Indeed, suppose that $c \Rightarrow d$ is a rule from Δ such that $c \subseteq \alpha$, but $d \cap u = \emptyset$. Then $D, u \vDash A$, for any $D \in d \cap \alpha$, and consequently $\bigvee(d \cap \alpha) \rightarrow A$ will belong to u. Also, we have that the rule $a, \neg(d \backslash \alpha) \Rightarrow \bigvee(d \cap \alpha)$ belongs to $N(\Delta)$, and consequently $\bigvee(d \cap \alpha) \in \mathcal{C}^n(\alpha)$. Then $\bigvee(d \cap \alpha) \in u$, and therefore $A \in u$—a contradiction.

Thus, (α, u) is a bitheory of $\Rightarrow_\Delta$, and $A \notin u$. Therefore $a \nRightarrow_\Delta A$. This shows that all singular rules from $\Rightarrow_\Delta$ are included into $\Rightarrow^n$.   □

Lemma (6.14) *A disjunctive causal relation is singular iff, for any world α, $\alpha \Rightarrow b, c$ only if either $\alpha \Rightarrow b$ or $\alpha \Rightarrow c$.*

Proof. If Δ is a set of singular rules and u a set of propositions, we will denote by $\Delta(u)$ the set $\{B \mid a \Rightarrow B \in \Delta \ \& \ a \subseteq \alpha\}$. Then we will show first that, for any u and v,

$$u \Rightarrow_\Delta v \text{ iff } v \cap \text{Th}(\Delta(\alpha)) \neq \emptyset, \text{ for any world } \alpha \text{ such that } u \subseteq \alpha$$

Indeed, bitheories of $\Rightarrow_\Delta$ are bitheories that are closed with respect to the rules of Δ. In our case, (α, u) is a bitheory of $\Rightarrow_\Delta$ if and only if $\Delta(\alpha) \subseteq u$. Consequently, minimal bitheories are pairs of the form $(\alpha, \text{Th}(\Delta(\alpha)))$, so the result follows from the completeness theorem.

Now if $\Rightarrow$ is a singular causal relation, and $\alpha \Rightarrow b, c$, then by the above result $(b \cup c) \cap \mathrm{Th}(\Delta(\alpha)) \neq \emptyset$, and hence either $b \cap \mathrm{Th}(\Delta(\alpha)) \neq \emptyset$, or $c \cap \mathrm{Th}(\Delta(\alpha)) \neq \emptyset$. In the first case we have $\alpha \Rightarrow b$ (by the same result), while in the second case $\alpha \Rightarrow c$.

Assume now that a disjunctive causal relation $\Rightarrow$ satisfies the above condition and $\Rightarrow_0$ is an arbitrary disjunctive causal relation that includes all singular rules from $\Rightarrow$. Suppose that $\Rightarrow \not\subseteq \Rightarrow_0$, and hence $a \Rightarrow b$ and $a \not\Rightarrow_0 b$ for some a, b. By the completeness theorem, there is a world α that includes a and such that $\alpha \not\Rightarrow_0 b$. Hence $\alpha \not\Rightarrow_0 B_i$ for any $B_i \in b$. But then $\alpha \not\Rightarrow B_i$ for any $B_i \in b$. Indeed, if $\alpha \Rightarrow B_i$, then there is $a_i \subseteq \alpha$ such that $a_i \Rightarrow B_i$, and therefore $a_i \Rightarrow_0 B_i$ since $\Rightarrow_0$ includes all singular rules from $\Rightarrow$. But the latter immediately implies $\alpha \Rightarrow_0 B_i$, contrary to our assumptions.

Now we can use the above condition to conclude that $\alpha \not\Rightarrow b$ (since otherwise α would cause at least one of B_i). But the latter contradicts our assumptions that $a \subseteq \alpha$ and $a \Rightarrow b$. The contradiction shows that $\Rightarrow$ is included in $\Rightarrow_0$. Since $\Rightarrow_0$ has been chosen to be an arbitrary disjunctive causal relation that includes all singular rules from $\Rightarrow$, we have shown, in effect, that $\Rightarrow$ is singular. $\qquad\square$

Theorem (6.15) *A disjunctive causal relation is singular if and only if it satisfies the following condition:*

If $a \Rightarrow b, c$, then $a, A \Rightarrow b$ and $a, \neg A \Rightarrow c$, for some proposition A.

Proof. Assume first that $\Rightarrow$ is a singular causal relation and $a \Rightarrow b, c$ for some a, b, c. Let us consider the following two sets of propositions:

$$u = \{\neg A \mid a, A \Rightarrow b\} \qquad v = \{\neg A \mid a, A \Rightarrow c\}.$$

It can be easily verified that u and v are deductively closed sets. Suppose now that $u \cup v \cup a$ is logically consistent and α is a world that contains this set. Then $\alpha \not\Rightarrow b$. Indeed, if $\alpha \Rightarrow b$, then by compactness $a, B \Rightarrow b$ for some $B \in \alpha$, contrary to the fact that $\neg B \in u$. Similarly, it can be shown that $\alpha \not\Rightarrow c$. Since the causal relation is singular, we infer $\alpha \not\Rightarrow b, c$ by corollary 6.14, contrary to our assumptions that $a \Rightarrow b, c$ and $a \subseteq \alpha$. Hence, $u \cup v \cup a$ is a logically inconsistent set. Then there must exist $B \in u$ and $A \in v$ such that $a, A \vDash \neg B$. As a result, we have $a, A \Rightarrow b$ and $a, \neg A \Rightarrow c$, as required.

Assume now that the above separation condition holds for a disjunctive causal relation $\Rightarrow$. If $a \Rightarrow B_1, \ldots, B_n$, then applying the separation condition a finite number of times, we obtain that there are n pairwise incompatible propositions $A_1, \ldots, A_n$ such that their disjunction is a tautology and $a, A_i \Rightarrow B_i$, for any $i \leq n$. Moreover, if for some disjunctive causal relation $\Rightarrow_0$ we have $a, A_i \Rightarrow_0 B_i$, for any $i \leq n$, and the disjunction of all A_i is a tautology, then $a \Rightarrow_0 B_1, \ldots, B_n$ (by Right Monotony and Or). An immediate consequence of these facts is that, if $a \Rightarrow B_1, \ldots, B_n$ holds, it should belong to any disjunctive causal relation that contains all the singular rules from $\Rightarrow$. In other words, $\Rightarrow$ is a least causal relation containing these singular rules, so it is singular itself. $\qquad\square$

Theorem (6.17) *For any disjunctive causal relation $\Rightarrow$ in a language $\mathcal{L}$, there exists a singular causal relation $\Rightarrow_s$ in an extended language $\mathcal{L}_s \supseteq \mathcal{L}$ such that $\Rightarrow$ coincides with the restriction of $\Rightarrow_s$ to the language $\mathcal{L}$.*

Proof. To begin with, note that the set of rules

$$a, \neg r_1, \ldots, \neg r_n \Rightarrow \mathbf{f} \quad \text{and} \quad a, r_i \Rightarrow B_i, \text{ for all } i \leq n$$

implies $a \Rightarrow B_1, \ldots, B_n$ by Right Monotonicity and Or. Consequently, $\Rightarrow_\Delta$ is included in $\Rightarrow_{\Delta_s}$.

Assume that $a \not\Rightarrow_\Delta b$, where $a, b \subseteq \mathcal{L}$. Then there exists a minimal bitheory (α, u) of $\Rightarrow_\Delta$ such that $a \subseteq \alpha$ and u is disjoint from b. We will transform this bitheory into a bitheory of $\Rightarrow_{\Delta_s}$. To begin with, we extend α to a world α_s in the extended language as follows: if r is a rule $c \Rightarrow d$ from Δ such that $a \subseteq \alpha$, then there must exist $D \in d \cap u$; we choose exactly one such D_i and add the corresponding atom r_i to α_s. After that, we add to α_s the negations of the rest of new atoms. It is easy to verify that α_s will be a causally consistent world for $\Rightarrow_{\Delta_s}$ since it is closed with respect to all the rules from Δ_s. Moreover, since the rules in Δ_s do not contain the new atoms in heads, we immediately obtain $\mathcal{C}_s(\alpha_s) \cap \mathcal{L} \subseteq u$, where $\mathcal{C}_s$ is the provability operator corresponding to $\Rightarrow_{\Delta_s}$. Therefore, b is disjoint from $\mathcal{C}_s(\alpha_s)$, and consequently $a \not\Rightarrow_{\Delta_s} b$. Thus, $\Rightarrow_{\Delta_s}$, restricted to $\mathcal{L}$, will coincide with $\Rightarrow_\Delta$. $\square$

Theorem (6.18) *The stable semantics of a disjunctive causal relation coincides with the nonmonotonic semantics of its normal subrelation.*

Proof. By lemma 6.6, a world α is causally explained with respect to a disjunctive causal relation $\Rightarrow$ if and only if it satisfies the following condition, for any proposition A,

$$A \in \alpha \quad \text{iff} \quad \alpha \Rightarrow \overline{\alpha}, A.$$

We will show now that $\alpha \Rightarrow \overline{\alpha}, A$ holds if and only if $\alpha \Rightarrow A$. Indeed, if $\alpha \Rightarrow \overline{\alpha}, A$, then by compactness $a \Rightarrow b, A$ for some finite $a \subseteq \alpha$, $b \subseteq \overline{\alpha}$. Consequently, $a, \neg b \Rightarrow A$ by Reduction. But since α is world, $b \subseteq \overline{\alpha}$ implies $\neg b \subseteq \alpha$, and therefore $\alpha \Rightarrow A$, as required.

Now the above condition can be rewritten as follows:

$$A \in \alpha \quad \text{iff} \quad \alpha \Rightarrow A.$$

But the latter condition is nothing other than the definition of explainable worlds with respect to the normal subrelation of $\Rightarrow$. Hence the result. $\square$

Lemma (6.19) *A disjunctive causal relation is stable if and only if it satisfies one of the following conditions:*

1. $A \Rightarrow B \vee A, \neg B \vee A;$
2. *If $a \Rightarrow b, C \vee D$, then $a, C, D \Rightarrow b, C, D;$*
3. *If $a \Rightarrow \bigvee b, c$, then $a, b \Rightarrow b, c;$*

4. *If $a, \neg b_1 \Rightarrow \bigvee b_2$, for any partition (b_1, b_2) of b, then $a \Rightarrow b$.*

Proof. (1). If Saturation holds, then $A \vee B, A \vee \neg B \Rightarrow A \vee B, (A \vee B) \rightarrow (A \vee \neg B)$, which is reducible to $A \Rightarrow A \vee B, A \vee \neg B$ by logical equivalence. In the other direction, (1) gives Saturation, if we substitute $A \wedge B$ for A.

(1 → 2). If $a \Rightarrow b, C \vee D$, then $a, C, D \Rightarrow b, C \vee D$. Now, Saturation gives

$$a, C, D \Rightarrow b, C, C \rightarrow D.$$

Combining the two, we immediately obtain $a, C, D \Rightarrow b, C, D$.

(2 ↔ 3). (3) can be obtained by repeated applications of (2).

(2 → 4). By induction on the number of propositions in b. If $b = \{B\}$, the claim is trivial. Suppose that the claim holds for all nonempty sets with less than n elements (where $n \geq 2$), and let $b_0 = b \cup \{B, C\}$ be a set with n elements. Then we can assume that the following rules hold, for any partition (b_1, b_2) of b:

$$a, \neg b_1 \Rightarrow (\bigvee b_2) \vee B \vee C \tag{i}$$

$$a, \neg b_1, \neg B \Rightarrow (\bigvee b_2) \vee C \tag{ii}$$

$$a, \neg b_1, \neg C \Rightarrow (\bigvee b_2) \vee B \tag{iii}$$

$$a, \neg b_1, \neg B, \neg C \Rightarrow \bigvee b_2. \tag{iv}$$

By the inductive assumption, (i) and (iv) imply $a \Rightarrow b, B \vee C$, (ii) and (iv) give $a, \neg B \Rightarrow b, C$ while (iii) and (iv) give $a, \neg C \Rightarrow b, B$. Now, the first rule implies $a, B, C \Rightarrow b, B, C$ by (2) while the last two give $a, \neg B \vee \neg C \Rightarrow b, B, C$ by Right Monotonicity and Or. Applying Or to the latter two, we obtain $a \Rightarrow b, B, C$.

(4 → 1). Since $\Rightarrow (A \vee B) \vee (A \vee \neg B)$ by Truth, and $A, \neg(A \vee B) \Rightarrow \mathbf{f}$ and $A, \neg(A \vee \neg B) \Rightarrow \mathbf{f}$ by Falsity, we immediately obtain (1) from (4). This completes the proof. □

Theorem (6.21) *A disjunctive causal relation is stable if and only if it is generated by a saturated causal semantics.*

Proof. It can be easily verified that Saturation is valid for saturated causal semantics. Hence any saturated causal semantics generates a stable causal relation. In the other direction, it is sufficient to show that the canonical causal semantics of a stable causal relation is saturated.

Let (α, u) be a bitheory of a stable causal relation $\Rightarrow$. If $u \neq \alpha$, then there exists a proposition $A \in \alpha$ such that $A \notin u$. By (1) from lemma 6.19, either $A \vee B \in u$ or $A \vee \neg B \in u$, for any proposition B. Consequently, $\text{Th}(u, \neg A)$ is a world, say β. Clearly, $u \subseteq \alpha \cap \beta$. Moreover, if $B \in \alpha \cap \beta$, then $B \in \alpha$ and $A \vee B \in u$. But we have also $A \rightarrow B \in u$ by Saturation, so $B \in u$. Thus, u coincides with $\alpha \cap \beta$, and therefore (α, u) is a saturated bitheory. □

Theorem (6.22) *The normal subrelation of a disjunctive causal relation $\Rightarrow$ coincides with the normal subrelation of $\Rightarrow^{st}$.*

Proof. To begin with, it can be verified that bitheories of $\Rightarrow^{st}$ are precisely saturated bitheories of $\Rightarrow$.

Now, if $a \not\Rightarrow B$, then there exists a bitheory (α, u) of $\Rightarrow$ such that $a \subseteq \alpha$ and $B \not\subseteq u$. Consequently, u can be extended to a world β such that $\neg B \in \beta$. Then $(\alpha, \alpha \cap \beta)$ will be a saturated bitheory of $\Rightarrow^{st}$, and $B \not\subseteq \alpha \cap \beta$. Hence, $a \not\Rightarrow^{st} B$, and therefore all singular causal rules in $\Rightarrow^{st}$ are included in $\Rightarrow$. $\qquad\square$

Theorem (6.25) *Two disjunctive causal theories are strongly equivalent if and only if they are stably equivalent.*

Proof. Note first that $\Delta \cup \Phi$ is not equivalent nonmonotonically to $\Gamma \cup \Phi$ only if $\Delta \cup N(\Phi)$ is not equivalent to $\Gamma \cup N(\Phi)$. In other words, only singular rules need to be added to verify strong equivalence. This implies, in particular, that two singular theories are strongly equivalent if and only if they are strongly equivalent in the sense used in chapter 4.

Now, by the preceding lemma, Δ and Γ are strongly equivalent if and only if $N(\Delta)$ is strongly equivalent to $N(\Gamma)$, while the latter holds iff $N(\Delta)$ is causally equivalent to $N(\Gamma)$. Now the result follows from corollary 6.23. $\qquad\square$

Theorem (6.29) *Disjunctive causal theories are strongly c-equivalent if and only if they are causally equivalent.*

Proof. The implication from right to left is obvious.

Suppose that Δ is not causally equivalent to Γ, and assume for certainty that there are finite sets propositions a, b such that $a \Rightarrow_\Delta b$ and $a \not\Rightarrow_\Gamma b$. The latter fact means that there is a minimal bitheory (α, u) of $\Rightarrow_\Gamma$ such that $a \subseteq \alpha$ and $b \cap u = \emptyset$. Let us consider two cases.

Suppose first that α is not causally consistent in $\Rightarrow_\Delta$. Then we choose $\Phi = \{C \Rightarrow C \mid C \in \alpha\}$. Clearly, α is a covered (even explained) world for $\Gamma \cup \Phi$, though it is still causally inconsistent with respect to $\Delta \cup \Phi$.

Assume now that α is causally consistent with respect to $\Rightarrow_\Delta$. Since $\alpha \Rightarrow_\Delta b$, we have $b \cap v \neq \emptyset$ for any bitheory (α, v) of $\Rightarrow_\Delta$. In addition, we have $\bigvee b \in \alpha$, and consequently u should be a proper subset of α since it is disjoint from b. We will proceed now in two steps.

First, we add to both causal theories the following set of causal rules:

$$\Phi_1 = \{C \Rightarrow C \mid C \in u\}.$$

Let $\Gamma_1 = \Gamma \cup \Phi_1$ and $\Delta_1 = \Delta \cup \Phi_1$. Clearly, u is now the only minimal model of α with respect to Γ_1. In addition, since $\alpha \Rightarrow_{\Delta_1}$, we still have $b \cap v \neq \emptyset$, for any bitheory (α, v) of $\Rightarrow_{\Delta_1}$. This implies, in particular, that u is a proper subset of any model of α in $\Rightarrow_{\Delta_1}$.

Let $B' = \bigwedge b'$, where b' is some minimal subset of b such that $\alpha \Rightarrow_{\Delta_1} b'$. As can be verified, for each $B \in b'$ there exists a bitheory (α, v) of $\Rightarrow_{\Delta_1}$ such that B is the only proposition from b' that belongs to v.

Now we add to both Δ_1 and Γ_1 the following set of rules:

$$\Phi_2 = \{C \Rightarrow B' \rightarrow C \mid C \in \alpha\}.$$

Let $w = \{B' \rightarrow C \mid C \in \alpha\}$. Note first that $\mathrm{Th}(u, w)$ will be the only minimal model of α with respect to $\Gamma_1 \cup \Phi_2$. Moreover, since $B' \notin u$, we have that $\mathrm{Th}(u, w)$ is a proper subset of α, and therefore α is not a covered world with respect to $\Gamma_1 \cup \Phi_2$. We will show, however, that α is a covered world with respect to $\Delta_2 = \Delta_1 \cup \Phi_2$. To this end, we will prove first that b' is still a minimal subset of b that is caused by α with respect to Δ_2.

Let $b_1 = b' \backslash B$ for some $B \in b'$, and suppose that $\alpha \Rightarrow_{\Delta_2} b_1$. Since $\alpha \not\Rightarrow_{\Delta_1} b_1$, there exists a bitheory (α, v) of $\Rightarrow_{\Delta_1}$ such that b_1 is disjoint from v. Let $v_1 = \mathrm{Th}(v, w)$. Then (α, v_1) will be a bitheory of $\Rightarrow_{\Delta_2}$ since it will be closed with respect to all the rules from Δ_2. Since $\alpha \Rightarrow_{\Delta_2} b_1$, it must be the case that $B_1 \in \mathrm{Th}(v, w)$ for some $B_1 \in b_1$. The latter means that, for some $C \in \alpha$, the proposition $(\bigwedge b_1 \rightarrow C) \rightarrow B_1$ belongs to v. But this implies $B_1 \in v$, contrary to our assumption that b_1 is disjoint from v. This shows that b' is a minimal subset of b that is caused by α with respect to Δ_2. Consequently, every proposition from b' belongs to at least one minimal model of α with respect to Δ_2. As result, B' will be implied by the union of all models of α.

Suppose now that $A \in \alpha$. Then the proposition $B' \rightarrow A$ belongs to w, and consequently it belongs to any model of α with respect to Δ_2. In addition, we have shown that B' is implied by the union of all models of α. Consequently, A is also implied by the union of all models of α. This shows that α is a covered world with respect to $\Delta \cup (\Phi_1 \cup \Phi_2)$, though it is not a covered world with respect to $\Gamma \cup (\Phi_1 \cup \Phi_2)$. This completes the proof. $\square$

Lemma (6.30) *A rule* $a \Rightarrow B_1, \ldots, B_n$ *is reducible to the set of singular rules* $\{a, B_i \Rightarrow B_i \mid i = 1, \ldots, n\}$ *and a constraint* $a, \neg B_1, \ldots, \neg B_n \Rightarrow \mathbf{f}$ *if and only if* $a, B_i, B_j \Rightarrow B_i$, *for any* $i \neq j$.

Proof. We will show that the set of rules $\{a, B_i \Rightarrow B_i\}$, taken together with the constraint,

$$a, \neg B_1, \ldots, \neg B_n \Rightarrow \mathbf{f},$$

are jointly equivalent to $a \Rightarrow B_1, \ldots, B_n$ and the rules $a, B_i, B_j \Rightarrow B_i$.

We show first that the singular rules of the reduction always imply the source disjunctive rule. Let b denote the set of all B_i. Then we have $a, B_i \Rightarrow b$ for any i, by Right Monotonicity, and hence $a, \vee b \Rightarrow b$ by Or. But the constraint is representable as $a, \neg(\vee b) \Rightarrow \mathbf{f}$, and hence $a \Rightarrow b$ by Or. In addition, the rules $\{a, B_i \Rightarrow B_i \mid i = 1, \ldots, n\}$ imply $a, B_i, B_j \Rightarrow B_i$ by Left Monotonicity.

In the other direction, the rule $a \Rightarrow b$ implies the constraint $a, \neg b \Rightarrow \mathbf{f}$ by Reduction. Also, it implies $a, B_i, \neg b_i \Rightarrow B_i$, where $b_i = b \backslash \{B_i\}$. Now a rule $a, B_i, B_j \Rightarrow B_i$ gives us

$a, B_i, B_j, \neg b_{ij} \Rightarrow B_i$, where $b_{ij} = b \backslash \{B_i, B_j\}$. Combining all such rules by Or, we finally obtain $a, B_i \Rightarrow B_i$. $\qquad \qquad \qquad \qquad \qquad \qquad \qquad \qquad \qquad \qquad \qquad \qquad \qquad \qquad \qquad \square$

Theorem (6.32) *If $\Rightarrow$ is a disjunctive causal relation in a finite language, then its covering semantics coincides with the stable semantics of its normalization.*

Proof. As before, for any world α, we will denote by u_α the set

$$\text{Th}(\bigcup \{u \mid (\alpha, u) \text{ is a minimal bitheory of } \Rightarrow \}.$$

In the finite case we have that, for any set u,

$$u \Rightarrow^n A \text{ iff } A \in u_\alpha, \text{ for any } \alpha \supseteq u.$$

As a special case, for any world α we have $C^n(\alpha) = u_\alpha$, and hence $\alpha = u_\alpha$ if and only if $\alpha = C^n(\alpha)$. Consequently, causally covered worlds of $\Rightarrow$ coincide with causally explained worlds of $\Rightarrow^n$. Hence the result. $\qquad \qquad \qquad \square$

Lemma (6.36) *If Δ is a literal causal theory, then for any causally consistent world α and any set b of literals, $\alpha \Rightarrow_\Delta b$ only if $d \subseteq b$ for some rule $a \Rightarrow d \in \Delta^\neg$ such that $a \subseteq \alpha$.*

Proof. Assume that α is a causally consistent world such that, for any causal rule $a \Rightarrow d$ from $\Delta^\neg$ for which $a \subseteq \alpha$, we have $d \not\subseteq b$. Let u_0 be the set of all literals in $\alpha \backslash b$ and u the logical closure of u_0. We will show that (α, u) is a bitheory of $\Rightarrow_\Delta$. Clearly, u is a deductively closed set included in α. Hence we need only to show that (α, u) is closed with respect to the rules from Δ (see lemma 6.3).

If $a \Rightarrow d \in \Delta$ and $a \subseteq \alpha$, then $a, \neg d_0 \Rightarrow d_1 \in \Delta^\neg$, where $d_1 = d \cap \alpha$ and $d_0 = d \backslash \alpha$. Now since $\neg d_0 \subseteq \alpha$, we have $d_1 \not\subseteq b$, and therefore there exists a literal $l \in d_1 \backslash b$. But then $l \in u$, and therefore $d \cap u \neq \emptyset$. This shows that (α, u) is a bitheory of $\Rightarrow_\Delta$. Note, however, that $b \cap u = \emptyset$, and hence $\alpha \not\Rightarrow_\Delta b$. This completes the proof. $\qquad \qquad \square$

Lemma (6.37) *If Δ is literal, then C_b is a causal cover of b in $\Rightarrow_\Delta$.*

Proof. If $C_b \in \alpha$, for some causally consistent world α, then there exists $a \subseteq \alpha$ such that $a \Rightarrow d \in \Delta^\neg$ and $d \subseteq b$. Consequently, $\alpha \Rightarrow_\Delta d$, and therefore $\alpha \Rightarrow_\Delta b$. In the other direction, if $\alpha \Rightarrow_\Delta b$, then by lemma 6.36 $d \subseteq b$, for some rule $a \Rightarrow d \in \Delta^\neg$ such that $a \subseteq \alpha$. Consequently, $C_b \in \alpha$, and hence the result follows from lemma 6.33. $\qquad \qquad \square$

Lemma (6.38) *If Δ is a causal theory, then for any causally consistent world α and any set b, $\alpha \Rightarrow_\Delta b$ only if $d \models^\vee b$, for some rule $a \Rightarrow d \in \Delta^\oplus$ such that $a \subseteq \alpha$.*

Proof. Given a causally consistent world α, let us assume that, for any rule $a \Rightarrow d \in \Delta^\oplus$, if $a \subseteq \alpha$, then $d \not\models^\vee b$. Now, if $a \Rightarrow d \in \Delta^\&$, the rule $a, \neg(d \backslash \alpha) \Rightarrow d \cap \alpha$ will belong to $\Delta^\oplus$, and consequently $a \subseteq \alpha$ only if $d \cap \alpha \not\models^\vee b$; in other words, $D \not\models^\vee b$, for some $D \in d \cap \alpha$.

Let $\mathcal{D} = \{ d \mid a \Rightarrow d \in \Delta \ \& \ a \subseteq \alpha \}$. Let us say also that a finite set of propositions c is *good* if $c \subseteq \alpha$ and $\wedge c \not\models^{\vee} b$. Then the above condition can be expressed as follows: *for any finite subset of $\mathcal{D}$ there exists a good incision set*. Now, since all $d \in \mathcal{D}$ are finite sets, a simple combinatorial argument (a variant of König lemma) shows that there exists an incision set u_0 for $\mathcal{D}$ itself such that any finite subset of u_0 is good. In other words, u_0 does not imply any proposition from b. Now let $u = \mathrm{Th}(u_0)$. Then b is disjoint from u. Moreover, by the definition of u_0, the bitheory (α, u) will be closed with respect to all the rules from Δ, and hence it will be a bitheory of $\Rightarrow_{\Delta}$. Consequently, $\alpha \not\Rightarrow_{\Delta} b$. This completes the proof. $\square$

Theorem (6.39) C_b *is a causal cover of b in* $\Rightarrow_{\Delta}$.

Proof. If $C_b \in \alpha$, for some causally consistent world α, then there exists $a \subseteq \alpha$ such that $a \Rightarrow d \in \Delta^{\oplus}$ and $d \models^{\vee} b$. Consequently, $\alpha \Rightarrow_{\Delta} d$, and therefore $\alpha \Rightarrow_{\Delta} b$. In the other direction, if $\alpha \Rightarrow_{\Delta} b$, then by lemma 6.38 $d \models^{\vee} b$ for some rule $a \Rightarrow b \in \Delta^{\oplus}$ such that $a \subseteq \alpha$ and consequently $C_b \in \alpha$ by the construction of C_b. Now the result follows from lemma 6.33. $\square$

III EXPLANATORY CAUSAL REASONING

7 Abduction

Abduction is yet another kind of commonsense reasoning from facts to their explanations that has been introduced into logic and philosophy by Charles Sanders Peirce. It is widely used now in many areas of artificial intelligence, including diagnosis, action theories, truth maintenance, knowledge assimilation, database updates, and logic programming. In this chapter we will show that this kind of reasoning can be given a uniform and syntax-independent representation in the causal calculus.

Causal considerations play an essential role in abduction. They determine, in particular, the very choice of abducibles as well as the right form of descriptions and constraints (even in classical first-order representations). As has been shown already in Darwiche and Pearl (1994), system descriptions that do not respect the natural causal order of things can produce inadequate predictions and explanations.

The intimate connections between causation and abduction have become especially vivid in the so-called abductive approach to diagnosis (see, e.g., Cox and Pietrzykowski 1987; Darwiche 1995; Poole 1994; Konolige 1992, 1994). As has been acknowledged in these studies, reasoning about causes and effects should constitute a logical basis for diagnostic reasoning. Unfortunately, the absence of an adequate logical formalization for causal reasoning has relegated the latter to the role of an informal heuristic background, with classical first-order logic serving as the main representation language. This naturally rises the question whether classical logic can always be used as an adequate and sufficiently general underlying logic for abductive reasoning and diagnosis. We will give in what follows grounds for a negative answer to this question.

In our approach, we will base abductive reasoning entirely on causal descriptions. As we will see, however, the resulting formalism will subsume the "classical" abductive reasoning as a special case, and we will give precise conditions when it is appropriate. The formalism will provide us, however, with additional representation capabilities that will encompass important alternative forms of abduction, namely, abductive reasoning suitable for theories of actions and change and for abductive logic programming. As a result, we obtain a generalized theory of abduction that covers in a single formal framework practically all kinds of abductive reasoning studied in the AI literature.

Systems of abductive reasoning provide a primary example of nonmonotonic formalisms that are based on the explanatory closure assumption, which is nothing other than a formal counterpart of the principle of sufficient reason. Such systems are designed to capture the relation between facts and their admissible explanations. In this section we will show that production inference relations, coupled with the corresponding nonmonotonic semantics, allow us to provide a formal representation of abductive reasoning. In addition to specific results, this will also pave the way to more realistic kinds of causal reasoning, such as theories of actual causality that we will discuss in subsequent chapters.

With a few exceptions, proofs of all the results in this chapter will be omitted, though they can be found in Bochman (2005, 2007a).

7.1 Abductive Systems and Abductive Semantics

We will begin with a general formal representation of abductive reasoning that subsumes the majority of abductive frameworks suggested in the literature. It will serve as a *logical* basis for our subsequent constructions.

A general *abductive system* can be defined as a pair $\mathbb{A} = (\mathrm{Cn}, \mathcal{A})$, where Cn is (a derivability operator of) a supraclassical Tarski consequence relation while $\mathcal{A}$ is a distinguished set of propositions called *abducibles*. A proposition A is *explainable* in an abductive system $\mathbb{A}$ if there exists a consistent set of abducibles $a \subseteq \mathcal{A}$ such that $A \in \mathrm{Cn}(a)$; in this case the set a is called *an explanation* of A. Clearly, A may have, in general, a number of different (and even incompatible) explanations.

In practical applications, the supraclassical consequence relation Cn is usually given indirectly by a generating conditional theory Δ, namely, by a set of inference rules of the form $a \vdash A$. The corresponding abductive system can be represented in this case by a pair $(\Delta, \mathcal{A})$. However, given a conditional theory Δ, we can identify Cn with Cn_Δ—the least supraclassical consequence relation containing Δ. Then $A \in \mathrm{Cn}(a)$ will hold if and only if A is derivable from a using the classical entailment and the rules from Δ. Accordingly, the pair $(\Delta, \mathcal{A})$ can be faithfully represented by an abductive system $(\mathrm{Cn}_\Delta, \mathcal{A})$ as defined above.

In abductive systems, acceptance of propositions depends on existence of explanations, and consequently, such systems sanction not only forward inferences determined by the associated consequence relation but also backward inferences from facts to their possible explanations as well as combinations of both. All these kinds of abductive inference can be captured formally by considering only theories of Cn that are generated by the abducibles. This suggests the following notion of an abductive semantics:

Definition 7.1 *The* abductive semantics $\mathcal{S}_\mathbb{A}$ *of an abductive system* $\mathbb{A}$ *is the set of theories* $\{\mathrm{Cn}(a) \mid a \subseteq \mathcal{A}\}$.

Note that a consequence relation Cn is uniquely determined by the set of all its theories. Accordingly, by restricting the set of these theories to theories generated by abducibles, we obtain a semantic framework containing more information. Generally speaking, all the information that can be discerned from the abductive semantics of an abductive system can be seen as abductively implied by the latter. Recall, however, that any set of deductively closed theories determines a unique supraclassical Scott consequence relation (see chapter 2). Moreover, in many cases the correspondence between sets of theories and Scott consequence relations is one-to-one (e.g., in a finite case). Accordingly, in most cases, the information embodied in the abductive semantics can be made explicit by considering the associated Scott consequence relation.

Definition 7.2 *An* abductive consequence relation *associated with an abductive system* $\mathbb{A}$ *is a supraclassical Scott consequence relation* $\vdash_{\mathbb{A}}$ *determined by the abductive semantics* $\mathcal{S}_{\mathbb{A}}$.

The abductive consequence relation associated with an abductive system can be defined explicitly as follows:

$$b \vdash_{\mathbb{A}} c \equiv (\forall a \subseteq \mathcal{A})(b \subseteq \mathrm{Cn}(a) \rightarrow c \cap \mathrm{Cn}(a) \neq \emptyset).$$

By the above description, $b \vdash_{\mathbb{A}} c$ holds if any set a of abducibles that explains b explains also at least one proposition from c. In other words, $b \vdash_{\mathbb{A}} c$ holds if any explanation of b is also an explanation of some $C \in c$.

The abductive consequence relation $\vdash_{\mathbb{A}}$ is an extension of Cn obtained by restricting the set of its theories. It describes not only forward explanatory relations, but also abductive inferences from propositions to their explanations. For example, if C and D are the only abducibles that imply A in an abductive system, then we will have $A \vdash_{\mathbb{A}} C, D$. Speaking generally, the abductive consequence relation describes the *explanatory closure* or *completion* of an abductive system and thereby allows a capturing of abduction by deduction (see Console, Dupre, and Torasso 1991; Konolige 1992).

Example 7.1 (Sprinkler) *The following abductive system describes a variant of the well-known example from Pearl (1987). Assume that an abductive system $\mathbb{A}$ is determined by the set Δ of rules*

$$Rained \vdash Grasswet$$

$$Sprinkler \vdash Grasswet$$

$$Rained \vdash Streetwet,$$

and the set abducibles

$$\mathcal{A} = \{Rained, \neg Rained, Sprinkler, \neg Sprinkler, \neg Grassswet\}.$$

By stipulating that both Rained and ¬Rained are abducibles, we make Rained an independent (exogenous) parameter and similarly for Sprinkler. However, only ¬Grassswet is an abducible, so non-wet grass does not require explanation, but wet grass does. Thus, *any theory of $S_{\mathbb{A}}$ that contains Grasswet should contain either Rained or Sprinkler, and consequently we have*

$$Grasswet \Vdash_{\mathbb{A}} Rained, Sprinkler.$$

Similarly, Streetwet implies in this sense both its only explanation Rained and a collateral effect Grasswet.

By the above construction, abductive reasoning with respect to a supraclassical consequence relation amounts to extending the latter to the corresponding abductive consequence relation.

Let us say that two consequence relations are $\mathcal{A}$-*equivalent* for a set $\mathcal{A}$ of abducibles if they have the same rules of the form $a \vdash A$, where $a \subseteq \mathcal{A}$. Clearly, two consequence relations are $\mathcal{A}$-equivalent if they provide precisely the same explanations. Consequently, $\mathcal{A}$-equivalence amounts to coincidence with respect to theories generated by abducibles. Moreover, the associated abductive consequence relation is precisely the greatest consequence relation that produces the same explanations as the source consequence relation.

Lemma 7.1 *Two consequence relations are $\mathcal{A}$-equivalent iff they determine the same abductive semantics. If $\mathbb{A} = (\mathrm{Cn}, \mathcal{A})$ is an abductive system, then $\vdash_{\mathbb{A}}$ is a greatest consequence relation among $\mathcal{A}$-equivalents of Cn.*

7.2   Abductive Production Inference

In order to translate abductive reasoning into the framework of the causal calculus, we will slightly extend the relevant notion of explanation and say that an arbitrary set u of propositions *explains* a proposition A in an abductive system if A is explainable by the abducibles that are implied by u. Then the set of propositions that are explainable by u will coincide with $\mathrm{Cn}(\mathrm{Cn}(u) \cap \mathcal{A})$.

Now, the main idea behind the following representation consists in viewing this latter notion of an explanation as a particular kind of production inference. In other words, u will be taken to cause A if it explains A.

Definition 7.3 *A production inference relation associated with an abductive system $\mathbb{A}$ is a production relation $\Rightarrow_{\mathbb{A}}$ determined by all bimodels of the form $(\mathrm{Cn}(u \cap \mathcal{A}), u)$, where u is a consistent theory of Cn.*

Note that the classical binary semantics described in the above definition is consistent, and therefore the associated production inference relation will always be regular.

To simplify the subsequent representation, we will assume that the set of abducibles $\mathcal{A}$ of an abductive system is closed with respect to conjunctions, that is, if A and B are abducibles, then $A \wedge B$ is also an abducible. In addition, to deal with limit cases, we will also assume that the constants $\mathbf{t}$ and $\mathbf{f}$ are abducibles. Then it turns out that the above defined production inference relation admits a very simple syntactic characterization. According to the description that follows, $A \Rightarrow_\mathbb{A} B$ holds if and only if A implies some abducible that explains B.

Lemma 7.2 *If* $\Rightarrow_\mathbb{A}$ *is a production inference relation associated with an abductive system* $\mathbb{A}$*, then*

$$A \Rightarrow_\mathbb{A} B \quad iff \quad (\exists C \in \mathcal{A})(C \in \mathrm{Cn}(A) \ \& \ B \in \mathrm{Cn}(C)).$$

As a consequence of the above description, abducibles of an abductive system correspond precisely to "reflexive" propositions of the associated production inference relation.

Corollary 7.3 *If* $\Rightarrow_\mathbb{A}$ *is a production inference relation associated with an abductive system* $\mathbb{A}$*, then* $C \Rightarrow_\mathbb{A} C$ *if and only if* C *is* Cn-*equivalent to an abducible from* $\mathcal{A}$*.*

Due to this correspondence, default (self-explanatory) propositions of a production relation can be seen as abducibles, and hence we introduce the following definition.

Definition 7.4 *A proposition A will be called an* abducible *of a production relation* $\Rightarrow$ *if* $A \Rightarrow A$*.*

Note that the set of abducibles of a production inference relation conforms to our earlier agreement: namely, it is closed with respect to conjunctions, and both **t** and **f** are abducibles (see definition 4.1 in chapter 4).

Another important fact about abducibles of a production relation is that any set of abducibles is obviously explanatory closed. As a result, we obtain the following corollary.

Corollary 7.4 *If a is a set of abducibles of a regular inference relation* $\Rightarrow$*, then* $\mathcal{C}(a) = \mathrm{Cn}_\Rightarrow(a)$ *and* $\mathcal{C}(a)$ *is an exact theory of* $\Rightarrow$*.*

It turns out that production inference relations corresponding to abductive systems form a special class that is described in the next definition.

Definition 7.5 *A production inference relation will be called* abductive *if it is regular and satisfies*

(Abduction) *If* $B \Rightarrow C$*, then* $B \Rightarrow A \Rightarrow C$*, for some abducible A.*

Thus, inferences in abductive production relations are always mediated by abducibles. Consequently, for such production relations, propositions are caused, in effect, by the derived set of abducibles: for any set u of propositions,

$$\mathcal{C}(u) = \mathcal{C}(\mathcal{C}(u) \cap \mathcal{A}),$$

where $\mathcal{A}$ is the set of abducibles of $\Rightarrow$.

Clearly, any abductive production relation is dense, and consequently its exact theories are sets of the form $\mathcal{C}(u)$ (see lemma 4.19 in chapter 4). Moreover, in our case we have that all exact theories are produced by abducibles:

Lemma 7.5 *The nonmonotonic semantics of an abductive production relation is the set of all sets of propositions of the form $C(u)$, for $u \subseteq \mathcal{A}$.*

The following result shows that abductive production relations are precisely production inference relations that are generated by abductive systems.

Theorem 7.6 *A production inference relation is abductive if and only if it is generated by an abductive system.*

Finally, the following key result shows that the abductive semantics of an abductive system coincides with the general nonmonotonic semantics of the associated abductive production inference relation.

Theorem 7.7 *If $\Rightarrow_\mathbb{A}$ is a production inference relation corresponding to an abductive system $\mathbb{A}$, then the abductive semantics of $\mathbb{A}$ coincides with the general nonmonotonic semantics of $\Rightarrow_\mathbb{A}$.*

Due to the above results, abductive production relations provide a faithful logical representation of abductive reasoning. Note that abductive production relations provide in this sense a syntax-independent description of abduction since the set of abducibles is determined as a set of propositions having certain logical properties with respect to the production relation (namely reflexivity).

Example 7.2 (Sprinkler, continued) *The following causal theory determines the abductive production relation corresponding to the Pearl's example, described earlier.*

$$Rained \Rightarrow Grasswet \quad Sprinkler \Rightarrow Grasswet \quad Rained \Rightarrow Streetwet$$

$$Rained \Rightarrow Rained \qquad \neg Rained \Rightarrow \neg Rained$$

$$Sprinkler \Rightarrow Sprinkler \qquad \neg Sprinkler \Rightarrow \neg Sprinkler$$

$$\neg Grasswet \Rightarrow \neg Grasswet \qquad \neg Streetwet \Rightarrow \neg Streetwet.$$

As follows from the preceding results, the general nonmonotonic semantics of this causal theory coincides with the abductive semantics of the original abductive system, and hence it determines the same abductive inferences.

7.2.1 Quasi-Abductive Causal Inference

Now we are going to show that the concept of abduction is in a sense implicit in the causal calculus itself. More precisely, any production inference relation includes an important abductive subrelation; in many regular situations the latter subrelation will determine the same nonmonotonic semantics.

Given a regular inference relation $\Rightarrow$, we will define the following production relation:

$$A \Rightarrow^a B \equiv (\exists C)(A \Rightarrow C \Rightarrow C \Rightarrow B).$$

Clearly, if $A \Rightarrow^a B$, then we have $A \Rightarrow B$ (due to transitivity), though not vice versa. Moreover, we have the following theorem.

Theorem 7.8 *If $\Rightarrow$ is a regular inference relation, then $\Rightarrow^a$ is the greatest abductive production relation included in $\Rightarrow$.*

It turns out that the above abductive subrelation of a production relation preserves many properties of the latter. For example, both have the same constraints, that is, $A \Rightarrow \mathbf{f}$ holds if and only if $A \Rightarrow^a \mathbf{f}$. Similarly, both have the same "axioms": $\mathbf{t} \Rightarrow A$ holds iff $\mathbf{t} \Rightarrow^a A$. Last but not least, both have the same abducibles. The latter fact implies, in particular:

Lemma 7.9 *The nonmonotonic semantics of $\Rightarrow^a$ is the set of all theories of $\Rightarrow$ of the form $\mathcal{C}(u)$, where u is a set of abducibles of $\Rightarrow$.*

Let $\mathcal{A}_\Rightarrow$ denote the set of abducibles of $\Rightarrow$ and $\mathrm{Cn}_\Rightarrow$ the least supraclassical consequence relation including $\Rightarrow$. Then the following result shows that the abductive system $(\mathrm{Cn}_\Rightarrow, \mathcal{A}_\Rightarrow)$ generates $\Rightarrow^a$ as its associated production relation.

Lemma 7.10 *If $\Rightarrow$ is a regular inference relation, then its abductive subrelation $\Rightarrow^a$ is generated by an abductive system $(\mathrm{Cn}_\Rightarrow, \mathcal{A}_\Rightarrow)$.*

The general nonmonotonic semantics is based on the law of causality, according to which any accepted proposition should be caused by other accepted propositions. But the latter should also be caused by accepted propositions, and so on. It should be clear that if our "explanatory resources" are limited (for example, if our language is finite), such an explanation process should stop somewhere. More exactly, it should reach abducible propositions (that are self-explanatory). This indicates that in many regular cases the nonmonotonic semantics of a production relation should coincide with the nonmonotonic semantics of its abductive subrelation. Below we will make this claim precise.

Definition 7.6 *A production relation will be called* quasi-abductive *if it has the same nonmonotonic semantics as its abductive subrelation.*

The next definition will give us an important sufficient condition for quasi-abductivity.

Definition 7.7 *A production inference relation $\Rightarrow$ will be called* well founded *if any infinite sequence $\{A_0, A_1, A_2, \dots\}$ of propositions such that $A_{n+1} \Rightarrow A_n$ for every $n \geq 0$ contains an abducible.*

The above definition describes a variant of a standard notion of well-foundedness with respect to the partial order determined by a production relation. Namely, it says that any infinite descending sequence of productions should always contain reflexive elements. The following lemma provides an alternative description of well-foundedness.

Lemma 7.11 *A production relation is well founded if and only if any explanatory closed set of propositions contains an abducible.*

It should be clear that any production relation in a finite language is well founded. Moreover, let us say that a regular production relation is *finitary* if it is a least regular production relation containing some finite causal theory. Then we have the following theorem.

Theorem 7.12 *Any finitary regular production relation is well founded.*

Finally, the following result shows that a well-founded production relation determines precisely the same nonmonotonic semantics as its abductive subrelation.

Theorem 7.13 *Any well-founded regular production relation is quasi-abductive.*

Since abductive production relations correspond exactly to abductive systems, the above result shows that in the well-founded case the nonmonotonic semantics of a regular inference relation is representable by some abductive system, and vice versa. In other words, in this case the nonmonotonic, causal reasoning of the causal calculus can be captured by an abductive reasoning in an associated deductive system.

7.3 Abduction in Literal Causal Theories

It turns out that a certain well-known class of abductive systems can be directly interpreted as causal theories. The description below will show, in effect, that the causal reading of abductive systems has long been present in the study of abduction and diagnosis.

By a *literal* inference rule we will mean a rule of the form $a \vdash l$, where l is a propositional literal and a a set of literals. In what follows, the constants $\mathbf{t}$ and $\mathbf{f}$ will also be treated as literals. A conditional theory Δ will be called a *literal* one if it consists only of literal rules. Finally, an abductive system $\mathbb{A} = (\Delta, \mathcal{A})$ will be called a *literal* one if Δ is a literal conditional theory and the set of abducibles $\mathcal{A}$ is also a set of literals.

The above simplified abductive framework has been extensively used in the theory of diagnosis under the name "causal theory" (see, e.g., Console, Dupre, and Torasso 1991; Konolige 1992, 1994; Poole 1994). The name has a different meaning in our study: namely, it denotes an arbitrary set of causal rules. Still, these two notions of a causal theory turn out to be closely related.

Recall that any conditional theory can also be viewed as a causal theory in our sense. Moreover, it has been shown earlier that abducibles can be incorporated into causal theories by accepting corresponding reflexive rules $A \Rightarrow A$. Accordingly, for an abductive system $(\Delta, \mathcal{A})$, we will introduce a causal theory $\Delta_{\mathcal{A}}$ which is the union of Δ (viewed as a set of causal rules) and the set $\{l \Rightarrow l \mid l \in \mathcal{A}\}$. As before in this chapter, $\Rightarrow_{\Delta_{\mathcal{A}}}$ will denote the least regular inference relation containing $\Delta_{\mathcal{A}}$.

To begin with, it is easy to verify that the abductive semantics of $\mathbb{A}$ is included in the nonmonotonic production semantics of $\Rightarrow_{\Delta_{\mathcal{A}}}$.

Lemma 7.14 *If $a \subseteq \mathcal{A}$, then $\mathrm{Cn}_{\Delta}(a)$ is an exact theory of $\Rightarrow_{\Delta_{\mathcal{A}}}$.*

However, the reverse inclusion in the above lemma does not hold, even in our present literal case, and it is important to clarify the reasons why this happens. First of all, the production relation $\Rightarrow_{\Delta_{\mathcal{A}}}$ is not well founded in general, so it may have exact theories that are not generated by abducibles. Second, even in the well-founded case, the causal theory $\Delta_{\mathcal{A}}$ may create new abducibles of its own if some of the propositions happen to be inter-derivable. Taking a simplest example, if we have that both $p \vdash q$ and $q \vdash p$ belong to Δ, then both p and q will be abducibles of $\Rightarrow_{\Delta_{\mathcal{A}}}$.

Both of the above reasons for a discrepancy will disappear, however, if Δ is an *acyclic* conditional theory. As a matter of fact, a restriction of this kind has been used extensively in the literature—see, for example, Console, Dupre, and Torasso (1991) and Poole (1994). Pearl's causal models can also be seen as belonging to this class, insofar as they are determined by directed acyclic graphs (DAGs) of causal dependencies.

By a (literal) *dependency graph* of a literal conditional theory Δ we will mean the directed graph such that its nodes are the literals occurring in Δ while the arcs are the pairs of literals (l, m), for which Δ contains a rule of the form $l, a \vdash m$. As usual, a literal conditional theory Δ will be called *acyclic* if its dependency graph does not contain infinite descending paths. However, in what follows we will use a weaker condition that will be sufficient for our purposes.

Definition 7.8 *A literal abductive system* $\mathbb{A} = (\Delta, \mathcal{A})$ *will be called* abductively well founded *if any infinite descending path in the dependency graph of* Δ *contains an abducible from* $\mathcal{A}$.

As can be seen from the above definition, an abductive system is abductively well founded if it does not have infinite descending chains of dependencies that consist of non-abducibles only. This implies, in particular, that non-abducibles do not form loops of dependencies. Clearly, any acyclic theory will also be abductively well founded. Note also that we do not require that abducibles should not appear in heads of the rules, a condition often imposed on such abductive systems.

The following result shows that in this case the causal theory $\Delta_{\mathcal{A}}$ captures the "abductive content" of the source abductive system.

Theorem 7.15 *If* $\mathbb{A}$ *is an abductively well-founded literal abductive system, then the abductive semantics of* $\mathbb{A}$ *coincides with the nonmonotonic semantics of* $\Rightarrow_{\Delta_{\mathcal{A}}}$.

The above result shows that, from the perspective of abductive reasoning, abductively well-founded literal conditional theories can be viewed directly as causal theories.

7.4 Abductive Causal Inference

Now we will consider abductive systems that correspond to stronger causal inference relations.

To begin with, the following result shows that the abductive subrelation of a causal inference relation is also causal.

Lemma 7.16 *If ⇒ is a causal inference relation, then its abductive subrelation ⇒^{a} is causal.*

It might be tempting to suppose that causal inference relations correspond to classical abductive systems (Cn, $\mathcal{A}$), where Cn is a classical Tarski consequence relation. Indeed, it is easy to verify that any abductive production relation generated by such an abductive system will already be causal. Still, the reverse claim turns out to be wrong since classical abductive systems correspond in this sense to a stronger, quasi-classical production inference described later in this chapter. An adequate abductive setting for causal inference is described in the next definition.

Definition 7.9 *An abductive system* $\mathbb{A} = (\text{Cn}, \mathcal{A})$ *will be called* $\mathcal{A}$-disjunctive *if* $\mathcal{A}$ *is closed with respect to disjunctions and* Cn *satisfies the following two conditions for any abducibles* $A, A_1 \in \mathcal{A}$ *and arbitrary* B, C:

- *If* $A \in \text{Cn}(B)$ *and* $A \in \text{Cn}(C)$, *then* $A \in \text{Cn}(B \vee C)$;
- *If* $B \in \text{Cn}(A)$ *and* $B \in \text{Cn}(A_1)$, *then* $B \in \text{Cn}(A \vee A_1)$.[1]

The following result shows that $\mathcal{A}$-disjunctive systems are precisely abductive systems that generate causal abductive production relations.

Theorem 7.17 *An abductive production relation is causal if and only if it is generated by an* $\mathcal{A}$-disjunctive abductive system.

7.5   Classical Abduction

Finally, we consider production relations that correspond to classical abductive systems.

Definition 7.10 *A production relation will be called* quasi-classical, *if it satisfies the rule:*

(Weak Deduction)   *If* $A \Rightarrow B$, *then* $\mathbf{t} \Rightarrow (A {\to} B)$.

Weak Deduction says, in effect, that material implications corresponding to causal rules can be seen as universally valid propositions in production inference. The following lemma provides two alternative descriptions of quasi-classicality.

Lemma 7.18 *A production relation is quasi-classical if and only if it satisfies one the following conditions:*

(Dual Cut)   *If* $A \Rightarrow B$ *and* $C \Rightarrow A \vee B$, *then* $C \Rightarrow B$.
(Contradiction)   *If* $\neg A \Rightarrow A$, *then* $\mathbf{t} \Rightarrow A$.

---

1. This rule corresponds to the rule Ab-Or in Lobo and Uzcátegui (1997).

Note that Dual Cut also implies Transitivity, so any quasi-classical production relation will already be transitive.

For regular inference relations, quasi-classicality becomes equivalent to the well-known *reductio ad absurdum* principle that can be found already in Aristotle's *Prior Analytics*:

(Reductio) If $A \Rightarrow \mathbf{f}$, then $\mathbf{t} \Rightarrow \neg A$.

A semantic interpretation of basic quasi-classical production relations can be obtained by requiring *right reflexivity* of the accessibility relation: if $\alpha R \beta$, then $\beta R \beta$.

Theorem 7.19 *A basic production relation is quasi-classical if and only if it is determined by a right reflexive possible worlds model.*

Taken together with quasi-reflexivity, right reflexivity gives a full reflexivity of the accessibility relation for quasi-classical causal inference:

Corollary 7.20 *A causal relation is quasi-classical if and only if it is determined by a reflexive possible worlds model.*

A constructive description of the least quasi-classical production relation containing a causal theory is given in the next lemma.

Lemma 7.21 *If $\Rightarrow^q_\Delta$ is the least quasi-classical production relation containing a causal theory Δ, then $C^q_\Delta(u) = \mathrm{Th}(\overrightarrow{\Delta} \cup \Delta(\mathrm{Th}(u)))$.*

The above description shows that quasi-classical production inference amounts to a general production inference which is based, in a sense, on a modified background logic Th that includes $\overrightarrow{\Delta}$ as the set of axioms. As a consequence, we obtain the following description of the associated causal nonmonotonic semantics.

Corollary 7.22 *A world α is an exact world of $\Rightarrow^q_\Delta$ iff $\alpha = \mathrm{Th}(\overrightarrow{\Delta}, \Delta(\alpha))$.*

Recall that causally consistent worlds are precisely the worlds that satisfy $\overrightarrow{\Delta}$. Accordingly, the above description says that α is an exact world of $\Rightarrow^q_\Delta$ if and only if it is the only *causally consistent* world containing $\Delta(\alpha)$.

Our interest in quasi-classical inference stems from the fact that it corresponds precisely to abductive reasoning based on classical consequence.

An abductive system $(\mathrm{Cn}, \mathcal{A})$ will be called *classical* if Cn is a classical consequence relation (see chapter 2 for a definition of the latter). A classical consequence relation can be seen as a classical entailment augmented with a set of nonlogical axioms. Consequently, a classical abductive system can be safely equated with a pair $(\Sigma, \mathcal{A})$, where Σ is a set of classical propositions (the domain theory) and $\mathcal{A}$ is a set of abducibles.

To begin with, the following result shows that the least supraclassical consequence relation $\vdash_\Rightarrow$ containing a quasi-classical regular relation $\Rightarrow$ is already fully classical.

Lemma 7.23 *If* $\Rightarrow$ *is a regular inference relation, then* $\vdash_{\Rightarrow}$ *is a classical consequence relation if and only if* $\Rightarrow$ *is quasi-classical.*

Moreover, the following result establishes precise correspondence between classical abductive systems and quasi-classical causal relations.

Theorem 7.24 *An abductive production relation is generated by a classical abductive system if and only if it is causal and quasi-classical.*

Despite apparent plausibility, the following example shows, however, that quasi-classical inference produces somewhat weaker conclusions.

Example 7.3 *A causal theory* $\Delta = \{\neg p \Rightarrow q, \ \neg q \Rightarrow \neg q, \ \neg p \Rightarrow \neg p\}$ *has a single exact world* $\{\neg p, q\}$.[2] *However,* $\neg p \Rightarrow q$ *implies* $\mathbf{t} \Rightarrow (\neg p \rightarrow q)$ *by Weak Deduction and then* $\neg q \Rightarrow (\neg p \rightarrow q)$ *by Strengthening. Coupled with* $\neg q \Rightarrow \neg q$, *this gives us* $\neg q \Rightarrow p$, *namely the contraposition of* $\neg p \Rightarrow q$. *As a result,* $\Rightarrow_{\Delta}^{q}$ *acquires also an additional exact world, namely* $\{p, \neg q\}$.

The above example shows, of course, that quasi-classical inference can change the causal nonmonotonic semantics. It should be noted, however, that since Weak Deduction is a classically valid rule, quasi-classical inference can only add new exact worlds and never eliminate them (cf. corollary 4.37). Accordingly, the nonmonotonic conclusions obtained by this inference will always be valid with respect to the "standard" causal nonmonotonic semantics of causal theories. Still, the upshot is that a fully classical abductive system turns out to be an informationally weaker system than the abductive formalism based on causal inference.

---

2. Note that this causal theory corresponds to a logic program $\{q \leftarrow \mathbf{not}\,p\}$—see section 4.6 in chapter 4.

8 Actual Causality

Actual causation (known also as "singular causation," "token causation," or "causation in fact") involves causal claims of the form *"C was a cause of E."* In other words, it deals with causal explanation of effects or, more precisely, with *post factum* attribution of causal responsibility for actual outcome.

In this chapter we are going to provide a formal definition of actual causality in the framework of the causal calculus. We compare our definition with other approaches on standard examples. On the way, we also explore general capabilities of our logical representation, both for representing structural equation models and beyond.

Much recent work on actual causation is conducted within the structural equation framework. As has been shown in chapter 5, however, structural equation models are representable in the causal calculus, and we will even make use of this representation for a logical translation of the examples of actual causation, described in this literature. We will not adopt, however, the dominant counterfactual approach to analyzing this notion. Instead, we will return to the traditional regularity approach, an approach "that has unjustly fallen into disfavor in some quarters" (Paul and Hall 2013). Our suggested definition of actual causality will be a particular instantiation of the INUS, or NESS, test, though elevated to causal language of the causal calculus.

Our plan in this chapter is as follows. First, we briefly review the counterfactual approaches to actual causation and point out some general problems that arise with them. Then we turn to the regularity approach and provide a formal definition of actual causality that is based on it. We will show, in particular, that the difference between general and actual causation could even be recast as a difference in their underlying logics. Then we proceed to some key examples and counterexamples of actual causation discussed in the literature. In usual cases, we will make use of existing representations of these examples provided by the structural approach. In some key examples, however, we will exploit the opportunities provided by the logical language of the causal calculus that will be shown to have representational capabilities that go beyond structural equation models.

During the exposition, we will also point out some discrepancies that will still remain between the predictions of our formal definition (and, for that matter, also of counterfactual approaches) and commonsense understanding of causation. These discrepancies will serve as one of the reasons for the necessity of an alternative, upgraded definition of causality that will be developed in the next chapter on relative, or proximate, causality.

8.1 An Overview

For a long period of time, a standard, if not the only, formal approach to analyzing causality has been the so-called regularity approach that originated in the "covering law" analysis of causation by John Stuart Mill. A large literature has been written about this approach in the past century,[1] but it has also accumulated a large list of problems that has been associated with it. As we mentioned in chapter 1, most of these problems stemmed from the classical logical understanding of the key notion of a lawlike conditional that lay at the basis of the regularity theory. The above studies have made it abundantly clear that regularity or, as Hume has termed it, constant conjunction, is not causality, though the distinction has turned out to be difficult to capture in logical terms.

The dominance of the regularity theory has ended (at least in the philosophical literature) with David Lewis's paper *Causation* (Lewis 1973a), where he argued that the regularity approach has numerous difficulties and apparently cannot be repaired. Following Lewis, an overwhelming majority of current approaches to actual causality today attempt to define it in terms of counterfactuals.

The starting point of all counterfactual accounts of causality is the but-for test (sine qua non) commonly used in both tort law and criminal law. The test asks, "but for C, would E have occurred?" If the answer is yes, then *C* is an actual cause of *E*. Viewed as a counterfactual assertion, "had C not occurred, E would not have occurred," sine qua non provides indeed a reliable test of causality in many regular situations. However, it breaks down, for instance, in cases of redundant causation (e.g., preemption or overdetermination), wherefore, by David Lewis's phrase, we need extra bells and whistles (Lewis 2000, p. 182).

A broad scheme of generalizing the but-for test that is in accord with numerous counterfactual approaches in the literature can be described along the lines of the notion of *"de facto dependence"* from Yablo (2004): E de facto depends on C just in case had C not occurred, and *the right things held fixed*, then E would not have occurred. The trick is to say what "the right things" are. The majority of counterfactual approaches, including the prominent HP definitions of Halpern and Pearl (Halpern and Pearl 2005; Halpern 2016a), could be viewed as particular instantiations of this general scheme. Many accounts have followed David Lewis in assuming that the relevant parameters are determined by a path of counterfactual dependencies that should exist between the cause and effect (see, e.g.,

---

1. Especially if we take into account the related deductive-nomological account of explanation (Hempel 1965)— see chapter 1.

Woodward 2003; Hitchcock 2001, 2007; Weslake 2015), though the HP definitions are more general.

It is difficult to adjudicate the advantages and shortcomings of the host of counterfactual accounts that have been suggested in the literature and, even more generally, the precise role of counterfactuals in assertions of actual causation (cf. Menzies 2011). To begin with, however, there is a broad consensus among psychologists that while causal and counterfactual reasoning are closely connected, they are distinct forms of reasoning. Thus, while counterfactual reasoning is useful in identifying the enabling conditions and preventative mechanisms, it is not useful in identifying causes proper—see Byrne (2005, chapter 5); cf. also Mandel (2003) for the corresponding judgment dissociation theory, and Menzies (2007) for a general discussion. Moreover, we argue that there are some general, "blanket" problems for the counterfactual approaches to actual causation that transcend the boundaries of specific definitions.

First of all, it appears that beyond the but-for test, there are no general principles, or rationality postulates, for the choice of the right counterfactual definition of actual causation. This creates an obvious trust problem for any potential definition of this kind.[2] In practice, most of this research is largely example-driven, but even on this score there are still doubts whether the current "empirical pool" of examples is sufficiently representative (see Glymour et al. 2010). All this leads to what seems to be an inevitable situation where the jury is still out on what the "right" counterfactual definition of causality is (see Halpern 2016a, p. 27).

In an important, though often overlooked paper (Maudlin 2004), Tim Maudlin has argued that there are no direct analytical connections between (actual) causation and counterfactuals in either direction. In his first "thought experiment," he suggested to consider a world in which all forces are extremely short range (within an angstrom) and in which there is a particle P that is at rest at t_0 and moving at t_1, and that in the period between t_0 and t_1 only one particle, particle Q, came within an angstrom of P. Then we know with complete certainty what caused P to start moving: it was the collision with Q. As Maudlin has argued, once we know the laws, we can make this causal claim without being certain about the validity of any associated counterfactuals. As a further confirmation, we will show in the next section that the old INUS test can be used to provide a natural definition of actual causality that does not use counterfactuals.

Maudlin's second example was purported to show that fixing truth values for all counterfactuals does not always fix the truth values of all causal claims. This example will turn out to be intimately related to our results in chapter 5 about impossibility of distinguishing basically equivalent causal theories using only counterfactuals.

---

2. "Consider some other relation, *schmausation*, which can be defined in terms of counterfactual dependence, adding drums and trumpets instead of bells and whistles. From the perspective of the counterfactual theory, schmausation is no less natural or distinctive" (Hitchcock 2011).

Example 8.1 (Game of Life) *In John Conway's Game of Life (see Gardner 1970), each square on a grid is either empty or occupied. The rules of the game specify for each three-by-three pattern of occupation on the greed whether the central square is or is not occupied at the next instant. As a result, the state of the grid evolves deterministically through time. Moreover, the rules thereby uniquely determine the validity of every counterfactual assertion about this game.*

Let us introduce some logical notation. Assume some fixed enumeration of the nine squares of a three-by-three grid, with zero being the central square, and let $p_i, i = 0, \ldots, 8$ denote the fact that the ith square is occupied while l_i will denote corresponding literals, that is, either p_i or $\neg p_i$. Then any rule of the Game of Life can be written as a causal rule of the form $A \Rightarrow l_0$, where A is a propositional formula in this language.[3]

Suppose now that there are two patterns of occupation that differ only on square 1, but which both yield that the central square 0 will be occupied. There are, however, two possibilities about how these transitions are generated by the rules. One possibility is that there is a *single* mechanism behind both these transitions which does not involve square 1; this mechanism can be encoded, for instance, by a causal rule of the form

$$l_2, \ldots, l_8 \Rightarrow p_0$$

that does not mention p_1 at all. Another possibility, however, is that these two transitions are governed by two *different* mechanisms depending on whether p_1 holds or not, so they are instantiations of two different laws, for instance:

$$p_1, l_2, \ldots, l_8 \Rightarrow p_0 \quad \text{and} \quad \neg p_1, l_2, \ldots, l_8 \Rightarrow p_0.$$

The above difference does not affect the associated counterfactuals, but it influences our causal judgments. Thus, for a transition from a pattern where square 1 is occupied, p_1 can be naturally viewed as one of the causes of p_0 for the case of alternative mechanisms, though not in the case of a single mechanism. As has been noted by Maudlin, where the laws are in dispute, the *causes* are in dispute, all while the truth values of the counterfactuals remain unquestioned.

As can be seen, the above two possibilities correspond to two basically equivalent causal theories, so our results in chapter 5 determine that these theories are indistinguishable by counterfactual means. However, the main lesson from the above example is that such theories may still support different claims about actual causation.

The problem of structural equivalents. It turns out that the above phenomenon could also be held responsible for the problem of "structural equivalents" in the structural approaches to actual causation. The problem can be broadly described as the problem that causally different situations that intuitively sanction different answers to the question of

---

3. To avoid clutter, we ignore that the antecedent and consequent of this rule refer to different time points.

what is a cause sometimes receive formally isomorphic representations in terms of structural equations (see, e.g., Menzies 2017). It can be illustrated on the following preemption example from Pearl (2000).

Example 8.2 (Desert Traveler) *Enemy 1 poisons T's canteen (p), and enemy 2, unaware of enemy 1's action, shoots and empties the canteen (x). A week later, T is found dead (y).*

An enriched causal model, appeared in Pearl (2000), included also variables C (for cyanide intake) and D (for dehydration) and contained the following equations:

$$c = p \land \neg x \quad d = x \quad y = c \lor d.$$

If we substitute c and d into the expression for y, we obtain a disjunction

$$y = x \lor (p \land \neg x).$$

Pearl has argued, however, that, though $x \lor (\neg x \land p)$ is logically equivalent to $x \lor p$, these two expressions are not "structurally equivalent," and it is this asymmetry that makes us proclaim shooting (x) and not poisoning (p) to be the cause of death.

There is a perceptible "anti-logical" overtone in the above Pearl's argument that appears to suggest that purely logical descriptions do not always provide an adequate representation of the relevant "structural" differences. We suggest, however, a somewhat different diagnosis. As we have argued in chapter 5, basic equivalence is built in Pearl's structural account due to the underlying assumption that any endogenous variable is determined by a single causal mechanism. Recall in this respect that the second example of Maudlin was based on a possibility that a variable (square occupation) might be determined by two different, *alternative* mechanisms. In the structural account, however, all such mechanisms are combined by disjunction into a single equation, so the relevant structural differences could be preserved only if we either sacrifice logical equivalence or use auxiliary variables (C and D in the above example). In fact, introduction of auxiliary variables has even been suggested in Halpern and Pearl (2005) as a modeling rule for such cases: "If we want to argue in a case of preemption that c is a cause of e rather than d, then there must be a random variable . . . that takes on different values depending on whether c or d is the actual cause." This dependence of causal ascriptions on the use of auxiliary variables produces, however, a seemingly problematic dimension of variability, or instability, of causal claims depending on the auxiliary variables we use (see Halpern 2016b).

The regularity approach that we are going to describe in the next section will allow us to provide a more succinct description of the above example without sacrificing classical equivalence.

8.2 Causal Regularity Approach

And on this rock every one must split, who represents to himself as the first and fundamental problem
of science to ascertain what is the cause of a given effect, rather than what are the effects of a given
cause. . . . If we discover the causes of effects, it is generally by having previously discovered the
effects of causes.
—Mill (1872) Bk. V, Ch. III, S. 7

According to the regularity approach, actual causality can and should be analyzed in terms of necessary *and* sufficient conditions. The requirement of sufficiency immediately makes a marked contrast with the counterfactual approaches which are based only on necessary conditions. A cause is something that determines, produces, implies or even necessitates its effect. As has been aptly mentioned by Elizabeth Anscombe (1981), this understanding of causes as sufficient for their effects is "a bit of *Weltanschauung*; it helps to form a cast of mind which is characteristic of our whole culture."

The first explicit formulation of the corresponding definition of (actual) causation has been developed in the first edition of Hart and Honoré (1985) (appeared in 1959) and reappeared as the well-known INUS condition[4] of Mackie (1974). A more perspicuous formulation has been suggested in Wright (1985) as the NESS test:[5]

> The NESS test: a condition c was a cause of a consequence e if and only if it was necessary for the sufficiency of a set of existing antecedent conditions that was sufficient for the occurrence of e.

As we already mentioned earlier, since Lewis's paper *Causation*, a standard opinion in the philosophical literature has become that regularity theories have insurmountable difficulties (see, e.g., Maslen 2012). This opinion, however, is largely unjustified. To begin with, a large part of the difficulties with the regularity theory can be met by adopting more stringent conditions on necessary and sufficient conditions (see Baumgartner 2013). However, a more radical amendment has been suggested, for example, in Strevens (2007), according to which the very notion of sufficiency (which has been assumed to be classical in the original regularity theory) should be given a causal interpretation. This view has been endorsed by the author of the NESS test himself:

> The required sense of sufficiency, which I call 'causal sufficiency' to distinguish it from mere lawful strong sufficiency, is the instantiation of all the conditions in the antecedent ('if' part) of a causal law, the consequent ('then' part) of which is instantiated by the consequence at issue. (Wright 2013)

According to Wright, a critical feature of causal laws and of the related concept of causal sufficiency is their successional or directional nature: the instantiation of the conditions in

4. Insufficient but Nonredundant part of an Unnecessary but Sufficient condition.
5. Necessary Element of a Sufficient Set.

the antecedent of the causal law causes the instantiation of the consequent, but not vice versa. A sequence of such causal laws that links the condition at issue with the consequence provides the required justification for the causal claim.

The above descriptions bring us very close to the formal definition of actual causation, given below. Basically, what we have to do is to explicate the relevant notion of causal sufficiency in terms of causal inference.[6]

Before we plunge into technical details, let us illustrate some key features of our subsequent definition on an extremely simple (and presumably highly confusing) formal example. On our view, this example highlights one of the main difficulties for any potential definition of actual causation (and especially for counterfactual accounts). This pattern will reappear a number of times in the examples discussed in the next section, often as part of a more complex picture.

Example 8.3 (Direct Switch) *Let us consider the following causal theory:*

$$\{p \Rightarrow q, \quad \neg p \Rightarrow q\}.$$

On a plausible understanding,[7] this theory provides a (very abstract) description of two causal mechanisms; the first produces q for the input p, whereas the second produces the same output for the input $\neg p$. These two mechanisms do not interact since they cannot both be active in the same situation.

If we take p to be an exogenous parameter while q an endogenous one, then this theory has two causal (exact) worlds, $\{p, q\}$ and $\{\neg p, q\}$, so q holds in both of them. Suppose now that $\{p, q\}$ is an actual world. Then it seems natural to hold p (as well as the first mechanism $p \Rightarrow q$) responsible for q holding in this world; otherwise q would be uncaused, and consequently the world $\{p, q\}$ would not satisfy the principle of sufficient reason. And indeed, p is a NESS condition of q, so according to the definition below, p is an actual cause of q in this world. Note that no counterfactual account could sanction this since if it were $\neg p$, then q would still hold (and there are no other variables to intervene on). However, things are not too simple also on the regularity side. Indeed, the above theory is causally (and even basically) equivalent to the following one: $\{\mathbf{t} \Rightarrow q\}$, so q is *causally guaranteed to hold* with respect to it. Worse still, though the latter theory has the same causal worlds as the original one, it already does not support the claim that p is an actual cause of q in the world $\{p, q\}$.

The first lesson we can learn from the above example is that the full logic of causal inference turns out to be too strong for dealing with actual causation since it does not distinguish

---

6. Such explications have already been attempted in the framework of structural equation models—see, for example, Baldwin and Neufeld (2004) and Halpern (2008).

7. "If I drink this cup of coffee, I'll die of poisoning. If I don't, I'll die of thirst." (This is a slight variation of an example suggested to me by Vladimir Lifschitz.)

causal theories that could produce different causal claims of this kind (cf. the problem of structural equivalents above). That is why the definition of actual causality that will be given below will be based on a weaker, regular inference, and we will show that the latter constitutes an adequate causal logic for actual causation.

Remark. Once more Aristotle weighs in and provides further support for our choice of the underlying causal logic. In chapter I.5 of the *Posterior Analytics* he argued that even if you demonstrate of each triangle that each has two right angles (that is, 180°)—separately of the equilateral and the scalene and the isosceles—you do not yet know of triangles that they have two right angles, except in the sophistical way (74a25–32). In other words, the logical rule of disjunction in the antecedent (alias reasoning by cases) is not a valid inference rule in Aristotle's apodeictic.

As a second lesson from the above example, we argue that what justifies our claim that p is an actual cause of q in the world $\{p, q\}$ in the original causal theory is that causal rules and mechanisms that are not "active" in the actual world should not influence our causal judgments about this world. In other words, actual causation should be *intrinsic* to the actual world where it is evaluated.[8] In the following exposition, this requirement will be implemented by restricting the set of causal rules that are used to evaluate actual causation to the rules that are active in the relevant world.

The third, and final, lesson from the above example is that, despite their close relations, general and actual causation should be carefully distinguished.

As a prerequisite of the suggested explication of the NESS test, we will severely restrict the form of the causal rules that constitute a causal theory because actual causation will turn out to be highly sensitive to this syntactic form. That is why we will require from the outset that the relevant causal theory should be a *literal* theory, namely, it should consist only of causal rules of the form $l_1, \ldots, l_n \Rightarrow l$, where l and all l_i are literals.

Remark. A reasonable explanation for the high sensitivity of actual causality to the syntactic form of causal rules that we will encounter in what follows could be found in the fact that, by their very definition, complex logical formulas are viewed as completely determined by the propositional atoms that appear in them. This naturally leads to the idea that their potential causal effect can always be "decomposed" into the effects of their constituent literals. This could explain why only plain literals are ordinarily considered as causes. Anyway, this syntax sensitivity is not specific to our definition or even to regularity accounts in general. Rather, it might well be a feature of the notion of actual causation itself. Thus, due to the stipulation that has been made already in Lewis (1973a), according to which causal relata are primitive "events," practically all counterfactual accounts of causation implicitly impose even more severe restrictions: there are profound differences

---

8. See, for example, Hall (2004a) and Lewis (2000) for a general discussion of intrinsicality.

in opinions even about whether "negations" of events (absences and omissions) can be causal relata, let alone "disjunctive" or more complex events. Some well-known difficulties involved in the use of negation, disjunction, and even conjunction, in causal rules and assertions of token causality will be discussed later in this chapter.

In Pearl's structural causal models, each endogenous variable is governed by a single causal mechanism. Such a sweeping unification is appropriate both for describing the causal worlds of a causal theory, and for interventions which cancel all relevant equations for a manipulated variable. However, as we have shown earlier, it becomes problematic for evaluating the claims of actual causation. In our approach to actual causation, the causal rules of a literal causal theory will be largely viewed as describing autonomous causal mechanisms.[9]

8.3   The Causal NESS Test

The terms "exact world" (as defined in chapter 4) and "causal world" from the structural account will be used indiscriminately in what follows (see theorem 5.1 in chapter 5).

On the account below, the claims of actual causality are always made in the context of a given causal theory Δ and an actual world α that is a causal (exact) world with respect to Δ. As before, $\Rightarrow^r_\Delta$ will denote the least regular inference relation that includes a causal theory Δ.

Definition 8.1 (Actual cause) *Let α be a causal world of a literal causal theory Δ. A literal $l_0 \in \alpha$ will be said to be an* actual cause *of a literal l in α wrt Δ if there exists a set of literals $L \subseteq \alpha$ such that*

- $l_0, L \Rightarrow^r_\Delta l$;
- $L \not\Rightarrow^r_\Delta l$.

The above description makes our definition of actual causation a straightforward formalization of the NESS test with regular inference as a logical explication of (causal) sufficiency. Indeed, l_0 is an actual cause of l by this definition if it is a necessary part of some set of existing conditions $\{l_0\} \cup L$ that is sufficient for l.

One of the first consequences of the above definition is that regular inference turns out to be adequate for reasoning about actual causation. In other words, any transformation of (literal) causal theories that is based on the derivation rules of regular inference will preserve actual causality:

Lemma 8.1 *Regularly equivalent literal causal theories support the same claims of actual causation.*

9. Cf. Vennekens, Bruynooghe, and Denecker (2010) for a similar approach.

Remark. In the first version of our approach that has appeared as Bochman (2018a), we have imposed an additional requirement on literal causal theories, namely the requirement of *parsimony* that roughly means that a causal theory should not contain redundant rules that are subsumed by other rules that belong to the theory. On a more careful analysis, this requirement has turned out to be unnecessary for the purposes of the definition.[10] Still, it is interesting to note that a somewhat weaker version of the parsimony requirement has been included in the original account of the NESS test by Richard Wright; it required that a properly formulated causal law should contain in its antecedent only those conditions that are necessary for the sufficiency of the set of conditions that causes its consequent (see Wright 2013).

On the suggested account, general and actual causality are conceptually different even on the logical scale. Namely, we can see general causation as a purely *logical* notion that is described by an appropriate formalism of causal inference. In contrast, actual causation is already an explicitly *nonmonotonic* notion since it depends on the absence (non-provability) of certain causal rules. Unlike general causation, claims of actual causation can be overridden with an addition of new causal rules to a causal theory. The following formal example illustrates this.

Example 8.4 *Suppose that $p, q \Rightarrow r$ is the only rule of a causal theory Δ that contains r in its head. Then any causal world of Δ where r holds should contain both p and q (by the law of causality). Moreover, both will be actual causes of r in such a world. However, if we add a rule $q \Rightarrow r$ to this causal theory, then p will no longer be an actual cause of r in the causal worlds of Δ.*

Yet another salient feature of the above definition is its high sensitivity to the syntactic form of causal rules. As a matter of fact, our definition "inherited" this sensitivity to the syntax from the original INUS condition. For instance, suppose that disjunctions of literals are allowed in the bodies of causal rules, and assume that l_0 is an actual cause of l in a world α by the above definition: for some set L of literals,

$$l_0, L \Rightarrow^r l \quad \text{and} \quad L \not\Rightarrow^r l.$$

Now let l' be an arbitrary literal from α. Then we have

$$l', (\neg l' \vee l_0), L \Rightarrow^r l \quad \text{and} \quad (\neg l' \vee l_0), L \not\Rightarrow^r l$$

by the logical properties of $\Rightarrow^r$. So, if we could add $\neg l' \vee l_0$ to the "witness" set L, we would obtain that l' is an actual cause of l, which is absurd. Kim (1971) was apparently the first to discover this problem; see also Strevens (2007).

The above "paradox" has occasionally been used as a decisive objection against taking facts and propositions as causal relata. In our approach, however, we view the notion

---

10. Thanks to Vladimir Lifschitz for pointing this out.

of actual causality itself as a prime suspect that is responsible for the necessity of the corresponding syntactic restrictions.

It has been shown in chapter 4 that causal reasoning in literal causal theories is much simpler than in the general case. Thus, the causal worlds of such theories can be identified with maximal consistent sets of literals that satisfy, in addition, the following fixpoint condition:

$$L = \Delta(L).$$

Moreover, regular causal inference in literal causal theories is quite similar to Horn inference for positive logical clauses, the only difference being that it admits both positive and negative literals indiscriminately. In particular, it has been shown that it can be reduced to the use of three literal inference rules, *Monotonicity*, *Cut*, and *Contradiction* (see theorem 4.23). Now, in the context of actual causality, we check only derivability for consistent sets of literals (since they belong to the same actual world), so we need only the first two inference rules. In other words, we obtain the following corollary.

Corollary 8.2 *If α is a causal world of a literal causal theory Δ, and $L_0 \subseteq \alpha$, then $L_0 \Rightarrow^r_\Delta l_0$ holds if and only if the causal rule $L_0 \Rightarrow l_0$ is derivable from Δ using the following two literal inference rules:*

(Literal Monotonicity)   *If $L \Rightarrow l$, then $L, L' \Rightarrow l$;*
(Literal Cut)   *If $L \Rightarrow l$ and $L, l \Rightarrow l'$, then $L \Rightarrow l'$.*

The above description implies, in particular, that checking claims of actual causality on our definition has the same complexity as verifying both derivability and nonderivability of propositional Horn clauses.

8.3.1 Forgetting

Horn inference is a well-developed part of a general logical theory, and it has been shown to have some nice logical features. One of these features involves *forgetting*—unfolding transformations that eliminate atoms from the clauses of a positive logical theory or logic program by (i) removing all clauses that contain this atom in heads, and (ii) replacing all occurrences of the atom in bodies with its "definition," namely the bodies of the removed clauses. It is well known that this unfolding transformation preserves derivability in Horn theories, which immediately raises the question whether this preservation property could be extended to actual causality. Unfortunately, the following counterexample shows that, in general, this does not hold.

Example 8.5 *Let us consider the following positive causal theory:*

$$p \Rightarrow q \quad q, r \Rightarrow s \quad p \Rightarrow s.$$

Then for the actual world $\{p, q, r, s\}$, *we obtain that* r *is an actual cause of* s *(due to the second rule). However, if we will eliminate* q *from this theory, we will obtain*

$$p, r \Rightarrow s \quad p \Rightarrow s,$$

and then r *will not be an actual cause of* s *in this world since any derivation of* s *from* r *would require* p *while* p *causes* s *by itself.*

Still, in many cases of interest, we will be able to eliminate intermediate links in a causal theory in such a way that will preserve causal assertions about remaining literals.

Let Δ be a literal causal theory and l some literal appearing in the heads of the rules in Δ. Assume that Δ_l is the set of all rules in Δ with the head l. Then Δ/l will denote a causal theory obtained from $\Delta \setminus \Delta_l$ by replacing every occurrence of l in the bodies of the remaining rules with each body of the rules from Δ_l. We will call this transformation *forgetting*, or *elimination*, of l in Δ.

The above example has shown that, in general, this forgetting transformation may change causal assertions. Still, we are going to describe some special conditions under which this transformation will preserve actual causality.

Let us say that a literal l_0 is a *direct cause* of a literal l in a literal causal theory Δ if Δ contains a rule of the form $l_0, L \Rightarrow l$.

Definition 8.2 *A literal* l *will be said to be* eliminable *with respect to a causal theory* Δ *if there is no literal that is a direct cause of both* l *and some other literal* l_0 *in* Δ.

To see the rationale behind this notion, note that in the above counterexample q is not eliminable since p is a direct cause of both q and s. Speaking more generally, the restriction is that if some literal is a direct cause of an eliminable literal, it cannot appear in causal rules for other literals. Though this constraint may appear to be overtly restrictive, it will turn out to be sufficient for the purposes of this book.

The following result shows that in such cases the forgetting transformation will preserve actual causality.

Theorem 8.3 *Let* Δ/l *be a forgetting transformation of* Δ *with respect to an eliminable literal* l, *and* α *is a causal world of* Δ. *Then, for any literals* l_0, l_1 *that are distinct from* l, l_0 *is an actual cause of* l_1 *in* α *with respect to* Δ *if and only if* l_0 *is an actual cause of* l_1 *with respect to* Δ/l.

Proof. Any causal rule of Δ/l either belongs to Δ or is derivable from Δ by regular inference. Therefore, $\Rightarrow^r_{\Delta/l}$ is included in $\Rightarrow^r_\Delta$. Moreover, just as for Horn inference, forgetting preserves derivability of causal rules with respect to regular inference, that is, if $l \notin L \cup \{l_0\}$, then $L \Rightarrow^r_\Delta l_0$ if and only if $L \Rightarrow^r_{\Delta/l} l_0$.

Suppose now that l_0 is an actual cause of l_1 in α with respect to Δ, that is,

$$l_0, L \Rightarrow^r_\Delta l_1 \quad \text{and} \quad L \not\Rightarrow^r_\Delta l_1,$$

for some set $L \subseteq \alpha$. If $l \notin L$, then clearly l_0 is an actual cause of l_1 with respect to Δ/l, and vice versa. Hence, assume that $l \in L$, that is,

$$l_0, l, L' \Rightarrow^r_\Delta l_1 \quad \text{and} \quad l, L' \not\Rightarrow^r_\Delta l_1,$$

for some L' that does not include l. Now, since $l \in \alpha$, and α is a causal world, Δ should contain a causal rule $L_0 \Rightarrow l$ such that $L_0 \subseteq \alpha$. Then we can replace l with L_0 in the first rule above and obtain $l_0, L_0, L' \Rightarrow^r_\Delta l_1$ by regular inference, and therefore $l_0, L_0, L' \Rightarrow^r_{\Delta/l} l_1$ (since L_0 does not contain l). Moreover, suppose that $L_0, L' \Rightarrow^r_\Delta l_1$. Note that $L' \not\Rightarrow^r_\Delta l_1$ due to the second rule above, and hence the supposed derivation should include (rules involving) at least one literal from L_0. However, any literal in L_0 is a direct cause of l, and therefore by eliminability of l there are no such rules. The obtained contradiction implies that $L_0, L' \not\Rightarrow^r_\Delta l_1$. Therefore, $L_0, L' \not\Rightarrow^r_{\Delta/l} l_1$, and consequently l_0 is an actual cause of l_1 with respect to Δ/l. $\square$

As we will see in a number of examples presented in what follows, the above result will allow us to eliminate intermediate links in causal theories in a way that preserves the causal structure for the remaining literals. This possibility can be seen as an important advantage of our approach to actual causality, especially because the corresponding unfolding transformation in the framework of structural equations (namely, replacement of variables with their determining functions) fails to preserve the claims of actual causality (see our discussion of *Desert Traveler* [example 8.2] above).

8.4   Actual Subtheories and Causal Inference

In this section we will show that full causal inference can also be used for checking claims of actual causality once the source causal theory is "pruned" by removing causal rules that are not active in the actual world.

Definition 8.3 *Let α be a causal world of a literal causal theory Δ. A rule $L \Rightarrow l$ from Δ will be called* active *in α if $L \subseteq \alpha$. The* actual subtheory *of Δ wrt α is the set Δ_α of all causal rules from Δ that are active in α.*

For a given actual causal world α, the actual subtheory of Δ with respect to α will also be called the *actual causal theory* of α. In what follows, $\Rightarrow_\alpha$ will denote the least *causal* inference relation that includes Δ_α.

Remark. The above notion of causal inference with respect to the actual causal theory can be naturally viewed as a formal counterpart of the Aristotelian notion of demonstration in his *Analytics*. A more detailed comparison of our theory with Aristotle's apodeictic should wait, however, for another occasion.

Since causal worlds of Δ are closed with respect to the rules of Δ, the heads of all causal rules that are active in α will also hold in α. Note also that any causal world is uniquely determined by the causal rules that are active in it.

There is a lot of similarity, both in content and purpose, between the actual causal subtheory and the notion of a *causal beam* from chapter 10 of Pearl (2000). This makes our definition much similar to the original definition of actual causation, described in Pearl (2000). Note, however, that our construction is immune to the counterexamples that have led to abandoning this idea by Pearl. Our definition has also much in common with the approach of Beckers and Vennekens (2018). In particular, their notion of production can be viewed as a counterpart of our logical notion of causal inference $\Rightarrow_\alpha$ in the actual subtheory.

The following key result shows that causal inference with respect to the actual subtheory $\Rightarrow_\alpha$ can replace regular inference in the definition of actual causality.

Theorem 8.4 *Let α be a causal world of a literal causal theory Δ. Then for any $L \subseteq \alpha$ and any literal l,*

$$L \Rightarrow_\alpha l \ \text{ iff } \ L \Rightarrow^r_\Delta l.$$

A proof sketch. Let C_α and C^r denote the causal operators corresponding, respectively, to $\Rightarrow_\alpha$ and $\Rightarrow^r_\Delta$. Then we have to show that $C_\alpha(L) = C^r(L)$ for any $L \subseteq \alpha$. The inclusion of right into left follows from the fact that $C_\alpha(L)$ is closed with respect to the rules of Δ while $C^r(L)$ is the least such Δ-closed set that contains L. In the other direction, let us rename all negated atoms $\neg p \in \alpha$ as new atoms $\bar{p}$. Then the actual causal subtheory Δ_α will become a positive literal theory in the new language, and it can be verified that the derivabilty of positive atoms from this positive theory with respect to causal inference amounts to Horn inference for positive propositional clauses. Now the inclusion of left into right follows from the fact that regular inference includes the inference rules of Monotonicity and Cut that are jointly sufficient for Horn inference (even with respect to classical inference). $\square$

As a consequence of the above theorem, we immediately obtain the following equivalent characterization of actual causation:[11]

Corollary 8.5 *Let α be a causal world of a literal causal theory Δ. Then a literal $l_0 \in \alpha$ is an actual cause of a literal l in α wrt Δ if and only if there exists a set of literals $L \subseteq \alpha$ such that*

- $l_0, L \Rightarrow_\alpha l;$
- $L \not\Rightarrow_\alpha l.$

---

11. This description has actually been used as a definition of actual causality in Bochman (2018a).

In the examples below, the notion of an actual subtheory will provide us with a convenient tool for reducing the set of causal rules that need to be taken into account in determining actual causality.

8.5 Examples and Counterexamples

The examples we are going to discuss below occupy a prominent place in the literature mainly because each of them constitutes a counterexample for some past approach to actual causality. For the majority of these examples, there are already established representations in structural equation models (see, e.g., Halpern 2016a), and we will often use them as an initial basis of our logical characterizations. This will not mean, of course, that our suggested definition will always produce the same answers, though there will indeed be a large area of agreement.

Remark. In the description of the examples below that already have an accepted representation in terms of structural equations, we will often employ a more or less systematic "translation procedure" for transforming structural equations into literal causal rules. The reader should be warned, however, that this procedure will not always be determinate, so it can only be viewed as an informal guide for obtaining corresponding literal causal theories; only the latter will be considered as our "official" representation. First, each structural equation $p := F$ can be translated into a pair of causal rules, $F \Rightarrow p$ and $\neg F \Rightarrow \neg p$ (see chapter 5). Then, in cases where F is a complex formula, we can transform both F and $\neg F$ into a disjunctive normal form (DNF) and use them to decompose each of these causal rules into corresponding sets of literal rules. It should be kept in mind, however, that the transformation to DNF is equivocal, in general, so it may produce different outputs. Worse still, in many cases these different representations will not be equivalent in our account! Actually, we will even "abuse" this equivocation by suggesting alternative rule-based causal representations for the relevant examples.

In order to make our formal descriptions more succinct, we will often use the possibility of literal causal completion (see definition 4.28). In other words, we will often write only the positive part of a relevant causal theory, assuming that a complete theory could be obtained by supplying causal rules for the corresponding negative endogenous literals, as well as explanatory rules $l \Rightarrow l$ for all exogenous literals. This completion construction could be viewed as a plain notational convention, and it will be moderated somewhat in chapter 9.

To begin with, the role of the restriction to the actual subtheory can be illustrated on the following example. This example has served as one of the main reasons for the change from the original HP definition, described in Halpern and Pearl (2001) to the updated definition of Halpern and Pearl (2005).

Example 8.6 (Loader) *(Hopkins and Pearl 2003) A firing squad consists of shooters B and C. It is A's job to load B's gun, C loads and fires his own gun. On a given day, A loads B's gun. When the time comes, only C shoots the prisoner.*

The initial definition in Halpern and Pearl (2001) wrongly made A an actual cause of D. The structural equation for this example is as follows:

$$D = (A \wedge B) \vee C.$$

It corresponds to the following literal causal theory:[12]

$$A, B \Rightarrow D \quad C \Rightarrow D$$

$$\neg A, \neg C \Rightarrow \neg D \quad \neg B, \neg C \Rightarrow \neg D$$

For the actual causal world $\{A, \neg B, C, D\}$, the associated actual causal theory is just $\{C \Rightarrow D\}$. Thus, C is clearly an actual cause of D, but A is not since A does not even appear in the latter.

8.5.1   Overdetermination

Examples of overdetermination present prima facie problems for counterfactual theories of causation because in such cases there is no direct counterfactual dependence between the effect and its cause.

Example 8.7 (Window) *Billy (B) and Suzy (S) both throw rocks at a window. The rocks strike the window at exactly the same time. The window breaks (W).*

The structural equation of this story is just $W = B \vee S$, so the corresponding causal theory is as follows:

$$B \Rightarrow W \quad S \Rightarrow W \quad \neg B, \neg S \Rightarrow \neg W.$$

The actual causal world is $\alpha = \{S, B, W\}$, and only the first two rules are active in it. We have both $B \Rightarrow_\alpha W$ and $S \Rightarrow_\alpha W$, though $\mathbf{t} \not\Rightarrow_\alpha W$, and therefore both S and B are actual causes of W in α.

The first two HP definitions (Halpern and Pearl 2001, 2005) gave the same answers as above. However, according to the last, modified definition of actual causation given in Halpern (2015), S and B are not actual causes of W, though they are parts of the actual cause $\{S, B\}$.

A somewhat less natural feature of the above representation should also be mentioned. It is that in another possible causal world, $\{\neg B, \neg S, \neg W\}$, an unbroken window also has actual causes, namely $\neg B$ and $\neg S$. A modified causal representation that takes into account

---

12. We could have written only the first two rules since the last two rules could be obtained by negative completion.

normality and the distinction between causes and conditions could help with this deficiency (see chapter 9).

Intransitivity of actual causation. One of the most discussed features of causation is transitivity (see, e.g., Hall 2000; Hitchcock 2001; McDonnell 2018), and the suggested theory allows us to explain at least some deliberations arising with respect to this controversial property. In our theory, general causal inference is transitive while actual causation is not. The following overdetermination example, which is a variation of an example from Ehring (1987), forms perhaps the simplest counterexample to transitivity on our definition:

Example 8.8 (Purple Flame) *Jones puts potassium salts (P) into a hot fire (F). Because potassium compounds produce a purple flame when heated, the flame changes to a purple colour (PF), though everything else remains the same. Both flames ignite some flammable material (I).*

The corresponding positive causal theory is

$$P, F \Rightarrow PF \quad F \Rightarrow I \quad PF \Rightarrow I.$$

This theory also happens to be an actual causal subtheory for the actual world $\alpha = \{P, PF, F, I\}$. Consequently, P is an actual cause of PF, and PF is an actual cause of I. However, P is not an actual cause of I in this world since it is not a necessary part of any sufficient condition for I.

The above example (as well as the corresponding causal pattern) is especially important for our approach because it does not "endorse" most of the other suggested counterexamples to transitivity (such as *Switch*, discussed in the next subsection).

The above counterexample could be made even more vivid if, instead of the causal rule $PF \Rightarrow I$, we would make use of a more basic *meaning postulate* stating that purple flame is a flame, that is, $PF \vDash F$ in the underlying logic. Then the original causal theory boils down to

$$F \Rightarrow I \quad P, F \Rightarrow PF,$$

which is also the actual causal theory with respect to the same actual world as before. Now we still have $PF \Rightarrow_\alpha I$ by regular inference (due to Strengthening), and hence P is an actual cause of PF, and PF is an actual cause of I. However, as before, P is not an actual cause of I since the inference of I from P requires F while the latter causes I by itself.

The following example has been discussed in Rosenberg and Glymour (1918) in connection with Halpern's approach to actual causality.

Example 8.9 (Transitivity) *Let us consider the following formal example of a causal theory (in the original notation):*

$$X \Rightarrow Y \quad Y \Rightarrow Z \quad X \Rightarrow Z.$$

According to our approach, in the world $\{X, Y, Z\}$, *both X and Y are actual causes of Z, unlike Halpern's modified definition in Halpern (2016a), according to which only X is an actual cause of Z. Note, however, that this would change if we would replace the rule $Y \Rightarrow Z$ with $X, Y \Rightarrow Z$. In this case we would also obtain that only X is an actual cause of Z; this is because, just as in the preceding example, any set of literals sufficient for causing Z would contain X.*

8.5.2 Switches

A large group of examples in the literature involves a causal pattern where some variable acts as a switch between two mechanisms or processes, both leading to the same result. Accordingly, this variable does not influence the result on the general causal level, though its actual instantiations are actual causes of this result in each particular situation. A vivid example is given below.

Example 8.10 (Switch) *(Pearl 2000) A two-state switch is wired to two lamps. If the switch is in one state (S), the first lamp is activated (L_1), and if it is in the other state ($\neg S$) the second lamp is activated (L_2). In both cases, the room is illuminated (I).*

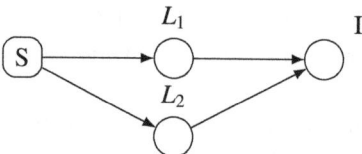

In the above picture, the rounded square labeled with S represents a *switch*; if S holds, it causes L_1, while if $\neg S$ holds, it causes L_2.

The corresponding positive causal theory for this setup is as follows:

$$S \Rightarrow L_1 \quad \neg S \Rightarrow L_2 \quad L_1 \Rightarrow I \quad L_2 \Rightarrow I.$$

On the general causal level, $S \Rightarrow L_1$ and $L_1 \Rightarrow I$ imply $S \Rightarrow I$ by transitivity, and similarly $\neg S \Rightarrow L_2$ together with $L_2 \Rightarrow I$ imply $\neg S \Rightarrow I$, so $\mathbf{t} \Rightarrow I$ by the Or postulate of causal inference. Accordingly, I is guaranteed to hold in all causal worlds of this causal theory.

For an actual causal world $\{S, L_1, \neg L_2, I\}$, however, the associated causal theory is

$$S \Rightarrow L_1 \quad L_1 \Rightarrow I,$$

and therefore S and L_1 are actual causes of I in this world (cf. our discussion of example 8.3, Direct Switch). Note also that actual causation is still transitive in this case: S is an actual cause of L_1, and L_1 is an actual cause of I.

The above causal description, according to which the position of the switch is an actual cause of the room being illuminated, has been widely disputed in the literature. Pearl himself, though accepting this description, has remarked that, although the position of the switch enables the passage of electric current through the lamp, it might not deserve the title "cause" in ordinary conversation. In fact, a number of approaches to actual causality have been designed with an explicit aim of preventing this conclusion. We will return to this example in chapter 9 where we will explore some additional representation possibilities for describing such situations.

Our next example usually serves as a paradigmatic case of *early preemption*.

Example 8.11 (Backup) *(Hitchcock 2007) Assassin poisons Victim's coffee (A). Victim drinks it and dies (D). If Assassin hadn't poisoned the coffee, Backup would have (B), and Victim would have died anyway.*

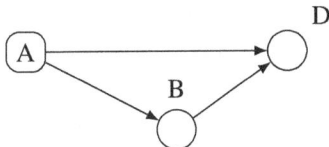

Note that this example is obviously similar to the previous *Switch* example. In fact, it can be obtained from the latter just by "forgetting" L_1. No wonder that our analysis will produce a similar result.

In the structural approach, Backup is usually represented using the following equations:

$$B = \neg A, \quad D = A \vee B.$$

These equations correspond to the following positive causal theory:

$$\neg A \Rightarrow B \quad A \Rightarrow D \quad B \Rightarrow D.$$

$\neg A \Rightarrow B$ and $B \Rightarrow D$ imply $\neg A \Rightarrow D$. Taken together with $A \Rightarrow D$, the latter implies $\mathbf{t} \Rightarrow D$ by Or. Thus, on the level of general causation, death (D) is "causally inevitable."

Still, for the original causal theory (obtained by negative completion from the above positive theory) and an actual world $\alpha = \{A, \neg B, D\}$, the actual causal theory is $\{A \Rightarrow \neg B, A \Rightarrow D\}$. We have $A \Rightarrow_\alpha D$, though $\mathbf{t} \nRightarrow_\alpha D$, and therefore poisoning A is an actual cause of death D in α, as is commonly expected in this case. However, a distinctive feature of our account is that in the situation where Assassin does not poison the coffee (that is, $\alpha = \{\neg A, B, D\}$), $\neg A$ is an actual cause of D (as well as B), since in this case the actual causal theory is

$$\neg A \Rightarrow B \quad B \Rightarrow D.$$

Surprisingly, the last, modified definition from Halpern (2016a) does not support this claim. Even more surprisingly, just as in the original counterfactual account of Lewis (1973a), the claim is restored if we add some intermediate event (say, "Victim drinks the poisoned coffee") on the causal path from A to D since in this case we would return to the preceding *Switch* pattern. We agree here with Hitchcock (2001) that we should be particularly troubled that we judge there to be a causal relationship on the basis of finding an intermediate event that is not made salient in the presentation of the example.

Now let us turn to more complex examples of this kind.

Example 8.12 (Push) *(M. McDermott 1995) I push (P) Jones in front of a truck (T), which hits (H) and kills him (D); if I had not done so, a bus would have hit (B) and killed him.*

Here is a corresponding structural model:

$$H = (P \wedge T) \vee (\neg P \wedge B), \quad D = H.$$

For a causal world $\{P, B, T, H, D\}$, the HP definition from Halpern and Pearl (2005) yielded P and T as causes of D, as we would expect. But, unfortunately, it also yielded B as a cause of D. Even the modified definition from Halpern (2016a), though it does not make B a cause of D, still makes it part of a cause $P \wedge B$—see example 4.1.1 in Halpern (2016a).

The corresponding positive causal theory is

$$P, T \Rightarrow H \quad \neg P, B \Rightarrow H \quad H \Rightarrow D.$$

Since the actual world is $\{P, B, T, H, D\}$, the actual causal theory is

$$P, T \Rightarrow H \quad H \Rightarrow D.$$

Thus, P and T are actual causes of D, but B is not.

The above example can also be used to illustrate once again that full causal inference is too strong for serving as an internal logic of actual causality. Thus, the first two rules imply the following rule by causal inference:

$$T, B \Rightarrow H.$$

Actually, this is an immediate consequence of a purely logical fact that $(P \wedge T) \vee (\neg P \wedge B)$ is equivalent to

$$(P \wedge T) \vee (\neg P \wedge B) \vee (T \wedge B)$$

in classical logic. However, if we would add the above causal rule to the source causal theory, it would appear also in the actual causal theory, and we would obtain that also B is an actual cause of D, contrary to our intuitions.

In our final "switch" example below, what is a cause on any counterfactual approach will not be an actual cause on our definition.

Example 8.13 (Inevitable Shock) *(M. McDermott 1995; Weslake 2015) Two switches are wired to an electrode. The switches are controlled by A and B respectively, and the electrode is attached to C. A flips her switch (A), which* forces *B to flip her switch (B) (B has no choice in the matter).*[13] *The electrode is activated and shocks C (C) if both switches are in the same position.*

The corresponding positive causal theory for this story is as follows:

$$A \Rightarrow B \quad A, B \Rightarrow C \quad \neg A, \neg B \Rightarrow C.$$

Again, $\mathbf{t} \Rightarrow C$ follows from the (negatively completed) theory by causal inference, so the shock is inevitable. Still, the actual world is $\{A, B, C\}$, so the actual causal subtheory is

$$A \Rightarrow B \quad A, B \Rightarrow C.$$

We obtain $A \Rightarrow C$ by Cut, so B cannot be an actual cause of C (since any set of literals that causes C will include A). This makes A the *only* actual cause of C. In fact, the source positive causal theory is *regularly* equivalent to the following one:

$$A \Rightarrow B \quad A \Rightarrow C.$$

In the above theory, B and C are just joint effects of the common cause A. Note, however, that B is a but-for cause of C in the original causal theory (due to the fact that a rule $A, \neg B \Rightarrow \neg C$ can be obtained by negative completion), so it is an actual cause of the latter on any counterfactual account. The causal regularity approach provides here more discriminate answers about actual causality than the counterfactual approach.

8.6 Negative Causality

In this section, we will provide an overview of some of the most prominent and difficult cases of negative causation in the literature and the problems associated with their interpretation. We will begin with cases of prevention and preemption and show that a deep representation that uses defeasible causal rules (see section 4.8 in chapter 4) will allow us to provide an adequate description of corresponding situations. However, when we turn to more complex cases of negative causation, it will be shown that the framework of actual causality described in this chapter still does not have sufficient representation capabilities for fully capturing such cases; this will pave the way to an alternative, "upgraded" representation we will describe in chapter 9.

8.6.1 Prevention and Preemption

Both natural language and commonsense reasoning maintain a clear separation between causing and preventing. Thus, the experiments of Walsh and Sloman (2005) have shown

---

13. Here we have slightly changed the original story in M. McDermott (1995) to make it more in accord with the equations used in Weslake (2015), specifically, $B := A$.

that "prevent X" is not equivalent to "cause not-X" for their participants. Depending on the exact story, participants would sometimes think that one or the other of these two constructions was appropriate, but they very rarely found them to be interchangeable. This distinction is also a natural consequence of singularist theories of causation which argue that prevention is not causation (see Moore 2009).

Traditional logical methods of handling preventive factors amount to placing the negations of such factors as additional conditions in the antecedent of a rule; Mill's conditionals could serve as a good illustration of this method (see chapter 1). Thus, if P is a condition that prevents the application of a causal rule $C \Rightarrow E$, then $\neg P$ should be added to the antecedent of this rule, which gives us

$$C, \neg P \Rightarrow E.$$

However, in the case when this is the only causal rule that causes E, it produces the following causal rules by negative causal completion:

$$\neg C \Rightarrow \neg E \quad P \Rightarrow \neg E.$$

One of the undesirable consequences of the second rule above is that P becomes an actual cause of $\neg E$ even when C does not hold, so there are no "real" causes for E to begin with. Of course, P *guarantees* the nonoccurrence of E in this situation, but being a guarantee of something is not being a cause of it. The following well-known example illustrates the problem.

Example 8.14 (Bogus Prevention) *(Hiddleston 2005) Assassin refrains from putting poison in Victim's coffee ($\neg A$). Bodyguard puts antidote in the coffee (B). Victim drinks the coffee and survives ($\neg D$).*

An initial description of this situation in the structural approach has been based on a seemingly appropriate equation $D = A \wedge \neg B$. Indeed, the equation apparently says that poison causes death in the absence of antidote. It corresponds to the following causal theory:

$$A, \neg B \Rightarrow D \quad \neg A \Rightarrow \neg D \quad B \Rightarrow \neg D.$$

However, this description (as well as the source structural equation) makes this case "isomorphic" to *Window* (symmetric overdetermination), with a counterintuitive conclusion that both the absence of poison $\neg A$ *and* the presence of antidote B are actual causes of survival $\neg D$ in the world $\{\neg A, B, \neg D\}$. This example has been used as one of the prominent instances of the problem of structural equivalents in the structural account.

As a replacement for this structural representation, we will suggest that a causal pattern of prevention should be formalized using defeasible causal rules (see section 4.8) and the following deep representation:

$$C, n \Rightarrow E \quad C, P \Rightarrow \neg n,$$

where n denotes the underlying causal mechanism. This causal pattern could be depicted using the following diagram:[14]

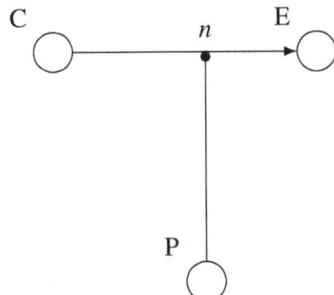

The second rule above, $C, P \Rightarrow \neg n$, is a *defeat (or cancellation) rule* that describes how the preventive condition P, being combined with a cause C, cancels the underlying mechanism n of the main causal rule $C, n \Rightarrow E$.

Now the negative completion of the source defeasible rule produces the following two rules:

$$\neg C \Rightarrow \neg E \quad \neg n \Rightarrow \neg E.$$

Combined with the above defeat rule, this gives us the following surface representation for $\neg E$:

$$\neg C \Rightarrow \neg E \quad C, P \Rightarrow \neg E.$$

On the other hand, default causal completion for n produces

$$\neg C \Rightarrow n \qquad \neg P \Rightarrow n.$$

Combined with the basic defeasible rule $C, n \Rightarrow E$, this reproduces the initial surface rule for E:

$$C, \neg P \Rightarrow E.$$

Summing up, the surface representation of prevention will amount to

$$C, \neg P \Rightarrow E \qquad \neg C \Rightarrow \neg E \qquad C, P \Rightarrow \neg E.$$

It is important to observe that the transition from the deep to surface representation in the above setting proceeds through "forgetting" of eliminable literals (see definition 8.2), and therefore it preserves all the claims of actual causation for the remaining surface literals.

As can be seen, the above representation is causally (and even basically) equivalent to the initial one, given earlier. Still, it is not regularly equivalent to the former, and this will imply that it will justify different claims of actual causality.

---

14. We should warn the readers that are familiar with Lewis's neuron diagrams that, from now on, our pictures will begin to deviate from the "laws" of the latter.

Now let us apply this modified representation to *Bogus Prevention*.

Example 8.15 (Bogus Prevention, continued) *The modified representation of this causal situation will be as follows:*

$$A, n \Rightarrow D \qquad A, B \Rightarrow \neg n.$$

As has been shown above, this deep representation produces the following surface one:

$$A, \neg B \Rightarrow D \quad \neg A \Rightarrow \neg D \quad A, B \Rightarrow \neg D.$$

In this causal theory, the last causal rule provides an implicit description of poison neutralization. For the actual world $\alpha = \{\neg A, B, \neg D\}$, the corresponding actual causal theory is just $\{\neg A \Rightarrow \neg D\}$, and consequently only $\neg A$ is an actual cause of $\neg D$.

Blanchard and Schaffer (2017) have suggested that the initial structural model for *Bogus Prevention* should be replaced with a more "apt" model that would take into account the underlying causal process of poison neutralization.[15] Indeed, if we add a new proposition N saying that the poison is neutralized, then the above story can be described using the structural equations

$$N = A \wedge B \qquad D = A \wedge \neg N$$

while the latter correspond to the following causal theory:

$$A, B \Rightarrow N \quad \neg A \Rightarrow \neg N \quad \neg B \Rightarrow \neg N$$

$$A, \neg N \Rightarrow D \quad \neg A \Rightarrow \neg D \quad N \Rightarrow \neg D.$$

Then the actual subtheory for the world $\alpha = \{\neg A, B, \neg N, \neg D\}$ will be

$$\neg A \Rightarrow \neg N \quad \neg A \Rightarrow \neg D,$$

so only Assassin not putting in poison is an actual cause of survival.

It can be easily seen, however, that the above theory can be obtained as a result of negative causal completion of the following positive theory:

$$A, B \Rightarrow N \quad A, \neg N \Rightarrow D,$$

while the latter theory is just a notational variant of our formalization of prevention:

$$A, B \Rightarrow \neg n \quad A, n \Rightarrow D.$$

In the framework of the causal calculus, however, we can "forget" n (see section 8.3.1) and unfold this theory to a more adequate surface causal theory in the source language, given above.

Short-circuit. Interesting (though confusing) situations arise when both causation and prevention are initiated by the same event.

---

15. This solution has also been mentioned in Halpern and Hitchcock (2015).

Example 8.16 (Boulder) *(Hall 2000) A boulder begins to roll down the hill toward the hiker's head, which causes the hiker to duck, which in turn causes the hiker to survive.*

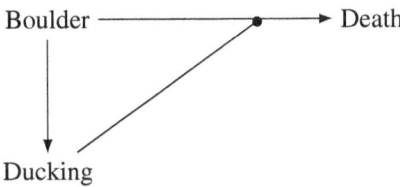

This example has often been used as yet another counterexample to transitivity of actual causality since the boulder's rolling down does not seem to be a cause of survival. Thus, Moore (2009) has argued that transitivity should not be expected in these cases because the ducking does not cause the hiker to survive; rather, the ducking prevents the hiker's death, and prevention should be distinguished from causality.

A similar problem is presented by the following variation.

Example 8.17 (Bogus Poisoning) *(Hitchcock 2007) Bodyguard puts an antidote in Victim's coffee. Assassin then poisons the coffee. Assassin would not have poisoned the coffee if Bodyguard had not administered the antidote first. Victim drinks the coffee and survives.*

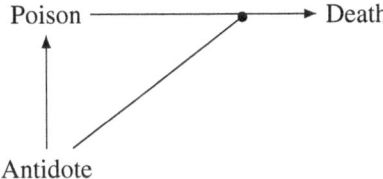

For this example, all variants of HP definitions from Halpern (2016a) make Bodyguard putting in the antidote a cause of Victim's surviving.

From the point of view of our representation, the above two examples are just special instances of a more general situation, called *short-circuit* in Hall (2007), that can be depicted as follows:

As was noted by Hall, this situation spells trouble both for the claim that causation is transitive and that counterfactual dependence suffices for causation.

Applied to this latter case, our representation of prevention produces the following causal theory:

$$C, n \Rightarrow E \qquad C \Rightarrow \neg n.$$

Note, however, that $C \Rightarrow \neg n$ implies $C, n \Rightarrow \mathbf{f}$ by Constraint (see section 4.1.2) while the latter implies $C, n \Rightarrow E$ (by Weakening). Thus, the above theory reduces to just

$$C \Rightarrow \neg n$$

for regular inference, so it does not provide causal explanations neither for E nor for $\neg E$; all it says is that C causes a cancellation of the main causal rule. We could even say that the causal rule $C, n \Rightarrow E$ is "causally incoherent" in such situations since its antecedent is causally inconsistent (there are no causal worlds where both C and n hold).

Remark. The above examples provide, however, a first indication (in the course of our exposition) that a "static" atemporal representation of causal rules does not always constitute an adequate description of causal situations. What we have, for instance, in the *Boulder* example is actually a consistent, though dynamic, chain of events which starts with boulder's rolling down that creates a *threat* to the hiker's life (which normally does not require causal explanation), followed by his ducking, which cancels this process and thereby the threat. This causal chain is similar to *abortion* that we will discuss later in this chapter and once again in chapter 9 in a more expressive framework.

Late preemption. The following well-known example of late preemption can actually be seen as a combination of two causal patterns—prevention and switch.

Example 8.18 (Bottle) *Suzy (ST) and Billy (BT) both throw rocks at a bottle. Suzy's rock arrives first and hits the bottle (SH), and the bottle shatters (BS); Billy's arrives second and so does not hit the bottle (BH). Both throws are accurate; Billy's would have shattered the bottle if Suzy's had not.*

The following structural equation model for this example has been suggested in Halpern and Pearl (2001):

$$SH = ST, \quad BH = BT \wedge \neg SH, \quad BS = BH \vee SH.$$

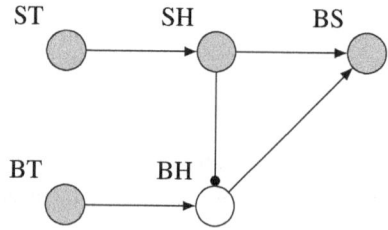

The positive causal theory corresponding to the above equations is:

$$ST \Rightarrow SH \quad BT, \neg SH \Rightarrow BH \quad BH \Rightarrow BS \quad SH \Rightarrow BS.$$

For general causation, we have an overdetermination, namely not only $ST \Rightarrow BS$ but also $BT \Rightarrow BS$! (Indeed, $BT \wedge ST \Rightarrow BS$ by Strengthening, but also $BT \wedge \neg ST \Rightarrow BS$ since $BT \wedge \neg ST \Rightarrow \neg SH$ by negative completion and $BT \wedge \neg SH \Rightarrow BH$. Consequently, $BT \wedge \neg ST \Rightarrow BH$ by Cut, and therefore $BT \wedge \neg ST \Rightarrow BS$ by transitivity. Taken together, $BT \wedge ST \Rightarrow BS$ and $BT \wedge \neg ST \Rightarrow BS$ imply $BT \Rightarrow BS$ by Or.)

However, for the actual world $\alpha = \{ST, BT, SH, \neg BH, BS\}$, the corresponding actual causal theory is just

$$ST \Rightarrow SH \quad SH \Rightarrow BS \quad SH \Rightarrow \neg BH.$$

We have $ST \Rightarrow BS$ by transitivity, so both SH and ST are actual causes of BS in α, as expected. It is clear also that BT cannot be an actual cause of BS in this world.

The above description of bottle shattering involved auxiliary variables SH and BH whose role consisted in enabling a counterfactual description of the difference between preempting and preempted cause. It has even been argued in Halpern (2016a) that if the model does not contain such variables, then it will not be possible to determine which one is in fact the cause, and the naive rock-throwing model which just has the endogenous variables ST, BT, and BS is insufficiently expressive in this sense because it cannot distinguish between the causal roles of Suzy and Billy.

We agree with Halpern that introduction of additional variables that describe the underlying mechanisms is often necessary for distinguishing situations that otherwise seem identical. However, unlike the structural account, our approach preserves causal claims under some simple forgetting transformations, and this gives us an opportunity to provide an adequate causal description without leaving the original language. Thus, in the present example, we can eliminate SH and TH by unfolding the above causal theory, and then it will be reduced to the following simpler theory:

$$ST \Rightarrow BS \quad BT, \neg ST \Rightarrow BS \quad \neg ST, \neg BT \Rightarrow \neg BS.$$

This "naive" theory could even be viewed, in turn, as a formal representation of the following simplified description.

Example 8.19 (Simplified Bottle) *Suzy (ST) and Billy (BT) both throw rocks at a bottle, but Suzy's rock arrives first and shatters the bottle (BS). Both throws are accurate: Billy's would have shattered the bottle if Suzy's had not throw.*

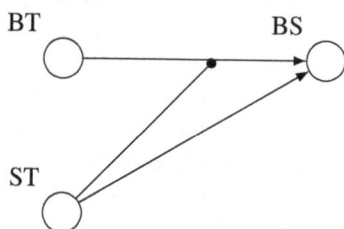

For the variables involved, this theory supports the same claims of actual causation. The adequacy of this causal theory in describing late preemption is based, however, on the fact that the set of causal rules $\{ST \Rightarrow BS,\ \neg ST, BT \Rightarrow BS\}$ is logically distinct from $\{ST \Rightarrow BS,\ BT \Rightarrow BS\}$ for regular inference, though they are equivalent with respect to basic inference (and therefore cannot be distinguished in the structural account).

As a final observation, we should note that even the simplified theory retains yet another problematic feature of the representation, namely that the absence of Suzy's throw ($\neg ST$) is an actual cause of bottle shattering in the situation when only Billy throws (namely in another causal world $\{\neg ST, BT, BS\}$). This intuitive discrepancy can actually be traced to our basic representation of prevention that includes the main positive rule:

$$C, \neg P \Rightarrow E.$$

This rule sanctions, in effect, that $\neg P$ is no less an actual cause of E than C itself. This shows that the above discrepancy is yet another symptom of a larger problem of negative literals. After all, a failure to prevent an outcome is in important respects distinct from actually causing it; this distinction becomes especially relevant in legal contexts. Therefore, an adequate representation should be capable of distinguishing the two, and this will be one of our objectives in chapter 9.

The above rock throwing example has actually the same structure as the *Desert Traveler* we have discussed earlier in this chapter in the context of the structural approach (see example 8.2).

Example 8.20 (Desert Traveler, revisited) *The causal model for the Desert Traveler story can be translated into the following causal theory:*

$$p \wedge \neg x \Rightarrow c \quad x \Rightarrow d \quad c \Rightarrow y \quad d \Rightarrow y$$

$$\neg p \Rightarrow \neg c \quad x \Rightarrow \neg c \quad \neg x \Rightarrow \neg d \quad \neg c, \neg d \Rightarrow \neg y.$$

The actual world is $\{p, x, y, \neg c, d\}$, *so the actual subtheory is*

$$x \Rightarrow d \quad d \Rightarrow y \quad x \Rightarrow \neg c.$$

Accordingly, shot (x) and dehydration (d), but not poison (p), are actual causes of death (y).

In our logical framework, the asymmetry between the preempting and preempted cause stems from the fact that basically equivalent sets of causal rules might be regularly non-equivalent, so they could support different assertions of actual causation. In the present case, the causal theory $\{x \Rightarrow c, \neg x \wedge p \Rightarrow c\}$ is not regularly equivalent to $\{x \Rightarrow c, p \Rightarrow c\}$, though they are equivalent with respect to basic inference. Moreover, in contrast to the structural account, the auxiliary variables c and d are not necessary for describing this example; as in the *Bottle* example, the following simple causal theory provides the same answers about actual causation among the salient variables $\{x, p, y\}$:

$$p \wedge \neg x \Rightarrow y \quad x \Rightarrow y \quad \neg x, \neg p \Rightarrow \neg y.$$

Desert Traveler is just one variant in a series of examples in the philosophical and especially in legal literature that originate in McLaughlin (1925). In the original version, the first enemy drains the water keg, replacing the water with salt, while the second enemy steals the keg full of salt, which he believes to be full of water. This latter version already falls into a category of preventive overdetermination that we will discuss in the next section. Still, all such variants are commonly considered to be cases of preemption rather than overdetermination, yet there has been no agreement on who is doing the preempting, the first enemy or the second (or even both). According to Moore (2009), legal theorists are all over the map in their proposed resolution of such cases—see, for instance, Sartorio (2015) for a recent attempt.

8.6.2 Preventive Overdetermination

As we have seen above, situations of overdetermination and preemption create challenging test cases for theories of actual causality. Still, even more challenging are cases of overdetermination that involve negative causation, and the main reason is related to the fact that it becomes much more difficult (if at all possible) to distinguish symmetric overdetermination from preemption when we are dealing with absences, omissions, and failures.

Let us begin with the following example from Moore (2009).

Example 8.21 (Joint Omission) *Betty (B) and Susan (S) both have to input their parts of a code in order to prevent a rocket launch (L), but each of them omits to input her part.*

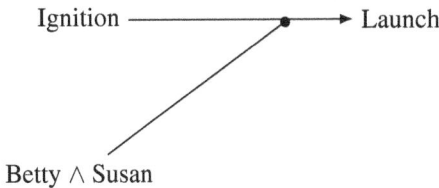

The following causal theory appears to provide a formal representation of this case of "joint" prevention:

$$I, n \Rightarrow L \quad B, S \Rightarrow \neg n.$$

Causal completion with respect to L and $\neg n$ will give us

$$\neg I \Rightarrow \neg L \quad \neg n \Rightarrow \neg L \quad \neg B \Rightarrow n \quad \neg S \Rightarrow n,$$

which produces the following surface representation (forgetting n):

$$\neg I \Rightarrow \neg L \quad B, S \Rightarrow \neg L \quad I, \neg B \Rightarrow L \quad I, \neg S \Rightarrow L.$$

If we consider now $\{I, \neg B, \neg S, L\}$ as an actual world, we obtain that both omissions, $\neg B$ and $\neg S$, are actual causes of the launch L. In other words, we obtain that this is a case of symmetric overdetermination.

The problem with the above representation, however, is that some alternative interpretations seem equally reasonable in such cases. Suppose, for instance, that Betty is actually unable to input her part of the code (for example, her hands are tied by an enemy). Then Susan cannot prevent the launch because her action or omission are *preempted* by the forced inaction of Betty, and so $\neg S$ already cannot be considered a cause of the launch in this case. Further, if both Betty and Susan are unable or unwilling to input their parts of the code, then each of these omissions could be viewed as preempting the other, so neither could be a cause of the launch![16] On the other hand, if we could manage to sequence these omissions as succeeding one another in time, the usual legal practice for dealing with multiple omissions is that the last omitter is considered as preempting the earlier one (see Fischer 1992; Wright 1985).

Finally, we should remember that in cases of ordinary, positive overdetermination, a decisive argument for making each of the concurrent causes an actual cause of an overdetermined effect has been that otherwise, by symmetry considerations, the effect would be uncaused. However, this obviously does not hold in the above case since the launch has its own positive causes.

Already the above example could be viewed as a clear sign that we may lack some expressive capabilities that would allow us to capture the relevant distinctions and variations in causal ascriptions. However, it still could be viewed as a limit, "esoteric" case, preserving the hope that in more standard cases of negative causation we would be able to systematically distinguish overdetermination from preemption based on the causal structure determined by causal rules. This naturally brings us to the well-known cases of double prevention where these problems will reappear in a more realistic setting.

---

16. This appears to be an opinion of Moore (2009).

8.6.3 The Problem of Double Preventions

An important class of situations involves positive causal assertions that arise as a product of double "causal negation." Richard Wright has even suggested calling just these situations instances of negative causation:

> A condition that causes the failure of a possible preventing cause, by negating a necessary positive condition for the occurrence of the preventing cause, is a *negative cause* of the consequence of the unprevented causal process. (Wright 2007)

Wright has argued that it is the distinction between positive causation and negative causation that is significant rather than the distinction between positive acts and negative omissions. He also rightly pointed out that questions about possible overdetermination or preemption of negative causes could not be treated in the same way as the corresponding questions about ordinary positive causation.

As a starting point in our exposition, let us begin with the following clear case of *triggering*. Actually, cases of this kind play a crucial role in arguments to the effect that negative causation is no less ubiquitous or important for our causal reasoning than the positive one.

Example 8.22 (Two Pillars) *(Paul and Hall 2013) Two pillars are leaning against each other. Suzy comes along and knocks one aside, thus causing the other to fall down.*

Let P_1 (respectively, P_2) denote the fact that pillar 1 (pillar 2) is lying down, S denote Suzy's knocking while G denote the presence of gravitation. Then the following causal theory can be used to provide a formal description:

$$G, n \Rightarrow P_2 \quad G, \neg P_1 \Rightarrow \neg n \quad S \Rightarrow P_1,$$

plus, of course, appropriate rules of negative causal completion.

By this formal description, $\neg P_1$ *prevents* activation of the first rule while Suzy's knocking action removes this hindrance (by causing P_1) and thereby triggers the rule. It appears to be perfectly natural to see both P_1 and S as causes of P_2 in this situation.

In accordance with the description, the actual world is $\{S, G, P_1, P_2\}$. Now, default causal completion with respect to n produces the rule

$$P_1 \Rightarrow n.$$

Then P_1 and S will indeed be actual causes of P_2. This is clearly reflected in the following surface formulation:

$$G, P_1 \Rightarrow P_2 \quad S \Rightarrow P_1.$$

The first thing that we should notice about this example (following Paul and Hall 2013) is that it can be viewed as an instance of double prevention: Suzy's action prevented the continued presence of the first pillar in an upright state which, had it occurred, would have prevented the second pillar from falling over.

It appears that our representation (as well as corresponding representations in the structural account) provides correct answers in such cases.

Such situations as firing a gun by pulling the trigger(!) and a host of other, similar situations have essentially the same causal structure. We can also add to this all the mirror situations of *abortion* (like death by strangulation) in which an unfolding causal process becomes aborted at some point by a preventing causal factor. Moreover, as have been rightly pointed out by Schaffer (2004) and Lewis (2000), we are not always in a position to know whether the causation we have in mind is an ordinary one, or some double prevention of this kind, for instance, when pressing a button on top of a black box makes a bomb explode. It appears that we should count it as a cause, no matter how the trick is done at the microlevel (see Maudlin 2004). Such considerations could also be used as a strong argument against enforcing a conceptual separation between positive and negative causation.

Unfortunately, the above harmony between our formal representations and common-sense intuitions ceases to hold when we turn to more complex and controversial cases of double prevention.

Let us turn to the following well-known example.

Example 8.23 (Bombing Mission) *(Hall 2004a) Suzy is piloting a bomber, and Billy is piloting a fighter as her escort. Billy shoots down an enemy plane, which would have otherwise shot down Suzy and prevented her from carrying out her bombing mission, which in fact she successfully completes.*

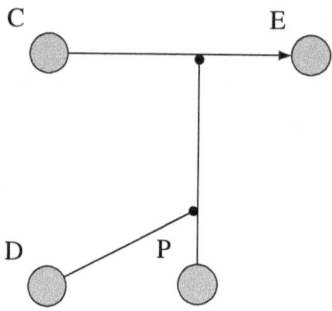

This is a typical case of double prevention: Billy prevents Enemy from preventing Suzy's mission. Accordingly, using our representation for preventions, we can write the following causal theory for this situation:

$$Suzy, n \Rightarrow Bombing \quad Suzy, Enemy, m \Rightarrow \neg n \quad Enemy, Billy \Rightarrow \neg m.$$

Now, it follows directly from the informal description that the bombing is counterfactually dependent on the firing, and many people report that Billy's firing is a cause of the bombing. In order to obtain this causal claim in our framework, it is sufficient to accept the causal rule

$$\neg m \Rightarrow n$$

that can be derived from the second rule above by default causal completion. Then we can immediately derive

$$Suzy, Enemy, Billy \Rightarrow Bombing$$

by regular inference, which will be sufficient for establishing that *Billy* is an actual cause of *Bombing* in this story. In fact, the surface positive theory for this example is

$$Suzy, Enemy, Billy \Rightarrow Bombing \quad Suzy, \neg Enemy \Rightarrow Bombing.$$

However, the above causal claim also raises some grave doubts. Notice first that the surface causal rule we used for establishing that *Billy* is an actual cause of *Bombing* provides also equal grounds for asserting that *Enemy* is also an actual cause of *Bombing*. More importantly, however, in this situation there is no natural, intrinsic causal process that connects Billy's shooting and Suzy's bombing; the shooting could happen, for instance, far away (both in time and location) from Suzy's flight (see Hall 2004b).

Hall has argued that what seems decisive for our causal judgment in this example is not existence of an intrinsic process but an existence of the requisite (counterfactual) dependence. However, Menzies (2006) has suggested the following modification of the story: Billy shoots down the enemy fighter exactly as before; but if he had not done so, the second escort, Ned, would have. Apparently, if we have counted Billy's firing as an actual cause of bombing in the original situation, we should still count it as a cause in this case also, though now there is even no counterfactual dependence between Billy's firing and Suzy's bombing.

An even more telling modification has been suggested by Hall himself (Hall 2004b, p. 244ff): If Enemy was about to receive instructions to shoot down Suzy, then Billy is a cause of the bombing (at least on a counterfactual account). If, however, Enemy was not about to receive instructions to shoot down Suzy, then Billy is not a cause of the bombing. But note that these two scenarios are intrinsically identical—because Billy shoots down Enemy before he would or would not have received the instructions—and yet their causal structures differ.

The above two modifications suggest somewhat conflicting lessons. The modification of Menzies indicates that even counterfactual dependence cannot serve as a test of causality in such cases, whereas Hall's modification clearly shows that what has actually happened (and not happened) is also insufficient for this purpose; we need to know what would happen in alternative circumstances where the actual preventions have not occurred.

The following famous case in the legal literature has the same causal structure of double prevention as *Bombing Mission*, but it focuses on a different causal world.

Example 8.24 (Saunders Case[17]**)** *A collision occurred when a motorist driving a rental car did not attempt to brake until it was too late to avoid the collision, but the brakes were defective due to a lack of proper inspection and maintenance by the rental car company and therefore would not have stopped the car in time even if the driver had applied them earlier.*

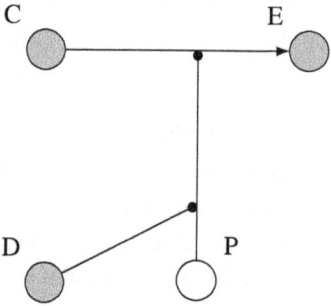

In the above picture, C denotes motorist's drive, E denotes collision, P denotes pressing the brakes, and D denotes the fact that the brakes are defective.[18] Using this labeling, the corresponding causal theory is as follows:

$$C, n \Rightarrow E \quad C, P, m \Rightarrow \neg n \quad P, D \Rightarrow \neg m.$$

As in the preceding case, if we apply causal completion, we will obtain the following surface representation:

$$C, \neg P \Rightarrow E \quad C, P, D \Rightarrow E.$$

Then, for the actual world $\{C, E, \neg P, D\}$, the actual causal theory is just $C, \neg P \Rightarrow E$, and therefore, in addition to car driving itself, the driver's failure to brake was the only actual cause of the collision. Defective brakes (D) would be a cause of the collision only in a different situation, $\{C, E, P, D\}$, where the motorist has actually used the brakes.

As a matter of fact, the above conclusions correspond precisely to the solution defended by Richard Wright for this case (see, e.g., Wright 2007). He acknowledged that intuitive evaluations of causation in these types of situations reportedly are mixed, and their proper theoretical resolution is controversial. Each of these noninstantiated conditions (not braking and defective brakes) independently *guarantees* the nonoccurrence or failure of the braking-stops-car causal process. Still, according to Wright, we should not confuse causes and conditions that merely guarantee the consequence. Only the failure to attempt to use the brakes had an actual negative causal effect, and this failure preempted the potential negative causal effect of the other non-instantiated conditions in this causal process.

---

17. Saunders System Birmingham Co v Adams 117 So 72 (Ala 1928).
18. We have omitted the rental company that was responsible for D.

Wright has actually suggested a certain methodology for dealing with the cases of negative causation, which amounts to focusing on the sequencing of the steps in the underlying positive causal process that failed, in order to determine at which step it failed (see, e.g., Wright 2013). This sequencing provides us with a key since the failure at some step preempts any potential failure at subsequent steps.

However, the above solution has not been universally accepted. Thus, Fischer (2006) has argued that Wright's reasoning could be reversed since failure to use the brakes will have an actual causal effect only if the brakes are in working order, and consequently the potential negative causal effect of this failure was preempted by the failure of the rental company to repair the brakes. In fact, our next example will provide a further support for this possible understanding.

Finally, let us turn to the following example that is also well known.

Example 8.25 (Catcher and Wall) *(M. McDermott 1995) A cricket ball is hit with substantial force (B) towards a window. I reach out and catch the ball (C). The next thing along in the ball's direction of motion is a solid brick wall (W). Beyond that is the window. Did my catch cause the ball to not hit the window (¬H)?*

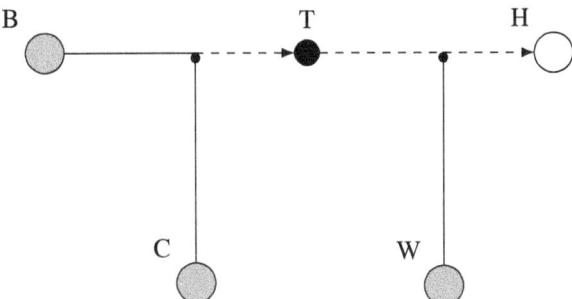

The above causal structure constitutes, in effect, a sequential "concatenation" of two preventions. To provide a formal representation for this structure, it is convenient to temporarily introduce an auxiliary transition state of the ball T between the two potential preventions (see the above picture). Then it can be described as follows:

$$B, n \Rightarrow T \quad B, C \Rightarrow \neg n \quad T, m \Rightarrow H \quad T, W \Rightarrow \neg m.$$

Now, the causal completion adds the following rules:

$$\neg B \Rightarrow \neg T \quad \neg n \Rightarrow \neg T \quad \neg B \Rightarrow n \quad \neg C \Rightarrow n$$

$$\neg T \Rightarrow \neg H \quad \neg m \Rightarrow \neg H \quad \neg T \Rightarrow m \quad \neg W \Rightarrow m.$$

As a next step, we will eliminate n and m, which will add the following surface rules:

$$B, \neg C \Rightarrow T \quad B, C \Rightarrow \neg T \quad T, \neg W \Rightarrow H \quad T, W \Rightarrow \neg H.$$

Finally, we will eliminate the auxiliary variable T and obtain the following surface representation:

$$B, \neg C, \neg W \Rightarrow H \quad \neg B \Rightarrow \neg H \quad B, C \Rightarrow \neg H \quad B, \neg C, W \Rightarrow \neg H.$$

If the actual world is $\{B, C, W, \neg H\}$, then the actual causal theory is just $\{B, C \Rightarrow \neg H\}$, and so the catch C will be an actual cause of the window being safe ($\neg H$) on this representation.

As has been rightly noted by McDermott, the first reaction of most people would be to reject this conclusion by arguing that the catch did not save the window because, due to the wall, the window has not been in danger anyway. Nevertheless, he has provided an interesting defense for this obviously controversial conclusion:

> "If the wall had not been there, and I had not acted, the ball would have hit the window. So between us—me and the wall—we prevented the ball from hitting the window. Which one of us prevented the ball hitting the window—me or the wall (or both together)?" And nearly everyone then retracts his initial intuition and says, "Well, it must have been your action that did it—the wall clearly contributed nothing."

Whether this intellectual trick is persuasive or not, however, it would be much easier for us to agree with it as well as with the original conclusion if we would replace the wall with, say, a second catcher, as has been suggested in Collins (2000).

In discussing this example, Maudlin (2004) has argued that our causal assertions depend on how we carve the situation into systems that can be assigned inertial behavior (behavior that can be expected if nothing interferes) along with a specification of things that can disturb it. Thus, we can see the situation as describing a cricket ball flying in the direction of an ordinary, fragile window, with two potential preventions on the way. But we can also carve up a "protected window" system that comprise the window plus brick wall. To count as a threat to this state, an object must be able to penetrate (or otherwise circumvent) the wall. But then the cricket ball will not be a threat to the inertial behavior of the window; only an artillery shell would be. So, in the given situation there are no threats to the window at all, and a fortiori *nothing* causes the protected window not to be hit.

In the formal framework of this chapter, such an alternative "carving" could be achieved by making the wall part of the background circumstances:

$$\mathbf{t} \Rightarrow W.$$

Given this rule, the last causal rule $T, W \Rightarrow \neg m$ of the original causal theory reduces to $T \Rightarrow \neg m$ while the latter rule already implies $T, m \Rightarrow H$ by regular inference. As a result, the causal theory boils down to

$$B, n \Rightarrow T \quad B, C \Rightarrow \neg n \quad T \Rightarrow \neg m \quad \mathbf{t} \Rightarrow W.$$

Note that the latter causal theory already does not include any causal rules for *H*. Consequently, it cannot be used for making any causal claims about window hitting.

Examples similar to *Catcher and Wall* have occupied a significant place in the legal literature, witness the following example from Hart and Honoré (1985):

Example 8.26 (Ship) *A ship is traveling down a river to deliver goods to Metropolis by a specific date. The ship is unable to arrive by that date, since its crew must and does stop when it reaches bridge A, which had collapsed into and blocked the river. The ship would not have been able to reach Metropolis on time even if bridge A had not collapsed due to another collapsed bridge, bridge B, located on the river between bridge A and Metropolis.*

As could be expected, Wright (1985) has provided and extensive defense of the claim that bridge A but not bridge B caused the ship's delay (see also Wright 2013). However, this view has not been universally accepted. And indeed, if we would make the collapse of bridge B part of the background (and well-known) circumstances, this would prevent us from making the collapse of bridge A an actual cause of not reaching Metropolis.

Now we should return to the Saunders case (example 8.24).

Example 8.27 (Saunders, reconsidered) *Recall that this example has been formalized using the following causal theory:*

$$C, n \Rightarrow E \quad C, P, m \Rightarrow \neg n \quad P, D \Rightarrow \neg m,$$

and our conclusion has been that, in the actual world $\{C, E, \neg P, D\}$, not pressing the brakes ($\neg P$) is an actual cause of collision E. However, as in the Catcher and Wall *example above, we could consider defective brakes as a preexisting (and even known to all) condition of this situation:*

$$\mathbf{t} \Rightarrow D.$$

Then the last rule will be reduced to $P \Rightarrow \neg m$ while the latter rule will already imply the second rule above by regular inference. Consequently, the causal theory will be reduced to

$$C, n \Rightarrow E \quad P \Rightarrow \neg m.$$

The latter theory will make the driving C the only actual cause of E. This could well be also a court decision in this case since knowingly driving with defective brakes is obviously an offense.

The problems discussed in this section clearly indicate that some crucial ingredients of our commonsense understanding of causality are still missing in our approach. We will return to these examples and its associated problems in chapter 9 where we will suggest an enhanced representation that goes beyond the pure NESS test.

8.7   On Composite Overdetermination

A large group of problematic examples in the literature, especially the legal one, concerns cases of overdetermination in which a set of quantitative causal factors add up to produce a combined effect. The problems with such cases can be illustrated by the following quite simple example.

Example 8.28 (Joint Poisoning) *Two assassins pour poison into Victim's coffee. Assassin1 pours a lethal dose (A_1) while Assassin2 adds half of the dose (A_2), insufficient by itself to cause death. As a result, Victim dies (D).*

The relevant causal theory appears to be just $\{A_1 \Rightarrow D; \ A_1, A_2 \Rightarrow D\}$, and hence by the NESS account, the action of Assassin1 is the only cause of death in this situation. But now let us replace Assassin1 with two assassins, Assassin3 and Assassin4, each pouring half a dose of poison, just as Assassin2. Then the relevant causal theory becomes

$$A_2, A_3 \Rightarrow D \quad A_2, A_4 \Rightarrow D \quad A_3, A_4 \Rightarrow D,$$

so in the world $\{A_2, A_3, A_4, D\}$, each of the three assassins becomes an actual cause of death, including Assassin2!

From a purely formal point of view, the above transformation could be viewed as resulting from a replacement of literal A_1 with an "equivalent" conjunction $A_3 \wedge A_4$, which shows once more high sensitivity of the NESS test (and our approach) to the syntactic form of causal rules, this time even with respect to conjunctions.

Richard Wright (see, e.g., Wright 2001) has used such examples, however, as an argument for a modification of his own NESS account that would allow us to count Assassin2 as a cause of death even in the original version of the story that involved Assassin1. Basically, the suggested modification amounted to using descriptions like "*at least* half of a dose" (for describing the contribution of Assassin1) in order to make the causal contribution of Assassin2 necessary for the sufficiency of the combined poisoning.

In fact, if we are looking just for technical solutions that would sanction such causal claims, then an easier option could be provided by a deep representation that uses names of underlying mechanisms. Thus, the original two-assassin situation could be viewed as a single causal process m jointly created by the two assassins, which is distinct from the poisoning process n initiated by Assassin1 alone:

$$A_1, n \Rightarrow D \quad A_1, A_2, m \Rightarrow D.$$

Then the actual world would be $\{A_1, A_2, m, n, D\}$, and therefore both Assassin1 and Assassin2 will become equally actual causes of death (using m in the witness set for showing that A_2 is a cause).

This kind of representation might even have legal precedents. Thus, Moore (2009) discusses the following case (People v. Lewis [1899]):

Example 8.29 (Two Wounds) *A victim suffers two wounds from two blows by two defendants acting independently of one another. One wound was sufficient to kill the victim by loss of blood while the other wound was not. The victim dies from loss of blood, both wounds bleeding.*

The California Supreme Court has provided the following causal intuitions for the decision that made both defendants responsible: "drop by drop" the blood flowed out of the victim "from both wounds."

This representation also appears to be especially suitable for an occasionally expressed position in the legal literature according to which there are no real overdetermination cases since combinations of concurrent causes always create different events in some respect or another. However, as was rightly noted by Moore, this fine-grained approach would generate enormous promiscuity in causal attributions.

Overdetermination examples of this kind represent a large portion of controversial legal cases, so no wonder that they occupy a prominent place in the legal literature (see, e.g., Fischer 2006; Moore 2009, for a detailed discussion as well as criticism of Wright's modification). The controversy extends even to court decisions, and the American Law Institute's Third Restatement of Torts has taken no position on whether the insufficient contribution is causal:

> For example, one actor's contribution may be sufficient to bring about the harm while another actor's contribution is only sufficient when combined with some portion of the first actor's contribution. Whether the second actor's contribution can be so combined into a sufficient causal set is a matter on which this Restatement takes no position and leaves to future development in the courts.

This book also will not venture a solution to this problem. We argue, however, that the above controversy should not to be construed as an argument against our approach or the NESS account itself, because it already spells even more troubles for rival counterfactual accounts. Rather, we believe that the controversy pertains to a prior *representation problem* whether and how causal factors can be decomposed and (re)combined.[19] More precisely, the controversy arises from a general problem of *individuation* for such composable factors, especially when they violate a fundamental principle of counting according to which a counting domain cannot contain overlapping individuals (see, e.g., Kratzer 2012, for a discussion of this latter problem).

In the philosophical literature on causation, this kind of overdetermination usually appears in the form of voting scenarios—see, e.g., Livengood (2013) and the discussion in Halpern (2016a). However, the phenomenon itself is not restricted to "conceptual" causal situations like voting and is present already in real "physical" situations such as when a

---

19. As was rightly noted by Fischer (2006), though Wright's modification allows us to combine parts of causal factors in order to explain a particular causal claim, it does not help us to decide whether to aggregate the factors in this way.

number of people are trying to push a door from both sides and the door opens as a result. Such situations create precisely the same problems of causal attribution as voting and the above legal cases. As we have already mentioned, however, we will leave such problems to future studies. Still, in the next section we will briefly consider a necessary prior step for representing such problems, which is important in its own right, namely the question of how such numerical, or multivalued, causal parameters could be represented in causal rules.

8.8 Causality among Multivalued Propositions

Under an overwhelming influence of the structural causal modeling in the analysis of actual causation, a lot of examples and counterexamples have been introduced that involved non-binary, multivalued variables. In order to represent such examples in our logical framework, two basic choices are available. The first consists in an appropriate generalization of the propositional causal calculus to multivalued propositions (this generalization has been used, e.g., in Giunchiglia et al. 2004) or even to a full-blown first-order causal calculus (as described in Lifschitz 1997, see also chapter 5). This move would complicate, however, the underlying causal logic(s) which has been taken as a logical basis of this book. Given our current modest purpose of explicating a few relevant counterexamples that involve non-binary variables, we will use yet another option. Namely, we will systematically rewrite the relevant causal conditions using appropriate binary variables. More precisely, given an n-valued variable X (where $n > 2$), we introduce n propositional atoms X_i that correspond to instantiations $X = x_i$ for every value x_i in the range of X. These propositional atoms, however, are not logically independent, so we should add the corresponding logical constraints to the background logic. Fortunately, this can be done without leaving the causal language of the causal calculus. Thus, the corresponding constraints can be incorporated by adding the following *nondeterminate* causal "laws" for each multivalued variable X:

$$\mathbf{t} \Rightarrow \bigvee X_i \qquad \mathbf{t} \Rightarrow \neg(X_i \wedge X_j)$$

for every $i \neq j$, where $i, j \leq n$. These causal rules will turn out to be sufficient for inferring all the relevant causal facts about possible values of X.

It turns out, however, that to provide an "apt" representation of the corresponding examples in this modified language, we have to deal with an emerging complication that stems from the fact that, unlike the case of binary variables, negations $\neg X_i$ of such new atoms implicitly describe disjunctive states (namely the disjunction of the rest of the values of X). In fact, we will see that the use of such "literals" may change the assertions of actual causation (see example 8.31 below).

To cope with the above problem, we will impose a further restriction on the literal causal rules forming a causal theory. Namely we will require that the new atoms representing multivalued variables should appear only in their pure (unnegated) form in these rules.

Note that this constraint also does not restrict our representation capabilities since any causal rule in this language can be logically reduced to literal rules of this kind.

The following example from Halpern and Pearl (2001) has created problems for the original definition of a causal beam from Pearl (2000).[20]

Example 8.30 (Vote Machine) *Two votes (V1, V2) are cast for a measure. The votes are summed by a machine (M), and the measure passes (P) if it receives at least one vote.*

The variable M (that counts the votes) is three-valued in this setting ($\{0,1,2\}$), so the corresponding causal theory is

$$V_1, V_2 \Rightarrow M_2 \quad \neg V_1, V_2 \Rightarrow M_1 \quad V_1, \neg V_2 \Rightarrow M_1$$

$$\neg V_1, \neg V_2 \Rightarrow M_0$$

$$M_1 \Rightarrow P \quad M_2 \Rightarrow P \quad M_0 \Rightarrow \neg P.$$

The actual causal world is $\{V_1, V_2, M_2, P\}$, so the corresponding actual causal theory is

$$V_1, V_2 \Rightarrow M_2 \quad M_2 \Rightarrow P.$$

Both V_1 and V_2 are clearly actual causes of P in this world.

The following "trumping" example from Schaffer (2000) illustrates, among other things, the role of our additional restriction on the form of causal rules.

Example 8.31 (Trumping) *Major (M) and sergeant (S) stand before the corporal, and both shout "Charge!" (M = 1, S = 1). The corporal charges (C = 1). Orders from higher-ranking soldiers trump those of lower rank, so if the major had shouted "Halt" (M = 2) the corporal would not have charged.*

This is one of the most controversial examples in the literature. The question is whether S is a cause of C in this situation and more generally whether trumping causation is a species of preemption or overdetermination.

The following structural equation provides a formal description:

$$C = (M = 1) \vee (S = 1 \wedge M \neq 2).$$

Note first that $(M \neq 2)$ is equivalent to the disjunction $M = 0 \vee M = 1$. Accordingly, by the suggested re-presentation of multivalued variables, the above equation produces three positive causal rules:

$$M_1 \Rightarrow C \quad S_1, M_0 \Rightarrow C \quad S_1, M_1 \Rightarrow C,$$

---

20. As noted in Halpern (2015), this is also a counterexample to the H-account in Hall (2007).

so by applying negative completion and removing (regularly) redundant rules, we obtain the following causal theory:[21]

$$M_1 \Rightarrow C \quad S_1, M_0 \Rightarrow C$$

$$M_0, \neg S_1 \Rightarrow \neg C \quad M_2 \Rightarrow \neg C.$$

The actual causal world is $\{M_1, S_1, C\}$,[22] so the associated actual causal theory is just

$$M_1 \Rightarrow C,$$

and therefore only M_1 is an actual cause of C. Thus, under the above representation, trumping is a species of preemption.

Remark. Suppose we ignore our additional constraint on the form of the causal rules and write the following (causally equivalent) causal theory:

$$M_1 \Rightarrow C \quad S_1, \neg M_2 \Rightarrow C$$

$$\neg M_1, \neg S_1 \Rightarrow \neg C \quad M_2 \Rightarrow \neg C.$$

Then the associated actual causal theory will be

$$M_1 \Rightarrow C \quad S_1, \neg M_2 \Rightarrow C.$$

On this representation, both S_1 and $\neg M_2$ would be actual causes of C (cf. Weslake 2015, that also rejects this solution).

Example 8.32 (Combination Lamp) *(Weslake 2015) A lamp (L) is controlled by three switches (A, B and C), each of which has three possible positions $(-1, 0, 1)$. The lamp switches on iff two or more switches are in the same position. In fact, the switches are in positions $A = 1, B = -1$ and $C = -1$.*

Under our suggested logical representation, this example is quite similar to *Loader* (example 8.6). The positive causal rules for L are as follows:

$$A_i, B_i \Rightarrow L \quad B_i, C_i \Rightarrow L \quad A_i, C_i \Rightarrow L$$

for every $i \in \{-1, 0, 1\}$.

The actual world is $\{A_1, B_{-1}, C_{-1}, L\}$, so the corresponding actual causal theory is just $B_{-1}, C_{-1} \Rightarrow L$. Therefore B_{-1} and C_{-1} are actual causes of L while A_1 is not, contrary to the HP definition from Halpern and Pearl (2005). However, the modified definition from Halpern (2015) handles it appropriately.

Still, contrary to Weslake (but in agreement with Halpern—see example 4.1.2 in Halpern 2016a), differences in *how* the effect is brought about matter in our causal judgments, even

---

21. To simplify the description, we treat S as a binary variable $\{Charge, \neg Charge\}$; our conclusions will not be influenced by this choice.

22. Note that we implicitly use the laws that determine the rest of the values for M_i in this world.

when this does not make a difference to *whether* it is brought about. The following variation of the above example, used in Weslake (2015), illustrates this.

Example 8.33 (Fancy Lamp) *(Weslake 2015) A lamp (L) is controlled by three switches (A, B, and C), each of which has three possible positions ($-1, 0, 1$). The switches are connected to detectors (N_{-1}, N_0, N_1), each of which is activated iff no switch is in position -1 or 0 or 1, respectively. The lamp switches on iff some detector is activated. In fact, the switches are in positions $A = 1, B = -1$, and $C = -1$, detector N_0 is activated, and $L = 1$.*

This time, the relevant causal theory is

$$A_i, B_j, C_k \Rightarrow N_l \text{ whenever } l \neq i, l \neq j, l \neq k$$

$$N_l \Rightarrow L$$

for every $i, j, k, l \in \{-1, 0, 1\}$.

The actual world is now $\{A_1, B_{-1}, C_{-1}, N_0, L\}$, so the actual causal theory is

$$A_1, B_{-1}, C_{-1} \Rightarrow N_0 \qquad N_0 \Rightarrow L.$$

As a result, not only B_{-1} and C_{-1} are now actual causes of L, but also A_1 is. Note, however, that we can remove the detectors from the above causal description (by forgetting) and obtain the following causal theory:

$$A_i, B_j, C_k \Rightarrow L \text{ whenever } \{i, j, k\} \text{ does not cover } \{-1, 0, 1\}.$$

Now it is easy to verify that the above causal theory is *basically* equivalent to the preceding theory for *Combination Lamp*. Still, the two theories are not regularly equivalent, and this explains why they provide different answers to our causal questions.

8.9 Interim Conclusions

One of the main objectives of this chapter has been to show that, once placed on proper logical and causal grounds, traditional regularity approach provides not only a natural but also a viable definition of actual causality. In some sense, the viability of this approach lends an additional support, and even new meaning, to the counterslogan from Pearl (2000): "Causation without manipulation? You bet!"

The suggested paradigm change (or, better, restoration) also appears to be in accord with the current trends in the legal theory where the NESS test is emerging as the new supplement to the but-for test for the twenty-first century (Fischer 2006). The significance of this renewed support for the regularity approach cannot be overestimated since legal practices provide a crucial test-bed for verifying various theories of causal attribution vis-à-vis our commonsense understanding of causality.

The problems of actual causality, described in this chapter, confirm that causation is a difficult and complex notion. In our logical approach, part of its complexity is reflected in the fact that the causal calculus is not a plain logical system with stipulated axioms but an essentially nonmonotonic formalism in which logic and nonmonotonic semantics are mutually interconnected. As we have seen, this representational complexity is even higher for actual causation. Still, our suggested definition of actual causality has produced reasonably simple algorithms for checking the relevant causal claims.

Of course, our approach does not "cancel" the counterfactual accounts; it only poses anew the old philosophical questions about the relation between (causal) laws and counterfactuals, questions that could be traced back to the famous double definition of causation by David Hume in his *Enquiry*. Still, we can't help to agree with Weslake (2015) that there is a nice irony in the fact that most plausible counterfactual theories of causation turn out to draw heavily from the resources of the regularity theories they were initially motivated by rejecting.

An important advantage of our general approach to causality is that both counterfactual and regularity accounts can be expressed in a single framework of the causal calculus. In fact, in chapter 9 we are going to describe a modification of the definition of actual causality that, without using an explicit language of counterfactuals, will incorporate some important insights of the counterfactual approaches, in particular the idea that causal claims are not determined only by what actually have happened, but often require comparison with what would happen in some alternative, "counterfactual" situations.

The best summary of our approach to actual causality as well as the hopes we associate with it can all be found in the following passage from Hall (2003):

> The philosopher interested in understanding causation would best spend her efforts by investigating the prospects for some nomological entailment account. And that is because even if she fails to come up with such an account, the lessons learned along the way—about such topics as varied as the nature of the causal relata, the varieties of causal pre-emption, tensions between basic theses involving causation such as Transitivity, Intrinsicness, and Dependence, the relation between ordinary causation and causation involving omissions, and much more—are rich enough to repay these efforts many times over. . . . No other approach has, as far as I can see, a prayer of providing the kind of account of causation that can set this important area of research on a sound conceptual footing.

We see the definitions and representations of this chapter only as a groundwork, a basic step toward a more comprehensive theory of actual causality. A further important step will be made in the next chapter, but the task will not be finished even there.

9 Relative Causality

In the course of chapter 8 we have repeatedly pointed out some remaining discrepancies between predictions of the formal definition of actual causality and our intuitive causal judgments. All the discrepancies we mentioned were "one-sided" in the sense that what has been claimed to be an actual cause according to the definition has not been considered a cause according to the commonsense understanding. We have suggested also that these discrepancies ought to be taken as an indication that some essential ingredients of our understanding of causality are still missing in our formalization.

In this chapter, we are going to argue that these discrepancies are symptoms of a larger and more systematic phenomenon according to which in ordinary cases of causal attribution only a few of the factors (typically just one) that are necessary for producing a given effect are called a (or even the) cause of the latter; the rest are treated as mere *conditions*. This will naturally lead us to a relativized notion of a *proximate cause* that will provide a formalization of this idea.

Though the but-for test is not viewed as a necessary condition of causality in modern counterfactual accounts, it is still generally considered to be a *sufficient* condition in these approaches. This makes, for instance, a birth of a person an ultimate cause of his death. Speaking more generally, already the adherence to the *sine qua non* principle leads to obliteration of the distinction between causes proper and mere conditions. Yet, even our regularity approach to actual causality described in the preceding chapter tends to retain this obliteration in many regular cases.

Though we will take the distinction between causes and conditions as a basis of our modified notion of proximate cause, we will see that our approach has close connections with a whole cluster of different concepts that have been used in the causal literature in attempts to capture much the same phenomenon, such as "causal field," "contrastive causation," "normal conditions," and "defaults."

In the framework of structural causal modeling, a wide consensus has been formed according to which structural models, taken by themselves, are still insufficient to deal with all the problems and counterexamples concerning actual causation, so these models need to be augmented with a *contextually determined* distinction between default and deviant variable values or, more generally, with considerations of normality. A number of approaches for accommodating these notions in the framework of structural equations has been suggested in Menzies (2004, 2007) and been developed by Hall, Hitchcock, and Halpern, to mention only a few. In a wider perspective, our formal approach in this chapter is also closely related to a recurrent idea in the causal literature that the very notion of a cause and that of a causal relation are not absolute but context dependent and contrastive by their nature—see Schaffer (2010, 2012) for an instructive discussion and references to the previous literature.

The contrastive feature of relative causality that will be introduced below will actually reproduce a characteristic property of counterfactual accounts, according to which the actual situation is not sufficient for determining claims of actual causation, and it needs to be supplemented with what is going on in alternative, background circumstances. Moreover, it will reaffirm an important role of interventions in our understanding of causality, so much stressed in Pearl's approach and manipulative accounts of causation such as Woodward (2003).

As a final correspondence, the difference between conditions and "true" causes is intimately related to yet another aspect of commonsense understanding, namely that causation is an inherently *dynamic* phenomenon that involves acts, events, changes, and happenings, in contrast to persistent conditions that only enable the latter—see, for example, Lombard (1990), Glymour et al. (2010), and Glymour and Wimberly (2007) and the experiments in Henne et al. (2019). Viewed from this perspective, the difference takes the form of the distinction between inertial (default) behavior of an object or system and deviations therefrom (see Maudlin 2004). In this respect, the notion of relative causation that will be developed in this chapter, though important in its own right, will occupy an intermediate position between a more abstract notion of actual causality described in chapter 8, and a fully dynamic notion that will be introduced in part IV of this book. In this dynamic setting, conditions will supply a description of the prior states, alias existing preconditions, of an actual causal event, where possible differences in our causal assertions could be attributed to different assumptions we implicitly make about such prior states.

9.1   Causes versus Conditions

If someone said that without bones and sinews and all such things, I should not be able to do what I decided, he would be right, but surely to say that they are the cause of what I do... is to speak very lazily and carelessly. To call those things causes is too absurd. [...] Imagine not being able to distinguish the real cause from that without which the cause would not be able to act as a cause.
—Plato, *Phaedo* 99a–b

Plato's remark above provides clear evidence that the distinction between causes and conditions has always been part of our pre-theoretic, commonsense understanding of causation. It seems that what Plato has suggested us to imagine could well be a promiscuous and counterproductive proliferation of causal information that could arise from ignoring this distinction as well as ensuing difficulties in conveying this information to others.

The distinction between causes and conditions has remained an undisputed part of the understanding of causality throughout the history until it was challenged by John Stuart Mill in his *System of Logic*. Mill argued that, although we usually refer to only one or some of the instantiated conditions in the antecedent of a causal rule as the cause(s) and treat the others as mere "conditions," philosophically and scientifically there is no basis for such discrimination: "We have no right to give the name of cause to one of the conditions exclusive of the others of them." According to Mill, nothing can better show the absence of any scientific ground for the distinction between the cause of a phenomenon and its conditions than the capricious manner in which we select from among the conditions that which we choose to denominate the cause.

It has been argued earlier in this book that it is this idea of Mill's logically complete, invariable conditional that could be seen as a key problem of the regularity approach and, even more generally, of the purely logical, deductive approach to our causal reasoning. As we have seen, the problems with such conditionals have been one of the main reasons behind the advance of nonmonotonic reasoning in artificial intelligence.

Nevertheless, Mill has clearly pointed out the main difficulty with the distinction between causes and conditions, namely that it is context-sensitive and theory-relative in ways that appear to be inconsistent with the idea that causal claims describe objective, mind-independent reality. Using an example from Hart and Honoré (1985), the cause of a great famine in India may be identified by an Indian peasant as the drought, but the World Food Authority may identify the Indian government's failure to build up food reserves as the cause and the drought as a mere condition. Though Hart and Honoré have actually suggested restoring the distinction between causes and conditions as an important part of our understanding of causation (see below), they have agreed with Mill that the distinction is not a "scientific" one in the sense that it is not determined by associated laws or generalizations. In fact, it even does not depend on the associated counterfactuals and probabilistic dependencies (see Menzies 2006).

To dispel these objectivity worries, however, it is instructive to compare causation with the notion of motion. Nobody seems to challenge the objectivity of this latter notion, though it is well known at least since Newton that motion descriptions are always made with respect to some frame of reference. In particular, the same physical motion can receive quite different descriptions in different reference frames. Now, in accordance with the formal definitions that will be introduced later in this chapter, the comparison between causation and motion is not just a vague analogy but a common generalization that has much the same conceptual roots. Just as with motion, the "perspectival" view of causation asserts that our causal judgments are always made in the context of existing background conditions, though this does not detract from the reality of the causal relations they describe (see Menzies 2007).

The relative notion of causation that will be introduced below will also give us an opportunity to reconsider some general properties of the causal relation, such as transitivity. Recall that general causal inference is transitive in our formalization. Moreover, except for some boundary cases, even actual causality is usually transitive on our definition of the latter. However, the supposition of transitivity leads almost inevitably to the idea that practically the whole "past cone" of any given event constitutes the set of its causes. For instance, such a view sanctions that birth is an actual cause of death (this is also the conclusion of the counterfactual sine qua non test). However, with the relative notion of causality, we will be able to express an alternative idea that such transitive causal chains can be broken, the idea that is well known in the legal literature under the name *novus actus interveniens*.

The distinction between causes and conditions has immediate practical implications for legal theory (see, e.g., Gardner and Honoré 2017). For instance, the very distinction between principle and accomplice could be explained on the basis of this difference, though the NESS test cannot distinguish them. Similarly, it could provide representation tools for distinguishing causing from occasioning, giving effect to, inflicting, failing to prevent, and enabling. There seems to be no "logical space" for all these legal distinctions on a "scientific" view of causation. Further, many legal systems divide the causal inquiry into two stages or phases. First there is what is known as the "cause-in-fact" phase. This phase can be roughly equated with the notion of actual causality, described in chapter 8. The second part of the inquiry, known as the "proximate cause" or "legal causation," treats some causal contributions as insufficient for legal responsibility. This phase, however, is usually infected with considerations of legal policy and so forth, so unsurprisingly, many are drawn to the position of "causal minimalism" according to which it is not truly part of the causal inquiry.

9.2   The Account of Hart and Honoré

Where both the regularity and counterfactual approaches have an abundant literature, the famous book (Hart and Honoré 1985) constitutes, in a sense, an exclusive source of information on the cause-condition distinction and its role in causal reasoning, especially in legal theory. With hundreds of examples drawn mainly from the legal literature and dozens of difficult problems, subtle distinctions, and complications discussed, this book supplies the main desiderata for any potential theory of causality that would attempt to account for this distinction. Our brief summary below will not do justice to the richness of this book and the wealth of its ideas. Rather, it will serve a more modest aim of providing reference to those key aspects of the approach of Hart and Honoré to causality (henceforth, the H&H approach) that we will attempt to capture in our formalization given later in this chapter.

The basic supposition of the H&H approach is that the contrast of causes with mere conditions is part of the meaning of causal expressions, in the same sense as the implicit reference to generalizations is (Hart and Honoré 1985, p. 12). This distinction is crucial for explaining events, controlling environment, and assigning praise or blame.

One of the central concepts by which "causes" are distinguished from "mere circumstances or conditions" is that of a contingency, usually a human *intervention*, in the natural course of events which *makes a difference* in the way these develop:

> The notion, that a cause is essentially something which interferes with or intervenes in the course of events which would normally take place, is central to the commonsense concept of cause, and at least as essential as the notions of invariable or constant sequence so much stressed by Mill and Hume. (Hart and Honoré 1985, p. 29)

If the waiter has placed a knife and fork on the table and the guest has plunged the knife into his hostess's breast, her death is not caused by the waiter, even though the waiter's action was a necessary *condition* of the stabbing.

According to Hart and Honoré, two contrasts are of prime importance in distinguishing between causes and conditions. These are the contrasts between what is abnormal in relation to any given thing or subject matter and what is *normal* and between a *free deliberate* human action and all other conditions (p. 33).

Generally, interventions created by a voluntary act or a conjunction of events amounting to *coincidence* in the causal chain (such as a suddenly falling bridge) operate as a limit on our notion of cause, in the sense that events subsequent to these are not attributed to the antecedent action or event as its consequence even though they would not have happened without it. In this sense, they "break the chain of causation." In other words, when seeking causal explanation for a given outcome, we work back along the causal chain till we find such a cause, which naturally completes our explanatory quest.

People (and courts) often use the term "proximate cause" for describing such causes. However, Hart and Honoré believed that this term is misleading rather than enlightening, apart from its solely negative force as a reminder that being a necessary condition is not sufficient for being a cause (pp. 86–87). They were equally dismissive about other "cloudy" terminology used by the courts, such as *novus actus interveniens* and "superseding cause" purported to single out those occurrences which break the chain of causation.

Unfortunately, numerous problems have arisen in attempts to apply this seemingly clear approach in particular legal cases, and a large portion of Hart and Honoré's book is devoted to discussing and clarifying such cases. For instance, sometimes we are responsible for a harm even if another agent "intervened" in the causal process—for example, if we left the car door unlocked in a high-theft area and the car was stolen. The idea that the provision of an opportunity may rank as the cause of the upshot when the opportunity is actually exploited is very important in both law and history (p. 2). In addition, problems arise in determining precisely when a human action is truly free and deliberate since if it is, for instance, enforced or coerced, then it does not break the chain of causation. Similarly, a coincidence could count as a cause only if it cannot be expected given the knowledge of the agent (or, better, what they should know).

Some key features of the H&H account of causation have been used by Peter Menzies in his suggested modifications of the structural approach (see, e.g., Menzies 2007). Thus, in order to capture context-sensitivity of causal claims, he has proposed to single out some values of variables as default ones and use them for representing *normal* states of a system, the latter serving as baselines for the calculation of difference-making produced by the causes. They can even be used for describing the distinction between causes and conditions. Menzies has also argued that a suitably enriched notion of counterfactual dependence that is based on such normal states would allow us to restore the connection between causation and counterfactual dependence that has been lost with respect to the standard definition of the latter.

On the opposite side, the H&H approach has been severely criticised by Richard Wright in a number of papers (see, e.g., Wright 2008). Wright has even argued that Hart and Honoré's account was overshadowed and distorted by their primary emphasis on elaborating supposedly "commonsense" principles for treating only some causally relevant factors as causes. Among other things, Wright has rightly mentioned, however, difficulties in using the notions of normality and abnormality for distinguishing causes from conditions. Although "what is abnormal" may often be identified as the "cause," so may what is normal: for example, a person's smoking causing cancer, a battery's running out of power causing electronic equipment to quit working, the sunlight's streaming into a window causing a person to wake up, a person's working hard causing her to become tired, and a person's becoming tired or bored causing her to fall asleep in class. Such examples clearly show that normality and abnormality in causal discourse is not a fixed feature of the relevant propositions or events, but something that is determined by and changes with the corresponding context.

9.3   Relative Models

Our formalization below will be based on formal definitions of background condition and proximate cause. These formal notions will be viewed, however, as explications of our intuitive, pre-theoretical understanding of these terms. In other words, by adopting a new formal meaning for the latter, we have implicitly accepted the widespread criticism of the existing legal interpretations of these notions as commonly infected with noncausal factors and considerations. Still, instead of throwing them to a legal waste bin, we have seen here an opportunity of restoring these key causal notions in their natural capacities.

In this section we are going to define the notion of an exact model of a causal theory that is relative to a given set of background conditions.

Recall that $\Delta(u)$ denotes the set of all propositions that are directly caused by a set of propositions u in a causal theory Δ, that is,

$$\Delta(u) = \{A \mid B \Rightarrow A \in \Delta, \text{ for some } B \in u\}.$$

Given this notion, an exact world of a causal theory from chapter 4 is a world that satisfies the fixpoint condition $\alpha = \mathrm{Th}(\Delta(\alpha))$. A relativization of this notion with respect to a given set of background conditions is given below.

Definition 9.1 *A causal scenario is a pair* $(\Delta, \mathcal{B})$, *where* Δ *is a causal theory, and* $\mathcal{B}$ *a set of propositions called* background conditions.

A world α *is an* exact model (or a causal world) *of a causal scenario* $(\Delta, \mathcal{B})$ *if*

$$\alpha = \mathrm{Th}((\mathcal{B} \cap \alpha) \cup \Delta(\alpha)).$$

Informally speaking, a world α is an exact model of a causal scenario,[1] if it is determined by its causal consequences, given the background conditions that remain to hold in α. An implicit effect of this modification of the notion of an exact model is that these background conditions are exempted from the need of causal explanation, which reflects an informal idea that these background conditions have been expected all along, so we do not need a cause for them. One of the important consequences of this exemption is that now it already becomes unnecessary for the source causal theory to contain causal rules for background conditions in order to produce a valid causal model.

As can be seen from the above description, we formalize the notion of a condition (and thereby a subsequent distinction between conditions and causes) as a semantic feature of propositions, a part of the new semantic framework for interpreting causal theories. In this respect our formalization is distinct, for instance, from a direct syntactic representation of causes and conditions as two kinds of literals in the bodies of causal rules that has been suggested in a recent approach of Denecker, Bogaerts, and Vennekens (2019).

Remark. Background conditions should be carefully distinguished from exogenous propositions, despite the fact that both are treated in our representation as propositions that do not require causal explanation. First, in contrast to conditions, exogeneity is a context-independent feature of propositions with respect to a given causal theory. Second, these two kinds of propositions roughly correspond to two kinds of boundary nodes of the "causal graph." Exogenous propositions correspond to initial nodes of the graph, facts that are not caused (with respect to a given causal theory) but can cause other propositions, whereas background conditions correspond in this sense to the end nodes—facts that can be caused (beforhand) but are not considered as causes but only as conditions of subsequent facts.

---

1. The term has been suggested by Vladimir Lifschitz.

In what follows, we will restrict our attention to scenarios where Δ is a determinate theory and $\mathcal{B}$ a set of literals. For this case, the above definition of a relative model admits a simpler and presumably more illuminating description:

Lemma 9.1 *If Δ is a determinate theory and $\mathcal{B}$ a set of literals, then a world α is an exact model of a causal scenario $(\Delta, \mathcal{B})$ if and only if α is closed with respect to the rules of Δ and any literal from α that is not a background condition is caused by α:*

$$Lit(\alpha)\backslash\mathcal{B} \subseteq \Delta(\alpha) \subseteq \alpha.$$

As before, $Lit(\alpha)$ above denotes the set of literals in the world α. The proof follows immediately from the above definition, given the properties of classical entailment.

Relative completion. Recall that the standard causal nonmonotonic semantics of a causal theory can be described "logically" using the notion of a propositional completion (see section 4.5.2 in chapter 4). It turns out that a similar description can be obtained also for the above relative nonmonotonic semantics.

Given a definite causal theory Δ, we can define its *relative propositional completion* with respect to a set $\mathcal{B}$ of background conditions, $comp(\Delta, \mathcal{B})$, as the set of all classical formulas of the forms

- $A \to l$, for any $A \Rightarrow l \in \Delta$;
- $l \to \bigvee\{A \mid A \Rightarrow l \in \Delta\}$, for any $l \notin \mathcal{B}$.

The following result shows that the classical models of $comp(\Delta, \mathcal{B})$ precisely correspond to exact worlds of the causal scenario $(\Delta, \mathcal{B})$. The proof follows readily from the preceding lemma.

Theorem 9.2 *The set of exact worlds of a causal scenario $(\Delta, \mathcal{B})$ coincides with the classical semantics of relative completion $comp(\Delta, \mathcal{B})$.*

9.4 Proximate Causes

As a final step in developing a relative notion of causality, the definitions below implement the idea that background conditions should not be treated as "real" or proximate causes, but only as mere conditions of what has happened in a causal situation.

Definition 9.2 (Proximate cause) *Let α be an exact world of a causal scenario $(\Delta, \mathcal{B})$. A literal $l_0 \in \alpha$ will be said to be a* proximate cause *of a literal l in α with respect to $(\Delta, \mathcal{B})$ if there exists a set L of literals such that $\mathcal{B} \cap \alpha \subseteq L \subseteq \alpha$ and*

- $l_0, L \Rightarrow^r_\Delta l;$
- $L \not\Rightarrow^r_\Delta l.$

Compared with our original definition of an actual cause, given in chapter 8, the above definition restricts the "evidence sets" L (used in checking the necessity of the potential causes) to sets that already include existing background conditions, and consequently it reduces the set of admissible actual causes. Note, in particular, that no background condition could be a proximate cause by this definition.

As before with the definition of actual causality, the restriction to regular inference can be lifted if the causal theory is reduced to an appropriate actual subtheory. Moreover, this will also simplify the corresponding definition.

Definition 9.3 *Let α be an exact world of a causal scenario $(\Delta, \mathcal{B})$. A relative actual subtheory of Δ wrt to α and $\mathcal{B}$ is the set of causal rules $\Delta_\alpha^\mathcal{B}$ obtained from the actual subtheory Δ_α of Δ by removing literals from $\mathcal{B}$ from the bodies of the rules.*

In what follows, $\Rightarrow_\alpha^\mathcal{B}$ will denote the least causal inference relation that includes $\Delta_\alpha^\mathcal{B}$. Then, as before, we can obtain the following description.

Theorem 9.3 *Let α be an exact world of a causal scenario $(\Delta, \mathcal{B})$. Then a literal $l_0 \in \alpha$ is a proximate cause of a literal l in α with respect to $(\Delta, \mathcal{B})$ iff there exists a set of literals $L \subseteq \alpha$ such that*

- $l_0, L \Rightarrow_\alpha^\mathcal{B} l$;
- $L \not\Rightarrow_\alpha^\mathcal{B} l$.

The above definition singles out proximate causes as a particular subset of actual causes in the sense of the preceding chapter (the NESS test). According to this definition, background conditions cannot be proximate causes. Note, however, that it also creates a possibility that an actual cause that is not a background condition may not be a proximate cause. The following formal example illustrates this.

Example 9.1 *Let Δ be a causal theory containing the following two causal rules as the only rules for r in heads:*

$$p, q \Rightarrow r \qquad q, s \Rightarrow r,$$

where s is a background condition. Suppose also that α is a causal world of Δ where all these atoms p, q, r, s hold. Then both p and q are actual causes of r in this world (due to the first rule). However, since s is a background condition, we have $q \Rightarrow_\alpha^\mathcal{B} r$, and therefore p is not a proximate cause of r, only q is.

Actually, the *Two fires* example discussed later in this chapter (see example 9.8) will provide an important illustration of this possibility.

There is an interesting relation between the notion of relative causality, introduced above, and counterfactual approaches. On the one hand, it agrees with the latter that the actual world is insufficient, taken by itself, for determining causality; we need to take into account some alternative (background) circumstances. On the other hand, the very distinction between causes and conditions cuts across counterfactual dependence since most of the conditions will still be sine qua non of the relevant effects, though they will not be considered (proximate) causes.

Proximate causation and normality. In order to show how our notion of proximate causality is related to causal descriptions that use normality and defaults as key concepts, we will introduce below an alternative, "default" way of describing proximate causes.

Definition 9.4 *A default update of a causal theory Δ with respect to a background $\mathcal{B}$ is the causal theory $\Delta_{\mathcal{B}} = \Delta \cup \{l \Rightarrow l \mid l \in \mathcal{B}\}$.*

The default update of a causal theory makes all the background conditions defaults (self-explainable) in it.

Let $\mathcal{D}_{\Delta}$ denote the set of defaults of a causal theory Δ. Then the following definition of a normal cause implements the well-known rule from Kahneman and Miller (1986) according to which a default value of a variable cannot be presented as a cause:

Definition 9.5 (Normal cause) *Let α be a causal world of a causal theory Δ. A literal $l_0 \in \alpha$ will be said to be a* normal actual cause *of a literal l in α if there exists a set of literals L such that $\mathcal{D}_{\Delta} \cap \alpha \subseteq L \subseteq \alpha$ and*

- $l_0, L \Rightarrow_{\alpha} l$;
- $L \not\Rightarrow_{\alpha} l$.

The above modification of the notion of an actual cause is obtained by requiring that the evidence set L should include all default conditions that hold in the actual world. As a result, defaults can never be normal causes. Consequently, no background condition could be a normal cause with respect to the default update $\Delta_{\mathcal{B}}$ of a causal theory. Note, however, that this definition excludes also all exogenous literals from being normal causes (at least according to our formal representation that makes all of them defaults). Moreover, it also exempts, in effect, all defaults from having normal causes. Still, beyond these boundary cases, the definitions of proximate and normal cause (with respect to the default update) provide much the same causal descriptions.

9.5 Further Examples

In this section we are going to use the notions of background condition and proximate cause for describing particular causal situations, including examples from chapter 8, in order to test whether and how the relative notion of causality can help in resolving the issues with actual causality that have been left behind.

But first, let us review the basic causal situations.

Example 9.2 (Oxygen and Fire) *Due to the oxygen in the air, a short circuit started fire in the house.*

A simplest representation of the above situation is provided by the following causal law:

$$Oxygen, Short\text{--}circuit \Rightarrow Fire.$$

Then, according to the definition of actual causation, in the causal world

$$\{Oxygen, Short\text{--}circuit, Fire\},$$

where all the relevant propositional atoms hold, both *Oxygen* and *Short--circuit* will be actual causes of *Fire*. However, in usual, or normal, circumstances *Oxygen* is viewed as a background condition, in which case only *Short--circuit* will be a proximate cause of *Fire* by our definition. Still, normality is not an inherent property of oxygen; as was noted already in Hart and Honoré (1985), in unusual circumstances where the presence of oxygen is abnormal, the statement "It was the presence of oxygen that caused the fire" makes perfect sense.[2]

The above causal rule can be viewed as a particular "surface approximation" of a more general, defeasible causal rule,

$$Short\text{--}circuit, n \Rightarrow Fire,$$

where *n* denotes the underlying causal mechanism. Thus, the original rule can be derived, in a sense, from the above rule and an auxiliary causal rule stating that absence of oxygen cancels the mechanism:

$$\neg Oxygen \Rightarrow \neg n.$$

Indeed, under the combined description, the absence of oxygen is the only factor that can cause $\neg n$, so by default causal completion (see definition 4.30) we obtain $Oxygen \Rightarrow n$. Combined with the main rule, this will allow us to derive $Oxygen, Short\text{--}circuit \Rightarrow Fire$, which will give us the corresponding surface representation.

As before, even under this more detailed representation, if *Oxygen* is a background condition, it will not be a proximate cause of *Fire*, but if it is not, then it will be a cause (along with *Short--circuit*).

Remark. Pearl (2019) has suggested that the explanatory distinction between these two causal factors can be accounted for using the distinction between necessary and sufficient causation, in particular by employing the technical notion of probability of sufficiency (PS). According to this suggestion, the difference is ultimately determined by the facts that short circuit (or even match-lighting) is a rare event, whereas the presence of oxygen is common. However, the example of Gardener and Queen below will show that such probabilistic considerations are not always sufficient for providing a general basis for the distinction between causes and conditions.

---

2. When there was a fire in an Apollo test capsule in 1967, the presence of (pure) oxygen was actually cited as the cause of the fire.

Considerations of normality are only a "default" way of determining background conditions. In many circumstances, the latter are determined by the *preceding situation*; in such cases background conditions of a current situation can be viewed as its *preconditions*. The following experiment from Henne et al. (2019) can be used as an illustration.

Example 9.3 (Motorboat) *The motorboat starts (S) when we turn the key (K) if the motor is in the lock position (L). Today, the motor is not in the lock position, so Ned puts it (P) in the lock position and turns the key. The motorboat starts.*

The causal theory for this example is

$$K, L \Rightarrow S \quad P \Rightarrow L.$$

According to the description, the actual world is $\{P, L, K, S\}$, whereas the preceding situation is $\{\neg P, \neg L, \neg K, \neg S\}$. Then both P and K will be proximate causes of S in this world. Henne et al. (2019) contrast this situation, however, with a different, inaction vignette in which the motor is already in the lock position, so Ned does not change it. Then the preceding situation is $\{\neg P, L, \neg K, \neg S\}$ while the actual world is $\{\neg P, L, K, S\}$. In this case $\neg P$ will not be a proximate cause of S but only its persistent background condition.

The notions of a background condition and proximate cause provide us with an immediate justification of why a birth of a person should not be considered a cause of his death (for instance, in a car accident) or why the absence of a meteor strike today is not among the causes of my writing this book (cf. Menzies 2007), despite the fact that, in both cases, they obviously pass the but-for test and hence will count as causes on counterfactual accounts.

According to the relational, or perspectival, view of causation that we advocate in this chapter, there is a significant degree of freedom in choosing the background conditions, so the latter are not determined solely by objective facts. For instance, a number of problematic examples, discussed in the preceding chapter, have (suspiciously often) involved an "event" of survival (absence of death) and its possible causes. In considering such cases, we should take into account that we normally consider continuation of life as something that is self-explanatory and hence a background condition, despite the fact that, from the point of view of physics, or even biology, life requires a good deal of supporting causes. By the same token, in such normal circumstances, death will always require causal explanation.[3]

Our next example has been widely discussed in the causal literature.

Example 9.4 (Gardener and Queen) *A plant requires regular watering during hot weather. A gardener whose job is to water the plant fails to do so during the hot spell, and the plant dies. If the gardener had watered the plant, it would have survived. But if the*

---

3. Cf. Hitchcock (2007): It is a plausible principle of folk biology that an individual will remain alive unless something causes them to die.

Queen of England had watered the plant, it would have survived. It is absurd, nonetheless, to say that the Queen's failure was a cause of the plant's death.

We can use the following causal theory to describe this situation:

$$Watered \Rightarrow \neg Death \quad Gardener \Rightarrow Watered \quad Rain \Rightarrow Watered \quad Queen \Rightarrow Watered.$$

The actual world is $\{\neg Gardener, \neg Rain, \neg Queen, \neg Watered, Death\}$. Now, by negative completion,

$$\neg Watered \Rightarrow Death \quad \text{and} \quad \neg Gardener, \neg Rain, \neg Queen \Rightarrow \neg Watered.$$

Still, *Gardener*, ¬*Rain*, ¬*Queen* and even *Watered* can be considered as *normal* background conditions in this situation, and consequently

$$\neg Gardener \Rightarrow \neg Watered$$

will belong to the relative actual causal theory, which will imply that only ¬*Gardener* is a proximate cause of *Death*.

It should be clear that the background conditions in this example are largely determined by normative considerations. It is the behavior of the gardener that is deviant in these circumstances (due to his job), and therefore it is his *omission*, even when rains are frequent in the area and even though the Queen often likes to water the plant by herself.

9.5.1 Negation as Default

A large part of the discrepancies between formal representations of chapter 8 and our causal intuitions can be "repaired" by assuming that negative propositions can *normally* be taken as background conditions.

Let us begin with overdetermination.

Example 9.5 (Window, reconsidered) *The* Window *example (example 8.7) has been formalized using the following causal theory:*

$$B \Rightarrow W \quad S \Rightarrow W \quad \neg B, \neg S \Rightarrow \neg W.$$

This theory provided appropriate answers for the situations (worlds) where W holds. As we have mentioned, however, for the causal world $\{\neg B, \neg S, \neg W\}$, *it has made Billy's and Suzy's not throwing rocks the actual causes of an unbroken window.*

In addition to intuitive inappropriateness, the above formalization has also a certain technical shortcoming, namely, it lacks "elaboration tolerance" (in the sense of McCarthy 1998): the last rule requires rewriting each time we add a new vandal throwing rocks at the window. Thus, if Ned has joined people throwing rocks at the window, we would have to add a new rule $N \Rightarrow W$, but we would also have to rewrite the last rule into

$$\neg B, \neg S, \neg N \Rightarrow \neg W.$$

This is just an instance of negative completion by which causal rules for negative literals are implicitly derivable from the corresponding positive rules.

But now let us suppose that we make an intact window ($\neg W$) a background condition for this situation. Then we still have that this causal scenario has the same causal worlds. Moreover, proximate causes of W in the worlds where it holds will coincide with the actual causes of W for the original causal description. So the only difference will be that in the causal world $\{\neg B, \neg S, \neg W\}$, the unbroken window will be excused from causal explanation (being a background condition). Moreover, the negative causal rule becomes obsolete, and the source causal theory can be safely reduced to its positive part

$$B \Rightarrow W \quad S \Rightarrow W$$

without any change in associated causal worlds and causal claims. This implies, in particular, that adding further rock-throwing actors to the situation can be achieved just by a modular addition of the corresponding positive rules like $N \Rightarrow W$.

The default interpretation of negation allows us also to further improve the representation of *late preemption*.

Example 9.6 (Simplified Bottle, revisited) *Recall the simplified formal description of bottle shattering (example 8.19):*

$$ST \Rightarrow BS \quad BT, \neg ST \Rightarrow BS \quad \neg ST, \neg BT \Rightarrow \neg BS.$$

This causal theory has justified the same causal claims as the representation of Halpern and Pearl (2001), including the problematic assertion that $\neg ST$ is a cause of bottle shattering when only Billy throws. This problem can now be immediately remedied by taking Suzy's not throwing a background condition in this situation. Moreover, as in the Window *example above, if we will take an intact bottle $\neg BS$ as yet another background condition, it will be exempted from the need of causal explanation, so neither $\neg ST$ nor $\neg BT$ will count as its causes. In this case the last, negative causal rule $\neg ST, \neg BT \Rightarrow \neg BS$ can also be dropped.*

The example of *Bogus Prevention* shows, however, that the default interpretation of negation has some limitations:

Example 9.7 (Bogus Prevention, revisited) *The following representation has been proposed for* Bogus Prevention *in chapter 8 (example 8.14):*

$$A, \neg B \Rightarrow D \quad \neg A \Rightarrow \neg D \quad A, B \Rightarrow \neg D.$$

To begin with, note that Victim's survival ($\neg D$) cannot unconditionally be taken as a background condition here since poison creates a threat to life, and hence in the world where A holds, $\neg D$ requires a causal explanation (which is provided by the last causal rule). Still, for another causal world $\alpha = \{\neg A, B, \neg D\}$, the actual causal theory reduces to

$\{\neg A \Rightarrow \neg D\}$, *and hence the absence of poison* $\neg A$ *is the only actual cause of* $\neg D$. *Now, in* this *world, we can consider both the absence of poison and survival as background conditions.*

9.5.2 Making a Difference

According to the account of Hart and Honoré, a cause is an intervention in the natural course of events that *makes a difference* in the way these develop. In this section, we will illustrate this idea on a number of important examples in the literature.

To begin with, the relational representation allows us to shed a new light on some peculiar cases of asymmetric overdetermination that have been extensively discussed in the legal literature (see Moore 2009, p. 427 and the references therein).

Example 9.8 (Two fires) *Two fires are burning their way toward the plaintiff's house. A defendant has negligently started one of the two fires, but the other fire is of natural origin caused by lightning. Either fire, by itself, will be sufficient both to reach the plaintiff's house and burn it to the ground. As it happens, the two fires join, and it is the larger, resultant fire that destroys the plaintiff's house.*

Some courts have denied liability of defendant to the plaintiff for his house in such cases, though as has been rightly noted by Moore, neither the regularity nor the counterfactual accounts could explain this asymmetry as compared with usual overdetermination cases.

As before with ordinary overdetermination, the following two causal rules provide a description:

$$Fd \Rightarrow H \quad Fl \Rightarrow H,$$

where *Fd* describes the fire started by the defendant while *Fl* stands for the fire caused by lightning. However, this time there is a natural inclination to see the lightning, and hence the second fire, as parts of a natural course of events and hence as background conditions. Thus, according Moore, the guiding intuition here is that what was going to occur naturally is a morally significant baseline. Now, given the actual world $\{Fd, Fl, H\}$, the background condition *Fl* makes valid $\mathbf{t} \Rightarrow_{\alpha}^{\mathcal{B}} H$ in the relative actual causal theory, and therefore neither *Fl*, nor *Fd* are proximate causes of *H* in this scenario. In a sense, the burning of the house *H* becomes part of the background conditions since it is a causal consequence of the latter.

Of course, taken by itself, the existence of such a relative causal representation cannot be considered a decisive reason for acceptance of the corresponding causal claims in legal cases. However, the very existence of deliberations in such cases clearly indicates that what is taken to be a background or preexisting condition plays an important role in our causal claims. A productive dispute in such cases could be reduced to the question of what facts could, or should, be considered as an appropriate background.

Now we are going to explore whether our enhanced representation capabilities could help in resolving the controversy involved in switch examples. It will turn out that the framework of relative causation naturally suggests an alternative interpretation of such cases that is more in accord with received views.

Example 9.9 (Switch, revisited) *The* Switch *example (see example 8.10) has been formalized using the following causal theory:*

$$\neg S \Rightarrow L_1 \quad S \Rightarrow L_2 \quad L_1 \Rightarrow I \quad L_2 \Rightarrow I$$
$$S \Rightarrow \neg L_1 \quad \neg S \Rightarrow \neg L_2 \quad \neg L_1, \neg L_2 \Rightarrow \neg I.$$

On this representation, the current position of the switch (S or $\neg S$) is an actual cause of the room being illuminated (I), a conclusion that has been rejected by many.

Suppose, however, that we single out one of the two possible situations in the above setup, say the one where the position of the switch is $\neg S$, as background circumstances. In other words, let us consider the world $\beta = \{\neg S, L_1, I\}$ as the set of background conditions for the second causal world $\alpha = \{S, L_2, I\}$ where the switch is flipped to S. Then S will be a proximate cause of L_2 in α, but neither S, nor L_2 will be proximate causes of the room being illuminated (I) since the latter, being a background condition, will be exempted from causal explanation! Thus, we have obtained an answer that is desirable by many, namely that position of the switch is not a (proximate) cause if the room being illuminated.

Still, a note of caution is in order here. The above representation is suitable for dynamic descriptions where we proceed from one situation to another by "flipping the switch." It is not suitable, however, for situations where this dynamic interpretation is not appropriate. Thus, in our example of *early preemption* (see the *Backup* example 8.11), we also have a switch pattern, but in the corresponding situation Assassin's poisoning act is viewed as fully voluntary (he chooses whether to poison or not at some specific time), so neither A, nor $\neg A$ can be taken as a background condition. Consequently, both poisoning and, respectively, refraining from poisoning by Assassin will be proximate causes of death in these circumstances, even according to our modified definition.

Nevertheless, when one of the states of a "switch" corresponds to a natural, or normal, course of events, the above representation appears to provide correct answers, as witnessed by our next example.

The following kind of examples has been used in the past as counterexamples to probabilistic approaches to causation. Recently, they have been employed for demonstrating that counterfactual reasoning is essential for our judgments of actual causality.

Example 9.10 (The Squirrel Case) *A golf ball is heading for a cup. A squirrel gets in the way and kicks the ball. However, the ball in fact ricochets against some trees and ends up in the cup after all.*

The following causal theory corresponds to this situation:

$$Ball, n \Rightarrow Cup \quad Ball, Squirrel \Rightarrow \neg n \quad Ball, Squirrel \Rightarrow Ricochet \quad Ricochet \Rightarrow Cup.$$

The actual world is $\{Ball, Squirrel, Ricochet, Cup\}$, and by the "plain" NESS account, both *Ball* and *Squirrel* are actual causes of *Cup* in this case. However, we can take into account the normal course of events for this situation; as suggested by the first sentence of the above informal description, it corresponds to the world $\{Ball, Cup, \neg Squirrel, \neg Ricochet\}$. If this latter world is taken as background circumstances, then *Cup* becomes a background condition, so it does not require causal explanation.

The above alternative representation appears to be somewhat controversial, but it seems to provide an explanation for the experiments of Gerstenberg et al. (2014) where (being translated to our vignette) people tended to reject that *Squirrel* is a cause of *Cup* in this situation. Anyway, such examples do not necessarily show the need in the language of counterfactuals for the analysis of causation; rather, they demonstrate that more basic features of difference making and contrast are important parts of our commonsense understanding of (relative) causality.

9.5.3 *Novus actus interveniens*

The following example provides perhaps a simplest illustration of what is known as *novus actus interveniens* in the legal literature.

Example 9.11 (Falling Tree) *(Hart and Honoré 1985) A defendant culpably pushes his victim to the ground, and the victim is killed on the ground by a falling tree.*

An appropriate causal theory for this case is as follows:

$$Push \Rightarrow Ground \quad Tree\text{–}fall, Ground \Rightarrow Death.$$

Now, the actual causal world of this story is $\alpha = \{Push, Ground, Tree\text{–}fall, Death\}$, and a straightforward application of the definition of actual cause for this world will produce that *Push* is an actual cause of *Death* (together with *Tree–fall*). However, our (pre-theoretical) intuition strongly suggests that *Push* is not a proximate cause of *Death*, though it is certainly a necessary *condition* of the latter. To obtain this result, we can consider *Push* and *Ground* as background conditions of this world; then our definition of a proximate cause will make *Tree–fall* the only proximate cause of *Death*.

The above solution might appear arbitrary or, using Mill's phrase, "capricious," but it could be made more justified if we will consider the original story as a *sequence* of two causal situations, or worlds, the first pushing the victim to the ground followed by a second one of the tree falling. The first situation corresponds to the world

$$\beta = \{Push, Ground, \neg Tree\text{–}fall, \neg Death\},$$

and in this world *Push* is a cause of *Ground* (even a proximate one). This world, however, can be considered a background condition of the world α (or its precondition), and this will make both *Push* and *Ground* "mere" conditions of *Death*. It is in this sense that the tree fall "broke the chain of causation" and justified the title *novus actus interveniens*.

The above sequential representation also brings to the fore the dynamic aspect of causal reasoning that will become our main subject in part IV of the book. This dynamic aspect is even more visible in the following example that has the same causal structure.

Example 9.12 (Collapsed Bridge) *Victim, stabbed by Assassin, will survive if he is taken to hospital by ambulance; alas a bridge collapses as the ambulance is crossing it and all inside, including Victim, are drowned in the river.*

In this example, the background situation is the world initiated by stabbing of Victim and normally (naturally) evolved to Victim's being taken to hospital by ambulance. However, the collapse of the bridge intervenes and creates a new situation, or world, in which the collapse causes drowning of Victim. On this representation, Assassin's action is a cause of Victim's being in the ambulance that crosses the bridge, but the latter is only a condition of Victim's drowning; the collapse of the bridge is the only proximate cause of the latter.

The above examples still represent only fairly simple cases of the distinction between conditions and (proximate) causes. What we will say, for instance, if the defendant from the first example sees (or knows) that the tree is going to fall, and deliberately pushes his victim toward it? What if Assassin in the second example has planted a bomb beforehand that has caused the collapse of the bridge? It seems obvious that in such cases we will say that the corresponding actions of the defendant and, respectively, Assassin are already full-fledged causes of death. The causal structure and causal rules remain the same, but these cases will be viewed already in alternative "frames of reference," namely as ordinary situations that involve a single causal world and a single, unbroken causal chain of events.

A detailed and subtle classification of such cases has been presented in Hart and Honoré (1985). From the perspective of the present study, however, these difficulties and possibilities of alternative representations show that our definitions and formal constructs, taken by themselves, *do not resolve* the problems of causality, just as the differential calculus does not resolve by itself the problems of physics; they only provide formal tools for a rigorous reasoning about them. Choosing an appropriate representation and, more generally, the "art of causal modeling" remain a crucial part of causal reasoning.

9.6   Negative Causality Revisited

Finally, in this section we are going to apply the framework of relative causation in an attempt to resolve the problems and controversies involved in interpreting negative causal assertions.

9.6.1 Triggering and Abortion

We have started our discussion of negative causation in chapter 8 with singling out two important patterns of negative causation where verdicts of the formal definition of actual causality are in full accord with our intuitive causal assertions, namely triggering and its reverse variant, abortion. Now, using the more elaborate formalism of the present chapter, we can give a more precise description of such cases. Among other things, this will allow us to distinguish such cases from other, more typical cases of double prevention, which will turn out to be more controversial.

Recall that our canonical representation of preventions consists of two rules:

$$C, n \Rightarrow E \qquad C, P \Rightarrow \neg n.$$

The first rule is the main, defeasible causal rule (where n denotes the underlying mechanism) while the second one is the cancellation rule.

Now, using this representation, *triggering* can be described as a transition from the background situation $\beta = \{C, P, \neg n, \neg E\}$ where the causal mechanism is canceled by P, to a situation $\alpha = \{C, \neg P, n, E\}$ where the mechanism is activated and produces the effect E.

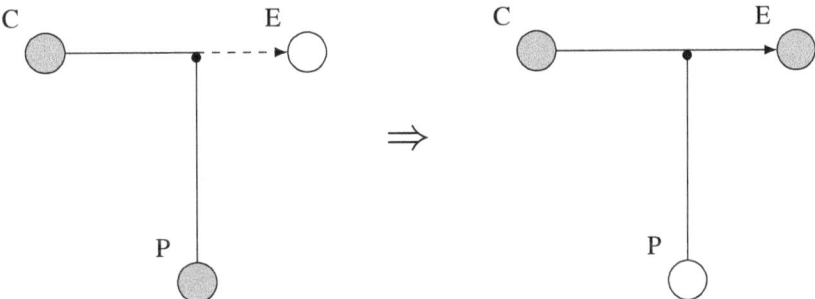

Two Pillars (see example 8.22) can be viewed as a paradigmatic example of triggering. If we consider the world β as a background of the world α, then $\neg P$ will be a proximate cause of E in this world while C will only be a background condition. For the *Two Pillars*, this means that Suzy's knocking is a proximate cause of the second pillar's falling down while gravity is only a background condition of this fall.

It is important to note what is unusual, or abnormal, in such triggering cases, beyond the use of negative literals as causes or effects. It is that in such cases what are usually considered causes and, respectively, conditions switch their roles. Thus, C, being normally a cause of E, is now only a background condition while $\neg P$, which is normally just a condition, is now a proximate cause of the effect E.

The second clear instance of negative causality is a mirror image of triggering, namely *abortion*.

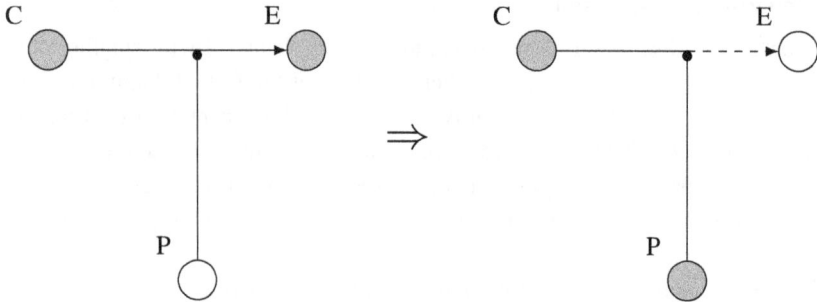

In this case we have a transition from the background situation $\alpha = \{C, \neg P, n, E\}$ where there is an active causal mechanism that produces the effect E to a situation $\beta = \{C, P, \neg n, \neg E\}$ where this mechanism becomes blocked by P, and as a result, the effect E ceases to hold. This time, P is a proximate cause of $\neg E$ in the world β, though as before, C is only a background condition. A simplest example of such an abortion is making a room dark by turning off a light switch, but death by decapitation also belongs to this category.

In what follows, we are going to argue, however, that the above two cases of negative causation are too special to serve as a standard for causal reasoning in more complex cases of double prevention. Fortunately, now we have an opportunity to clarify what are the differences, and our next example can be used as an illustration.

Example 9.13 (Unprevented Smash) *(Walsh and Sloman 2011) Frank kicks a ball, Sam moves out of its path, and the ball smashes a window.*

Formally, this example is quite similar to the preceding triggering case. In particular, the corresponding causal theory is isomorphic to that for *Two Pillars*:

$$Kicking, n \Rightarrow Window \quad Kicking, Blocking \Rightarrow \neg n \quad Moving \Rightarrow \neg Blocking.$$

The actual world is $\{Kicking, n, Window, Moving, \neg Blocking\}$. According to the definition of actual causality, both Frank's kicking a ball and Sam's move are actual causes of smashed window. Moreover, if we were taken Frank's kicking as a background condition for this situation (similarly to what we have done in *Two Pillars*), we would still obtain that Sam's move is a proximate cause of the smashing. However, in the test of Walsh and Sloman (2011), only fifth of the participants said that Sam's move is a cause of smashed window.

Since the syntactic descriptions for these cases are identical, the relevant differences should be semantic, and they amount to the fact that in triggering (*Two Pillars*) the underlying mechanism n has been disabled (canceled) beforehand, and then activated by the relevant action, whereas in the above example, the corresponding ball's flight n, once activated by Frank's kicking, has continued uninterruptedly till its "natural" end. Consequently, unlike the triggering case, n can be considered a background condition in this situation, which will immediately exclude Sam's move from the set of proximate causes of the smashed window, leaving only Frank's kicking as the proximate cause. Summing up, despite the syntactic similarity with triggering, the above story corresponds to a usual causal situation with an ordinary cause and absence of prevention as a background condition.

9.6.2 Double Prevention Reconsidered

"Begin at the beginning," the King said, very gravely, "Then go till you come to the end: then stop."
—Lewis Carroll, *Alice in Wonderland*

The solution we have suggested above for *Unprevented Smash* immediately brings us back to double prevention and, in particular, to *Bombing Mission* (see example 8.23).

Example 9.14 (Bombing Mission, revisited) *The following causal theory has been suggested for this situation in chapter 8:*

$$Suzy, n \Rightarrow Bombing \quad Suzy, Enemy, m \Rightarrow \neg n \quad Enemy, Billy \Rightarrow \neg m.$$

The relevant actual causal world is $\alpha = \{Suzy, n, Bombing, Enemy, Billy, \neg m\}$. Now, in this example also, the bombing process n, once initiated by Suzy, has continued uninterruptedly till its "natural" end. Accordingly, it can be considered a background condition for this situation, and if so, the absence of Enemy's shooting and a fortiori Billy's firing cannot be considered causes of it. But then Billy's firing will not be a proximate cause of the bombing, only its (necessary) condition.

It seems that even hesitations surrounding such cases could be explained in our framework as being created by an alleged alternative background "reference frame" that corresponds to the situation that would occur in the absence of Billy's shooting, namely, the possible world $\beta = \{Suzy, \neg n, \neg Bombing, Enemy, \neg Billy, m\}$.[4] Indeed, given such a background, it is easy to verify that Billy's shooting will be a proximate cause of *Bombing* in the world α.

Thus, we are apparently facing a choice between alternative backgrounds. However, we argue that, in contrast to the case of triggering, situation β cannot be taken as a *precondition* of the actual situation α. Once Suzy's plane has been shot down, she will be unable to resume the flight (to say the least), even if Billy will shoot the Enemy plane afterward. In other words, once β has actually occurred up to the moment of shooting down Suzy's

---

4. Thus, Menzies (2006) argued that it is precisely this situation that reflects typical default assumptions we make in this case.

bomber ($\neg n$), Billy's shooting the Enemy is not an intervention that could enable transition to situation α.

At this point, an attentive reader should have noticed that the above informal, dynamic descriptions have actually exceeded our current formal representation capabilities, meaning that our formalism requires a further upgrade to fully accommodate such a dynamic behavior. That is what we will attempt to do in part IV of the book.

IV DYNAMIC CAUSAL REASONING

10 Causal Dynamic Formalisms

The world we reason about and act in is essentially dynamic, and this is the ultimate reason why a logical analysis of the world (and information) dynamics occupies today an important place in general logical theory. There exists, however, a lot of possibilities and choices about how such a dynamics could be captured in a logical representation, each with advantages and problems of their own. It should be obvious that both an independent and comparative study of such alternatives is essential for a proper analysis and handling of the dynamic reality.

Our aims in this chapter will be twofold. Our general, though distant, objective amounts to singling out and exploring the ultimate logical ingredients of dynamic reasoning. In this respect, however, we will make only a few initial steps in this general quest. The second, more tangible aim will consist in demonstrating that *dynamic reasoning can be viewed as an instantiation of causal reasoning* as it is conceived in this book.

To begin with, we will presume that, at the most fundamental level, an adequate logical description of a dynamic world can be achieved by defining an appropriate dynamic consequence relation among (propositions denoting) temporally extended events and processes. In this respect, the logical formalism, described below, can be viewed as a particular implementation of the "dynamic turn in logic" that has been advocated in van Benthem (1996). We argue, however, that this dynamic consequence relation should reflect the basic *causal* relationships among events and processes. In other words, we argue that a natural and systematic description of dynamic domains can be given once we settle what causes or enables what in these domains; the rest of the properties and facts about the domain or situation should follow (at least in principle) by logical means. Moreover, the rules and postulates that characterize such a causal consequence relation cannot be viewed as *a priori* entailments that follow logically from the meaning of the consequence relation and propositions involved but rather as nontrivial claims, or assumptions, about the (causal) structure of the dynamic world. This understanding could actually be viewed as a particular implementation of Hume's approach to causation (see section 1.2.3 in chapter 1).

The idea that causality should constitute the basis of representation and description of dynamic universes is not new; as a matter of fact, it is widely used as a guiding principle in the main fields that deal with such dynamic descriptions, such as linguistic semantics, artificial intelligence (AI), and even general theory of computation. Taking only a few examples, it has been persuasively argued in Steedman (2005) that the so-called temporal semantics of natural language should not primarily deal with time at all but rather with representation of causality and goal-directed action (see also van Lambalgen and Hamm 2004, for a similar approach). Similarly, a number of causal approaches to reasoning about action and change have been developed in AI—see, for example, Giunchiglia et al. (2004), as well as chapter 12 of this book. In fact, our approach could also be seen as a logical counterpart of a qualitative causal modeling in AI developed a long time ago in Forbus (1984). Finally, so-called event theories of computation (see, e.g., van Glabbeek and Plotkin 2009) are based on the notion of *enabling* that could be viewed as a computational instantiation of our notion of dynamic causal inference that will be described in this chapter. The fundamental role of causation in the general theory of computation has also been pointed out in Hoare, Struth, and Woodcock (2019).

The majority of traditional logical approaches to dynamic evolution, including temporal and dynamic logics, stem from a fundamental modal logic paradigm, according to which the dynamic universum is a set of ordinary, "static" worlds structured by interworld accessibility relations. A syntactic counterpart of this view is that dynamic logic, broadly understood, is a conservative *extension* of (static) classical Boolean logic with (dynamic) modal operators that describe "accessible static options" for each static world.[1] By its very language, this paradigm presupposes that the essential information about the dynamics of the universe can be expressed, ultimately, in terms of such static accessibility assertions.

There is no need to argue these days that the above modal paradigm has provided (and still constitutes) a powerful and versatile tool for a logical analysis of dynamic discourses. There is something to be said, however, about discrepancies that exist between this modal representation and the ways dynamic reality is represented in ordinary language discourse. One of the most salient discrepancies of this kind is that the natural language seems to treat static and dynamic descriptions much on the same footing. In particular, language copulas and expressions corresponding to logical connectives of conjunction, disjunction, implication, and negation seem to be freely applicable to clauses of both kinds. In contrast, in propositional dynamic logic (PDL), a sentence "Event E brings about an effect A" should be "compartmentalized" as either a modal statement $[E]A$ (if E is an action/program) or as $E \to A$ when E is a static proposition.[2]

The above-mentioned discrepancies transcend the boundaries of small inessential syntactic variations and provide an initial justification for an alternative logical paradigm

---

1. This has been called an *explicit* design of logical systems in (van Benthem 2019) as opposed to what van Benthem has termed an *implicit* design that will be our choice in what follows.
2. Or even as $[E?]A$ with an additional syntactic sugar for tests.

according to which a logical description of a dynamic universe could be obtained by a direct *generalization* of classical logic to the dynamic setting.

Of course, we have to pay for this generalization by sacrificing some principles of Boolean logic. It is well known, for instance, that the language copula "and" is not commutative in linguistic discourse—"He drank a lot and drove home" is dangerously different from "He drove home and drank a lot." Still, the idea is that it is possible to single out a natural logical core of dynamic reasoning by assigning a more dynamic meaning to the ubiquitous logical connectives in a way that would give us a comprehensive basis for representing the dynamic world.

Again, the above idea is also not new. A first precise formalization of this idea has been suggested in the framework of dynamic predicate logic (DPL) described in Groenendijk and Stokhof (1991). The syntax of DPL included the usual logical connectives of conjunction, disjunction, implication, and negation but was based on a semantic interpretation in which propositions were directly represented as relations among worlds. In this way DPL instantiated another important idea, namely that meaning of linguistic expressions should be viewed, in general, as a capability of context change. This dynamic view of meaning occupies today a prominent place in natural language semantics (see, e.g., van Eijck and Stokhof 2006).

It should be noted, however, that our formalism of sequential dynamic logic (SDL), described below, is purported to capture general dynamic reasoning, so it is not restricted to specific (mainly linguistic) interpretations of DPL (see Dekker 2008, for an overview of the latter). Thus, we will show that our suggested generalization of classical logic is capable of (at least) reaching the level of sophistication and expressivity of propositional dynamic logic (PDL). Moreover, we hope the reader will see that our alternative, causal representation will actually provide a significant simplification of the corresponding formal development.

10.1 Dynamic Causal Inference

In this section we will introduce a basic, structural calculus of dynamic causal inference that is defined on sequences of propositions (called processes) and satisfies certain "sequential" variants of familiar inference rules such as monotonicity and cut (though not reflexivity!). This calculus will be assigned first an abstract monoid-based dynamic semantics, and the completeness results will be proved. Then we will show that the latter semantics can be transformed into more familiar relational or transition models. Moreover, it will be shown that our dynamic inference constitutes in this respect a generalization of the inference relation of DPL from Groenendijk and Stokhof (1991).

As a second step, we will introduce a two-sorted extension of the basic formalism in which the underlying set of propositions will be split into transitions (actions) proper and static propositions (also known as fluents, or tests). This extension of the formalism will be called a state-transition calculus. It will be shown that (under some additional conditions) the state transition calculus is also complete for the corresponding relational semantics.

Proofs of all the results in this section that do not appear in the text can be found in the appendix to this chapter.

10.1.1 Structural Dynamic Calculus

The language of the calculus will be a set $L = \{A, B, C, \dots \}$ of propositions that will denote primitive transitions or *events*. Finite sequences of events will be called *processes*, and we will use small letters $a, b, c, \dots$ to denote such processes. The set of all processes will be denoted by L^*. It will include, in particular, an empty sequence denoted by ϵ. As usual, ab will denote the concatenation of sequences a and b from L^* (and similarly for aA, aAb, and so on).

A *dynamic consequence relation* is a set of rules or sequents of the form $a \Rightarrow b$, where $a, b \in L^*$. The intended meaning of such rules is that a process a *causes* (or enables) a process b. This consequence relation will be required to satisfy the following postulates:

(Identity) $\epsilon \Rightarrow \epsilon$.
(Left Monotonicity)   If $a \Rightarrow b$, then $Aa \Rightarrow b$.
(Cut)   If $a \Rightarrow b$ and $ab \Rightarrow c$, then $a \Rightarrow bc$.

Already the very form of our causal sequents that involve *sequences* of propositions (instead of usual sets) indicates that the dynamic consequence relation is *substructural*, that is, it does not satisfy the usual structural rules for consequence relations such as contraction, permutation, and weakening. Note, in particular, that a minimal antecedent or a consequent of a dynamic sequent is not an empty set but an empty sequence ϵ.[3] Still, we will show in what follows that the analysis and representation of our dynamic calculus can proceed along quite the same lines as the usual theoretical development for standard consequence relations.

Though all the above postulates constitute a certain weakening of well-known structural rules for a (classical) sequent calculus, we argue that these postulates should rather be viewed as informed claims about the structure of processes and their interactions. In other words, we see the above postulates as assertions that have a nontrivial content that jointly determine the structure of the dynamic universe (more exactly, of its representation). Varying or extending these postulates (as will actually be done in subsequent sections) means changing the structure and relations of this universe.

In this respect, it is interesting to note that our postulates preserve locality (contiguity) and continuity of processes. Thus, if the premises of some postulate describe relations among contiguous processes, then its conclusion will also have this property. This feature

---

3. The first postulate, Identity, expresses in this sense the ancient causal principle *Ex nihilo nihil fit* (Nothing comes from nothing).

is especially evident in the form of the Cut postulate, which does not allow us to infer, for example, $a \Rightarrow c$ from $a \Rightarrow b$ and $ab \Rightarrow c$; such an inference would break contiguity and create nonlocal influences.

Due to the Horn form of the rules characterizing a dynamic consequence relation, the intersection of a set of consequence relations is again a consequence relation. This implies, in particular, that for any set of causal sequents, there exists a least dynamic consequence relation containing it.

Now we introduce the basic notion of a theory of a dynamic consequence relation. Intuitively, theories characterize admissible sets of processes (with respect to a given dynamic consequence relation).

Definition 10.1 *A set U of processes will be called a* theory *of a dynamic consequence relation $\Rightarrow$ if, whenever $a \Rightarrow b$ holds and s is some process such that $sa \in U$, then $sab \in U$.*

Speaking informally, a theory is a set of processes that is closed with respect to the dynamic causal rules: whenever it includes a process s_0 that ends with a (as its end segment), and a causes b, then the extended process $s_0 b$ should also belong to the set.

An important property of dynamic theories (common with ordinary Tarski theories) is that intersections of theories are again theories of a consequence relation. This immediately implies that, for any set of processes, there exists a least theory containing it. In other words, we have a natural closure operator $\mathrm{Th}(V)$ that assigns any set V of processes a unique least theory containing it.

Dynamic monoid semantics.   We will introduce first a very abstract notion of a dynamic semantics.

By a *dynamic frame* we will mean an arbitrary monoid $(P, \cdot, 1)$. In other words, $\cdot$ is an associative binary operation on P, and $u \cdot 1 = 1 \cdot u = u$, for any $u \in P$. Elements of P will be called *paths*. The operation $\cdot$ on P can be viewed as a concatenation of paths. It can be canonically extended to sets of paths as follows: if $U, V \subseteq P$, then

$$U \cdot V = \{uv \mid u \in U \ \& \ v \in V\}.$$

This extension is also an associative operation.

In what follows, we will often omit the operation sign $\cdot$ and write ab instead of $a \cdot b$. In addition, $x \preceq y$ will denote the fact that $y = xz$, for some z. It is easy to verify that $\preceq$ is a partial order. We will call it a *prefix relation* on paths.

Definition 10.2 *An abstract* dynamic model *is a tuple $D = (P, U, \cdot, 1, \mathcal{V})$, where $(P, \cdot, 1)$ is a dynamic frame and $U \subseteq P$ (called the set of* allowable paths*) while $\mathcal{V}$ is a valuation function assigning each proposition from the language a subset of P, that is, $\mathcal{V}(A) \subseteq P$, for any $A \in L$.*

A set of dynamic models will be called a dynamic (monoid) semantics.

A dynamic model can be seen as a restriction of a dynamic frame to the set of allowable paths. In other words, it can be viewed as a *partial* monoid in which the concatenation operation · is defined only if it produces an element of U.[4] Note, however, that the valuation function $\mathcal{V}$ also becomes a partial function on this view.

As a preparation, we will extend the valuation function to sequences of propositions (i.e., processes) from L^* as follows:

$$\mathcal{V}(A_1 \ldots A_n) = \mathcal{V}(A_1) \cdot \ldots \cdot \mathcal{V}(A_n).$$

We will extend $\mathcal{V}$ also to an empty sequence by stipulating $\mathcal{V}(\epsilon) = \{1\}$.

Every dynamic semantics $\mathcal{D}$ determines a dynamic consequence relation $\Rightarrow_\mathcal{D}$ defined as follows: $a \Rightarrow_\mathcal{D} b$ holds if and only if,

for any $D \in \mathcal{D}$ and any $u, x \in P$, if $ux \in U$ and $x \in \mathcal{V}(a)$, then $uxy \in U$ for some $y \in \mathcal{V}(b)$.

The following lemma verifies that this relation satisfies all the postulates of a dynamic consequence relation.

Lemma 10.1 $\Rightarrow_\mathcal{D}$ *is a dynamic consequence relation.*

If a consequence relation $\Rightarrow$ coincides with $\Rightarrow_\mathcal{D}$, for some dynamic semantics $\mathcal{D}$, we will say that $\Rightarrow$ is *generated by* $\mathcal{D}$.

It turns out that, similarly to ordinary consequence relations, any dynamic consequence relation can be generated by a canonical dynamic semantics constructed from the set of its theories.

Given a theory U of a consequence relation $\Rightarrow$, we can construct an associated dynamic model $D_U = (L^*, U, \cdot, \epsilon, \mathcal{V})$, where the dynamic frame is the set of all processes L^* with an operation of concatenation and ϵ as a unit while $\mathcal{V}$ is a trivial function $\mathcal{V}(A) = \{A\}$. Let $\mathcal{T}_\Rightarrow$ be the set of such dynamic models, for all theories of $\Rightarrow$. Then we have the following theorem.

Theorem 10.2 (Representation Theorem) *If* $\Rightarrow$ *is a dynamic consequence relation, then*

$$\Rightarrow \; = \; \Rightarrow_{\mathcal{T}_\Rightarrow} .$$

As a result, we obtain that dynamic consequence relations are complete with respect to the dynamic monoid semantics.

Corollary 10.3 $\Rightarrow$ *is a dynamic consequence relation if and only if it is generated by a dynamic monoid semantics.*

---

4. This notion also coincides with the notion of a *partially associative operation*, used in Barwise, Gabbay, and Hartonas (1995, definition 4.5).

Actually, the above result can be strengthened by making use of the fact that the canonical dynamic semantics $\mathcal{T}_\Rightarrow$, described above, is a dynamic semantics of a very special kind. Its specific properties are reflected in the following definition.

Definition 10.3 *A dynamic semantics* $\mathcal{D}$ *will be called* homogeneous *if:*

- *all its dynamic models involve the same dynamic frame* $(P, \cdot, 1)$ *and the same valuation function* $\mathcal{V}$.
- *The function* $\mathcal{V}$ *is singular:* $\mathcal{V}(A) \in P$, *for any* $A \in L$.

Thus, the variation of models in a homogeneous semantics reduces to the variation of admissible sets U of paths.

Now the representation theorem 10.2 immediately implies the following:

Corollary 10.4 $\Rightarrow$ *is a dynamic consequence relation iff it is generated by some homogeneous dynamic semantics.*

The above result will imply a number of consequences important for our study.

Another consequence of the above representation theorem is that dynamic consequence relations are uniquely determined by their theories. Moreover, for more expressive languages the above representation theorem can serve as a basis of constructing full-fledged semantics. In this case theories of a consequence relation will play, eventually, the role of its canonical models. In particular, just as for ordinary inference relations, inclusion among consequence relations amounts to inverse inclusion for their sets of theories. Accordingly, imposing further requirements on our consequence relations will amount to imposing additional properties on the associated theories.

Remark. To conclude our discussion in this section, we should mention an important dimension of generality that is embodied in the above notion of a dynamic monoid semantics and, in particular, in the very idea of allowable paths. As the reader may have noticed, the set of allowable paths is not required to be closed with respect to concatenation or subpaths, so concatenations of allowable paths are not always allowed and not all parts of allowable paths constitute allowable paths by themselves. Now, an important way to see this consists in viewing a compound process as possibly proceeding "in one leap" in which we cannot *temporally* (or sequentially) separate its constitutive parts. As an important extreme case, all the transitions in the process may be performed *concurrently*. For instance, a compound process of carrying a table by two people could still be described as a path $a \cdot b$ consisting of two concurrent actions of carrying, respectively, the left and right edge of the table. The causal and inference relations of such a process with other processes cannot be reduced directly to corresponding relations for its two components. Speaking more generally, the general dynamic monoid semantics allows us to accommodate the basic idea of Pratt (2003) that, unlike the sequential modeling, the evolution of time and information need not be synchronized.

Relational semantics.   Now we are going to show that our dynamic consequence relations can also be given a (suitably generalized) relational semantics.

We begin with the following well-known notion of a relational semantics.

Definition 10.4 *A relational model is a triple* $M = (S, L, \mathcal{F})$, *where S is a set of states, L is a set of propositions, and* $\mathcal{F}$ *is a function assigning each proposition from L a binary relation on S:*

$$\mathcal{F} : L \rightarrow \mathcal{P}(S^2).$$

For any binary relation R, $dom(R)$ will denote its domain, and $range(R)$ – its range. The function $\mathcal{F}$ can be canonically extended to sequences of propositions:

$$\mathcal{F}(A_1 \dots A_n) = \mathcal{F}(A_1) \circ \dots \circ \mathcal{F}(A_n),$$

where $\circ$ denotes the composition of binary relations. We will extend $\mathcal{F}$ also to an empty sequence by stipulating

$$\mathcal{F}(\epsilon) = \{(s, s) \mid s \in S\}.$$

Now, to provide a semantics for our dynamic inference, the above notion of a relational model has to be generalized as follows.

Definition 10.5 *A generalized relational model is a quadruple* $M = (S, \checkmark, L, \mathcal{F})$, *where* $(S, L, \mathcal{F})$ *is a relational model and* $\checkmark$ *is a subset of states called* termination states.

The notion of a termination state is actually known in the computational literature. In the latter, it denotes states in which a program or its parts successfully terminate as opposed to other states that do not have computational meaning. In our setting, we will stretch this notion to a general distinction between *real* and *virtual* states, guided by an idea that atomic transitions in a compound process are not necessary separated by real states. As we already mentioned at the end of the preceding section, we allow for a possibility that a process b may proceed in one leap, in which we cannot observe, or single out, sequential parts and associated intermediate states. Of course, this idea conflicts somewhat with the very notion of a relational model where transitions are *defined* as pairs of states. Still, the suggested way out consists in distinguishing between states that are actual (real) and all other, virtual states that may be viewed as purely theoretical constructs without "physical" meaning. This distinction will also serve as a preparation for the state-transition calculus that will be introduced later.

Now we are ready to formulate the following definition of validity for dynamic inference rules in this semantics:

Definition 10.6 *A dynamic rule* $a \Rightarrow b$ *will be said to be* valid *in a generalized relational model M if, for any states* $s \in S$ *and* $t \in \checkmark$ *such that* $(s, t) \in \mathcal{F}(a)$, *there exists a state* $r \in \checkmark$ *such that* $(t, r) \in \mathcal{F}(b)$.

$\Rightarrow_M$ *will denote the set of sequents that are valid in a relational model M.*

Let $\vec{\checkmark}$ denote pairs of states that end in $\checkmark$, that is, $\vec{\checkmark} = \{(s,t) \mid t \in \checkmark\}$. Then the above definition can be compactly written as follows:

$$a \Rightarrow_M b \equiv range(\mathcal{F}(a)) \cap \checkmark \subseteq dom(\mathcal{F}(b) \cap \vec{\checkmark}).$$

To begin with, it is easy to verify the following lemma.

Lemma 10.5 *If M is a generalized relational model, then $\Rightarrow_M$ is a dynamic consequence relation.*

In order to show that this relational semantics is also adequate for dynamic consequence relations, it is sufficient to demonstrate that any dynamic monoid semantics can be transformed into an equivalent relational model.

For the purposes of the construction that follows, we will use an expression $x \in U \in \mathcal{D}$ as a shorthand for $x \in U$ for an admissible set U of some dynamic model from $\mathcal{D}$. Now, given such an admissible set U, let $U{\downarrow}$ denote the set of prefixes of the elements of U, that is,

$$U{\downarrow} = \{x \mid x \preceq y, \text{ for some } y \in U \in \mathcal{D}\}.$$

Then the set of states of the relational model corresponding to a monoid semantics $\mathcal{D}$ will be a disjoint union of all $U{\downarrow}$, for every admissible set U. As usual, to achieve disjointness, each path $x \in U{\downarrow}$ will be indexed with the corresponding set U to become a new object x_U. As a result, we obtain the following construction of a generalized relational model $M_\mathcal{D}$:

- $S = \{x_U \mid x \preceq y, \text{ for some } y \in U \in \mathcal{D}\}$;
- $\checkmark = \{u_U \mid u \in U \in \mathcal{D}\}$; and
- for any $A \in L$, $\mathcal{F}(A) = \{(x_U, (xy)_U) \mid y \in \mathcal{V}(A) \text{ \& } (xy)_U \in S\}$.

The following theorem shows that the resulting relational model validates the same rules as the source dynamic semantics.

Theorem 10.6 *If $M_\mathcal{D}$ is the relational model corresponding to the dynamic semantics $\mathcal{D}$, then $\Rightarrow_\mathcal{D} = \Rightarrow_{M_\mathcal{D}}$.*

The above result immediately implies that dynamic consequence relations are also complete for the above notion of validity in generalized relational models.

Corollary 10.7 $\Rightarrow$ *is a dynamic consequence relation iff it coincides with $\Rightarrow_M$, for some generalized relational model M.*

As a matter of fact, the above completeness result can be strengthened by exploiting the fact that the dynamic semantics can be safely restricted to homogeneous one (see corollary 10.4).

Recall that a relational model is *deterministic* if $\mathcal{F}(A)$ is a partial function, that is, for any $A \in P$, if $(s,t) \in \mathcal{F}(A)$ and $(s,r) \in \mathcal{F}(A)$, then $t = r$. Now, it is easy to verify that the relational model corresponding to the canonical dynamic semantics $\mathcal{T}_\Rightarrow$ by the above construction is actually deterministic. Consequently, we immediately obtain that, at this stage, imposing determinism does not produce new valid rules of dynamic inference.

Corollary 10.8 $\Rightarrow$ *is a dynamic consequence relation if and only if it is generated by a generalized deterministic relational model.*

This result will no longer hold, however, when we will impose some additional conditions on dynamic consequence relations.

10.1.2 Sequentiality

A dynamic consequence relation will be called *sequential* if it satisfies

(Right Anti-Monotonicity)   If $a \Rightarrow bB$, then $a \Rightarrow b$.

Remark. When compared with Left Monotonicity, the above postulate makes vivid another essential difference between our consequence relations and common sequent calculi: namely the fact that both antecedents and consequents of our rules are interpreted conjunctively, unlike the usual, disjunctive understanding of succedents in sequents.

The postulates of sequential consequence relations almost coincide with the axiomatization of the calculus $\mathbf{G}_\mu$ presented in Kanazawa (1994) with the only exception that the latter does not have the Identity postulate (though appropriately restricts the discussion to non-empty sequences). As has been shown by Kanazawa, such consequence relations can be given a relational semantics of the kind used in the dynamic predicate logic (DPL) of Groenendijk and Stokhof (1991). In what follows we will reproduce this result as an "unfolding" of the corresponding representation theorem.

To begin with, note that sequentiality can be alternatively characterized by strengthening the Cut rule to the following "contextual" Cut:

(C-Cut)   If $a \Rightarrow bd$ and $ab \Rightarrow c$, then $a \Rightarrow bc$.

Indeed, the original Cut rule is a special case of C-Cut (when $d = \epsilon$). Moreover, the latter implies Right Anti-Monotonicity: Identity implies $ab \Rightarrow \epsilon$ by Left Monotonicity, and hence if $a \Rightarrow bc$ holds, then $a \Rightarrow b$ follows by C-Cut. In the other direction, given Right Anti-Monotonicity, C-Cut is derivable from Cut, since $a \Rightarrow bd$ implies $a \Rightarrow b$.

Sequentiality corresponds to the following restriction of the dynamic monoid semantics:

Definition 10.7 *A dynamic semantics $\mathcal{D}$ will be called* sequential *if, for any model $D \in \mathcal{D}$ with an admissible set U, if $u, v \in U$ and $u \preceq v$, then $w \in U$, for any w such that $u \preceq w \preceq v$.*

Thus, in a sequential dynamic model the set of admissible paths is interval-closed with respect to the prefix relation $\preceq$. The next lemma shows that such a semantics generates a sequential consequence relation.

Lemma 10.9 *If $\mathcal{D}$ is a sequential dynamic semantics, then $\Rightarrow_{\mathcal{D}}$ is a sequential consequence relation.*

Moreover, the main representation theorem 10.2 can be immediately extended to sequential inference, and we obtain the following completeness result:

Corollary 10.10 *A dynamic consequence relation $\Rightarrow$ is sequential if and only if it is generated by a sequential dynamic semantics.*

As before, we can use corollary 10.4 to conclude that the sequential dynamic semantics can be safely restricted to a homogeneous one.

Corollary 10.11 *A dynamic consequence relation $\Rightarrow$ is sequential if and only if it is generated by a homogeneous sequential dynamic semantics.*

This strengthening of the general representation result will be used in the next section.

Relational dynamic inference. It turns out that sequential consequence relations can be given a simpler relational semantics. More precisely, sequential dynamic inference amounts to validity in plain relational models, that is, models in which all states are termination ones.

Definition 10.8 *A dynamic rule $a \Rightarrow b$ will be said to be* valid *in a plain relational model M if*

$$range(\mathcal{F}(a)) \subseteq dom(\mathcal{F}(b)).$$

The above notion of validity corresponds to what has been called Update-to-Domain Consequence in van Benthem (1996), and it describes the inference relation adopted in the DPL of Groenendijk and Stokhof (1991). Unfolding the definition, it says that $a \Rightarrow b$ holds iff, for any $(s, t) \in \mathcal{F}(a)$ there exists $r \in S$ such that $(t, r) \in \mathcal{F}(b)$.

To begin with, we have the following lemma.

Lemma 10.12 *If M is a relational model, then $\Rightarrow_M$ is a sequential consequence relation.*

The proof amounts to a simple verification of Right Anti-Monotonicity. In order to show that the relational semantics is also adequate for sequential consequence relations, we will make use of corollary 10.4 and show that any *homogeneous* sequential dynamic semantics can be transformed into an equivalent relational model. Note, however, that the construction below is essentially different from the translation of the general dynamic semantics, described in the preceding section. Most importantly, the construction below creates an *indeterministic model* even from a homogeneous dynamic semantics.

Given a homogeneous sequential dynamic semantics $\mathcal{D}$ with a common frame $(P, \cdot, 1)$, we will construct a relational model $M_{\mathcal{D}}$ as follows:

- The set of states S is the set of all paths P together with all labeled paths from the the disjoint union of $\mathcal{D}$:
$$S = P \bigcup \{u_U \mid u \in U \in \mathcal{D}\};$$
- For any $A \in L$, $\mathcal{F}(A)$ is a set of all pairs (s, t) from S such that, for some paths $x, y \in P$, $y = x \cdot \mathcal{V}(A)$ and one of the following cases holds:

  1. $s = x$ and $t = y$;
  2. $s = x$ and $t = y_U$, if $y \in U \in \mathcal{D}$;
  3. $s = x_U$ and $t = y_U$, if $x, y \in U \in \mathcal{D}$.

The following theorem shows that the constructed relational model validates the same rules as the source sequential semantics.

Theorem 10.13 *If $M_\mathcal{D}$ is a (plain) relational model corresponding to the homogeneous sequential semantics $\mathcal{D}$, then $\Rightarrow_\mathcal{D} = \Rightarrow_{M_\mathcal{D}}$.*

As an immediate consequence of corollary 10.11, we conclude with the following completeness result.

Corollary 10.14 *A dynamic consequence relation is sequential if and only if it coincides with a dynamic inference generated by some plain relational model.*

As we noted in the course of the above construction, the relational models obtained by transforming sequential dynamic models are in general not deterministic. As we are going to see in the next section, this liberty is essential, because restriction of arbitrary relational models to deterministic ones will make valid an additional postulate of dynamic inference.

10.1.3   Deterministic Inference

A dynamic consequence relation will be called *deterministic* if it is sequential and satisfies the following postulate:

(Cumulativity)   If $a \Rightarrow Bc$, then $aB \Rightarrow c$.

It should be clear that a repeated application of Cumulativity generates the following structural rule for processes:
$$\frac{a \Rightarrow bc}{ab \Rightarrow c}.$$

Now, in deterministic consequence relations, any rule $a \Rightarrow bc$ is reducible to a pair of simpler rules:

Lemma 10.15 *If $\Rightarrow$ is a deterministic consequence relation, then $a \Rightarrow bc$ if and only if $a \Rightarrow b$ and $ab \Rightarrow c$.*

Proof. The direction from right to left follows directly by Cut, while the opposite direction follows from Cumulativity and Right Anti-Monotonicity. $\square$

As a consequence, any rule can now be reduced to a set of rules $a \Rightarrow A$ involving only singular conclusions (including ϵ). Moreover, this property can be viewed in some sense as a characteristic property of deterministic dynamic relations. In order to show this, let us introduce the following notions. Let us say that a sequent $a \Rightarrow b$ is *singular* if b is either an atom B or an empty sequence ϵ. Then a singular consequence relation, defined below, can be viewed as a restriction of a general dynamic consequence relation to singular sequents.

Definition 10.9 A singular dynamic consequence relation *is a set of singular sequents that satisfies Left Monotonicity and Identity.*

It can be easily verified that a set of singular sequents belonging to an arbitrary dynamic consequence relation forms a singular consequence relation. Moreover, it can be shown that the above two postulates exhaust the structural properties of such an inference relation.[5] So a singular consequence relation is actually a very simple inference relation.

Now, given a singular consequence relation $\Rightarrow$, we can inductively extend it to arbitrary sequents $a \Rightarrow b$ as follows: if $a \Rightarrow b$ is already defined, then we stipulate that $a \Rightarrow bA$ holds iff both $a \Rightarrow b$ and $ab \Rightarrow A$ hold. Let $\Rightarrow^m$ denote the resulting consequence relation. Then the next result shows that any deterministic consequence relation can be viewed as a definitional extension of a singular consequence relation.

Theorem 10.16 $\Rightarrow$ *is a deterministic consequence relation iff* $\Rightarrow = \Rightarrow_0^m$ *for some singular consequence relation* $\Rightarrow_0$.

Now let us turn to a semantic description. A dynamic monoid semantics of deterministic consequence relations can be defined as follows.

Definition 10.10 A *dynamic model* $D = (P, U, \cdot, 1, \mathcal{V})$ *will be called* deterministic *if the valuation function* $\mathcal{V}$ *is singular (that is,* $\mathcal{V}(A) \in P$, *for any* $A \in L$) *and the admissible set* U *is closed with respect to the prefix relation: if* $ab \in U$, *then* $a \in U$. *A dynamic monoid semantics will be called* deterministic *if all its dynamic models are deterministic.*

The next result verifies that any deterministic dynamic semantics generates a deterministic consequence relation.

Lemma 10.17 *If* $\mathcal{D}$ *is a deterministic dynamic semantics, then* $\Rightarrow_{\mathcal{D}}$ *is a deterministic consequence relation.*

A theory of a dynamic consequence relation will be called *deterministic* if it is closed with respect to the prefix relation. Clearly, the canonical dynamic model corresponding to a deterministic theory will be a deterministic model. Moreover, the corresponding canonical dynamic semantics is fully adequate for deterministic consequence relations (see the appendix), and consequently we obtain the following theorem.

---

5. A similar fact (though without ϵ) has been mentioned in van Benthem (1996).

Theorem 10.18 (Representation Theorem) *If $\mathcal{T}_{\Rightarrow}^d$ is a set of deterministic theories of a dynamic consequence relation $\Rightarrow$, then $\Rightarrow$ is deterministic iff $\Rightarrow \, = \, \Rightarrow_{\mathcal{T}_{\Rightarrow}^d}$.*

As a result, we can conclude with the following corollary.

Corollary 10.19 *A dynamic consequence relation $\Rightarrow$ is deterministic iff it is generated by a deterministic dynamic semantics.*

Note now that any deterministic dynamic semantics is already sequential and hence corresponds to some relational model. Moreover, we are going to show that deterministic semantics correspond precisely to deterministic relational models in which the accessibility relations are partial functions.

To begin with, we have the following lemma.

Lemma 10.20 *If M is a deterministic relational model, then $\Rightarrow_M$ is a deterministic consequence relation.*

The proof amounts to a straightforward verification of Cumulativity.

As before, to show that deterministic consequence relations are complete with respect to deterministic relational models, we will show that any deterministic dynamic semantics can be transformed into an equivalent deterministic relational model. The construction is actually a simplification of the transformation described earlier for general relational models.

Given a deterministic dynamic semantics $\mathcal{D}$, we will construct the corresponding deterministic relational model $M_{\mathcal{D}}$ as follows:

- $S = \{a_U \mid a \in U \in \mathcal{D}\}$;
- For any $A \in L$, $\mathcal{F}(A) = \{(a_U, b_U) \mid b \in U \in \mathcal{D} \ \& \ b = aA\}$.

It is easy to see that the above relational model is deterministic. The following theorem shows that it validates the same rules as the source dynamic semantics.

Theorem 10.21 *If $M_{\mathcal{D}}$ is the relational model corresponding to the deterministic dynamic semantics $\mathcal{D}$, then $\Rightarrow_{\mathcal{D}} \, = \, \Rightarrow_{M_{\mathcal{D}}}$.*

Combined with corollary 10.19, the above correspondence immediately gives us the following corollary.

Corollary 10.22 *A dynamic consequence relation is deterministic iff it coincides with a dynamic inference generated by some deterministic relational model.*

This result concludes our study of the basic variety of dynamic inference.

10.1.4   On Dynamic Inference with States and Transitions

The paper Bochman and Gabbay (2012a) contains also a generalization of the structural formalism of causal dynamic inference to a two-sorted language containing two kinds of events: proper *transitions* (or actions) and *states* (or conditions). This generalization (which has been called an ST-calculus) can be viewed as a structural counterpart of the sequential dynamic logic (SDL) that will be described in the next section.

Formally, the language L of events will become a union $Tr \cup St$ of the set $Tr = \{p, q, \dots\}$ of transitions and a disjoint set $St = \{A, B, C, \dots\}$ of state propositions (also called conditions, fluents, or tests). Still, as before, finite sequences of events (from L^*) will be called *processes*, and we will use the letters $a, b, c, \dots$ to denote such processes. Moreover, such processes will be assumed to satisfy all the postulates of dynamic inference, stated earlier, despite the fact that processes correspond now to mixed sequences of transitions and states. The actual difference with our basic, uniform setting will amount to a stipulation that states form a special kind of events that possesses some additional properties.

The distinction between states and transitions is actually much more subtle than it appears. Intuitively, states correspond to temporally extended occurrences with a relatively stable temporal behavior. As was argued in Pratt (2003), though physical states are never completely stationary, we usually assume that during a given state time passes while information remains fixed. Accordingly, states on our understanding include relatively static properties and facts (though extended in time) as well as cases of inertial change.

In a commonsense structuring of the dynamic universe, states are usually seen as boundaries, or limits of transitions and change, in the same sense as points play the role of boundaries of linear geometric segments. And just as in geometry, this role is twofold. First, boundaries *separate* and thereby single out parts of the continuum. But on the other hand, they are *links* or junctions that combine these parts into larger pieces. This natural, *mereological* view of the continuum and its boundaries (see Bochman 1980) is not too familiar in our modern times, though it can be traced back again at least to Aristotle. We owe this situation to the predominant alternative representation, namely the point-based model of the continuum where these two functions are less transparent (though definable).

Now we will turn to a syntactic description. The above understanding of states and transitions is embodied in the following extension of our basic dynamic calculus.

Definition 10.11 *A dynamic consequence relation in a two-sorted language will be called an* ST-consequence *relation if it satisfies the following additional rules for static propositions (below, A denotes a static proposition):*

(S-Identity) $A \Rightarrow A$.
(S-Monotonicity) *If $ab \Rightarrow c$, then $aAb \Rightarrow c$.*
(S-Cut) *If $a \Rightarrow A$ and $aAb \Rightarrow c$, then $ab \Rightarrow c$.*
(S-Expansion) *If $a \Rightarrow bc$ and $ab \Rightarrow A$, then $a \Rightarrow bAc$.*
(S-Reduction) *If $a \Rightarrow bAc$, then $a \Rightarrow bc$.*

It has been shown that static propositions in such inference relations satisfy already all the usual structural rules, such as contraction, permutation, and dilution.

Relational models for sequential ST-consequence relations can be defined as a straight-forward extension of general relational models.

Definition 10.12 *A relational ST-model is a relational model* $M = (S, L, \mathcal{F})$, *in which the valuation function* $\mathcal{F}$ *satisfies the following constraint: for any state proposition* $A \in St$, $\mathcal{F}(A) \subseteq Id$, *where* $Id = \{(s, s) \mid s \in S\}$.

A completeness of the above axiomatization with respect to this semantics has been established in Bochman and Gabbay (2012a).

Remark. Already the above axiomatization contains clear indications that the dynamic generalization of causality should be accompanied by a proper understanding of the difference between static conditions and dynamic actions or events. According to both the commonsense view of the world and Newtonian physics, continuation (persistence) of static conditions and, more generally, inertial motion do not require causal explanation, which is reflected in the above formalism by accepting the rule of S-Identity, though only for static conditions. Only true actions and events, including creation of conditions, can and should be caused. This view should remind the reader the relative notion of causality that has been described in chapter 9, where the causal assertions has been made relative to background conditions.

10.2   Sequential Dynamic Logic

Our aim in this section will consist in augmenting the underlying language of dynamic causal inference with logical connectives. This will raise our abstract structural formalism to a full-fledged logic.

We introduce below sequential dynamic logic (SDL) that will subsume a propositional part of dynamic predicate logic (DPL), and will be shown to be expressively equivalent to propositional dynamic logic (PDL). Completeness of SDL with respect to the intended relational semantics will be established.

The language of sequential dynamic logic will include propositional atoms of two kinds: dynamic transitions (or actions) and static atoms (or tests). The syntax of the language will be formed, however, by using the usual connectives of conjunction, disjunction and negation. The semantics of this language will be given in the framework of labeled transition models that are Kripke structures in which each propositional atom is interpreted as a binary relation. These models will determine also the intended notion of dynamic inference for our language.

We will show first that the resulting formalism of sequential dynamic logic is equivalent in expressive power to the modal formalism of propositional dynamic logic. This will be established by defining back and forth translations between the two systems.

As a next step, we will introduce a substructural sequent calculus that will provide a logical formalization of SDL and prove the corresponding completeness theorem.

Finally, we will extend the language of SDL with an iteration connective and thereby obtain a sequential counterpart of the full PDL. We will prove the corresponding completeness result by using (a simplified version of) a filtration method.

10.2.1 The Language and Semantics

The vocabulary of SDL contains propositional atoms of two kinds, *transitions* (alias actions, or programs) $\Pi = \{\pi, \rho, \dots\}$ and *static* atoms (also called fluents or tests) $P = \{p, q, \dots\}$. Logical formulas $\{A, B, C, \dots\}$ will be constructed from the set of propositional atoms (of either kind) using three connectives: (noncommutative) *conjunction* $\wedge$, (dynamic) *negation* $\sim$, and (ordinary) *disjunction* $\vee$. Semantic interpretation of these formulas will be given in the framework of a relational semantics.

The following definition describes a variant of the well-known notion of a relational model (see, e.g., van Benthem 1996).

Definition 10.13 *A transition model is a triple $M = (S, \Pi, P, \mathcal{F})$, where S is a set of states, Π and P are sets of atomic propositions, and $\mathcal{F}$ a valuation function assigning each atomic proposition a binary relation on S, subject to the constraint that, for any static atom $p \in P$, $\mathcal{F}(p) \subseteq Id$, where $Id = \{(s, s) \mid s \in S\}$.*

According to the above description, static propositions can be seen as a particular kind of propositions that possess some specific properties. This difference will be captured in the calculus described later by assigning special structural rules to static propositions.

Propositional atoms are interpreted in a transition model as relations among states. In this sense our language can be viewed as a particular language of (binary) relations, immediately raising the question of what is an appropriate *logical language* for describing dynamic domains. The trade-off between expressivity and complexity is especially pressing for relational languages since "maximal" choices in expressivity (such as a full relational algebra) lead to undecidability.

As a reasonable guidance in the choice of the logical operators in our setting we can make use of the notion of *safety for bisimulation* (see van Benthem 1996). Roughly, an operation on relations is safe for bisimulation if it preserves the relation of bisimulation between models that are bisimilar with respect to its argument relations. Now, the following safety theorem has been proved by van Benthem (see van Benthem 1996, theorem 5.17).

Theorem 10.23 (Safety Theorem) *Any first-order relational operation is safe for bisimulation iff it can be defined using atomic relations and atomic tests, using three operations—composition, dynamic negation, and union.*

The above three operations correspond, respectively, to the three connectives of our logic, namely conjunction $\wedge$, dynamic negation $\sim$, and disjunction $\vee$. These connectives can be interpreted in a transition model as follows:

- $s, t \models \kappa$ iff $(s, t) \in \mathcal{F}(\kappa)$, for any propositional atom κ;

- $s, t \vDash A \wedge B$ iff   there exists r with $s, r \vDash A$ and $r, t \vDash B$;
- $s, t \vDash \sim A$ iff $s = t$ and there is no r such that $s, r \vDash A$;
- $s, t \vDash A \vee B$ iff $s, t \vDash A$ or $s, t \vDash B$;

The above definitions extend, in effect, the valuation function $\mathcal{F}$ to all formulas of the language. The following properties of the connectives are immediate from these definitions.

The definition for conjunction $\wedge$ amounts to the equality $\mathcal{F}(A \wedge B) = \mathcal{F}(A) \circ \mathcal{F}(B)$, where $\circ$ is the usual composition of binary relations. This conjunction is of course non-commutative, but it is an associative connective, so we can safely omit parentheses in writing multiple conjunctions. Note also that the conjunction of static propositions collapses to the usual classical conjunction (including commutativity).

The condition for disjunction $\vee$ amounts to the equality $\mathcal{F}(A \vee B) = \mathcal{F}(A) \cup \mathcal{F}(B)$. Thus, it is a fully standard disjunction—it is associative, commutative, and idempotent.

The dynamic negation $\sim$ is a most specific connective of our language. Thus, negated proposition is always a subset of *Id*. In other words, negation determines a static proposition by its very definition, independently of its argument. This property will be reflected in our calculus by extending the notion of a static proposition to all negated formulas. Note also that, when restricted to static propositions, $\sim$ behaves exactly as a classical negation. In particular, this implies that, like an intuitionistic negation, the dynamic negation satisfies the triple negation law: $\sim\sim\sim A \equiv \sim A$.

Speaking more generally, from the semantic point of view, any logical combination of static propositions in our language will be a static proposition. Moreover, restricted to static propositions, the three connectives of our language generate the classical Boolean logic. In this sense the classical logic can be justifiably viewed as a "static fragment" of sequential dynamic logic.

In accordance with this, we will (recursively) define a *static proposition* as either a static atom, or a negated formula, or else a logical combination of static propositions. A static proposition that is not a logical combination of other static propositions will be called a *static literal*. Note, however, that a static literal can be an arbitrary complex (negated) formula.

Now we are going to define the dynamic entailment relation determined by a transition model.

For any binary relation R, $dom(R)$ will denote its domain, and $range(R)$ – its range. In addition, we will canonically extend the valuation function $\mathcal{F}$ to sequences of propositions:

$$\mathcal{F}(A_1 \dots A_n) = \mathcal{F}(A_1) \circ \cdots \circ \mathcal{F}(A_n),$$

where $\circ$ denotes the composition of binary relations.

The entailment relation of our logic will be based on dynamic causal rules of the form $a \Rightarrow A$, where A is a formula and a a sequence of formulas. The intended informal meaning of such rules will be taken to be that a process a *causes* an event A.

Definition 10.14 *A dynamic rule $a \Rightarrow A$ will be said to be* valid *in a relational model M if* $range(\mathcal{F}(a)) \subseteq dom(\mathcal{F}(A))$.

$\Rightarrow_M$ *will denote the set of dynamic rules valid in a relational model M.*

Unfolding the above definition, $a \Rightarrow A$ is valid in M if and only if, for any states $s, t \in S$ such that $(s, t) \in \mathcal{F}(a)$, there exists a state r such that $(t, r) \in \mathcal{F}(A)$. This notion of validity corresponds precisely to the inference relation adopted in the dynamic predicate logic of Groenendijk and Stokhof (1991).

It should be noted, however, that our language is slightly more expressive than DPL in that it contains a general disjunction connective $\vee$. In fact, (the propositional part of) DPL can be identified with the sublanguage of our language determined by $\{\wedge, \sim\}$. Still, DPL has a "static" counterpart of our disjunction, defined as $\sim(\sim A \wedge \sim B)$, and it can be easily verified that the latter coincides with our disjunction on static propositions.

In a few works, DPL has been studied from a logical point of view. Thus, Hollenberg (1997) provided a finite equational axiomatization of the corresponding relational algebra (including the extensions with disjunction and tests) while Blackburn and Venema (1995) gave a modal analysis of the corresponding dynamic implication (described below). The paper (van Eijck 1999) is more closely related to our representation, since it provides an axiomatization of dynamic predicate logic (for the first-order language) in a format very similar to ours (see below).

In what follows, we will occasionally use a few derived connectives definable in our language:

- *Dynamic implication $A \to B$, defined as $\sim(A \wedge \sim B)$.*

  A direct semantic definition of dynamic implication is as follows:

  – $s, t \vDash A \to B$ iff $s = t$ and for every r with $s, r \vDash A$, there exists r_0 with $r, r_0 \vDash B$.

  Dynamic implication constitutes a propositional counterpart of our dynamic inference in the same sense as the classical, material implication forms a propositional counterpart of classical inference. In other words, it satisfies the deduction theorem:

  $$aA \Rightarrow_M B \text{ iff } a \Rightarrow_M A \to B.$$

  Consequently, it can be used, if desired, to transform our rule-based sequential logic into a Hilbert-type propositional calculus.
- *Inconsistency $\perp$, defined as $\sim A \wedge A$, where A is an arbitrary proposition.*[6] As can be seen from the semantic interpretation, $\perp$ corresponds to the empty relation $\emptyset$. Note that since $\emptyset \subseteq Id$, the constant $\perp$ can be safely viewed as a static *falsity* constant.
- *Truth $\top$, defined as $\sim\perp$.* As follows from the definition, $\top$ denotes *Id*, so it is an ordinary static truth constant.

---

6. Note that the order is important here since $A \wedge \sim A$ may well be consistent.

10.2.2 SDL versus PDL

In this section we are going to show that the language of sequential dynamic logic is expressively equivalent to the language of propositional dynamic logic. More precisely, we will establish (polynomial) translations from each of the languages to another that preserve the respective entailment relations.

The proofs of all the results in this section that are not given in the text can also be found in the appendix to this chapter.

Just as the language of our SDL, the language of PDL (see, e.g., Harel, Kozen, and Tiuryn 2000) is based on two sets of atomic expressions, *programs* $\Pi = \{\pi, \rho, \dots\}$ and *tests* $P = \{p, q, \dots\}$. The language itself, however, involves a construction of two separate kinds of expressions, *formulas* and *programs*, each with connectives of its own, that are defined by mutual recursion:

Formulas Any atomic test is a formula, and if ϕ, ψ are formulas and α is a program, then
$\phi \& \psi$, $\neg \phi$ and $[\alpha]\phi$ are formulas.

Programs Any atomic program is a program, and if α, β are programs and ϕ a formula, then $\alpha; \beta$, $\alpha \cup \beta$ and $?\phi$ are programs.

The entailment relation is defined in PDL only for formulas, so the logical properties of programs are established only indirectly by their functioning as parts of formulas.

The models of PDL can be described as follows:

Definition 10.15 *A PDL model is a tuple* $M = (S, P, \Pi, V, R)$, *where S is a set of states, P is a set of atomic propositions, while* Π *is a set of atomic programs, V is a valuation function assigning each proposition from P a subset of S, and R is a function assigning each atomic program* α *a binary relation* $R(\alpha)$ *on S.*

The valuation V is extended to all *formulas* of PDL using the following definitions:

- $s \vDash p$ iff $s \in V(p)$;
- $s \vDash \phi \& \psi$ iff $s \vDash \phi$ and $s \vDash \psi$;
- $s \vDash \neg \phi$ iff $s \nvDash \phi$;
- $s \vDash [\alpha]\phi$ iff $t \vDash \phi$, for any t such that $(s, t) \in R(\alpha)$,

where the function R is extended to all *programs* as follows:

$$R(\alpha; \beta) = R(\alpha) \circ R(\beta)$$

$$R(\alpha \cup \beta) = R(\alpha) \cup R(\beta)$$

$$R(?\phi) = \{(s, s) \mid M, s \vDash \phi\}.$$

Finally the entailment relation of PDL has the form $\Gamma \vDash \phi$, where ϕ is a formula and Γ a finite set of formulas.

Definition 10.16 *A rule* $\Gamma \vDash \phi$ *is* valid *in a model M (notation* $\Gamma \vDash_M \phi$*) if for every state s,* $s \vDash \psi$ *for every* $\psi \in \Gamma$ *implies* $s \vDash \phi$.

$\vDash_M^{pdl}$ *will denote the set of all such rules that are valid in M.*

As a first step in the comparison between the two languages, let us note that the respective descriptions of transition models for SDL and PDL are actually notational variants of each other. Indeed, given an SDL model $(S, \Pi, P, \mathcal{F})$, we can split the valuation function $\mathcal{F}$ into two functions R and V_0 obtained by restricting the domain of $\mathcal{F}$ to atomic transitions and state atoms, respectively. Moreover, due to the constraint on $\mathcal{F}$, the function V_0 is uniquely determined by a function $V : P \mapsto \mathcal{P}(S)$, defined as $V(p) = \{s \mid (s, s) \in V_0(p)\}$. As a result, we obtain a PDL model (S, P, Π, V, R). Conversely, given a PDL model (S, P, Π, V, R), the corresponding SDL model is obtainable by defining a valuation function $\mathcal{F}$ that coincides with R on Π, while, for every $p \in P$, $\mathcal{F}(p)$ is taken to be $\{(s, s) \mid s \in V(p)\}$. Clearly, $(S, P, \Pi, \mathcal{F})$ will be an SDL model according to our definition.

Taking for granted the above correspondence between SDL models and PDL models, we will establish now translations between the two languages that will preserve the respective entailment relations.

From SDL to PDL. To begin with, the following translation δ transforms any SDL formula into a program of PDL:

$$\delta(\pi) = \pi$$

$$\delta(p) = ?p$$

$$\delta(A \wedge B) = \delta(A); \delta(B)$$

$$\delta(\sim A) = ?([\delta(A)]\bot)$$

$$\delta(A \vee B) = \delta(A) \cup \delta(B).$$

The following lemma shows that the translation δ is faithful with respect to the respective semantic interpretations in transition models:

Lemma 10.24 *For any SDL-formula A, and for every model M,*

$$\mathcal{F}(A) = R(\delta(A)).$$

Proof. By an easy induction on the complexity of A. $\qquad\square$

As a final step, we will adopt a variant of a "global" translation for dynamic inference rules described in van Benthem (1996). A dynamic causal rule $A_1 A_2 \ldots A_n \Rightarrow A$ of SDL will correspond to the following formula of PDL:

$$[\delta(A_1); \delta(A_2); \ldots; \delta(A_n)]\langle\delta(A)\rangle\top.$$

The next result shows that this transformation preserves the dynamic inference relation of SDL:

Theorem 10.25 $A_1A_2 \ldots A_n \Rightarrow_M A$ *holds if and only if the formula*

$$[\delta(A_1); \delta(A_2); \ldots; \delta(A_n)]\langle \delta(A) \rangle \top$$

is PDL-valid in M.

Proof. The formula $[\delta(A_1); \delta(A_2); \ldots; \delta(A_n)]\langle \delta(A) \rangle \top$ is valid in M iff for any states s, t such that $(s, t) \in R(\delta(A_1); \delta(A_2); \ldots; \delta(A_n))$, there exists r such that $(t, r) \in R(\delta(A))$. By the preceding lemma, this is equivalent to the claim that for any $(s, t) \in \mathcal{F}(A_1 \ldots A_n)$, there exists $(t, r) \in \mathcal{F}(A)$, which amounts to $A_1A_2 \ldots A_n \Rightarrow_M A$. Hence the result. $\qquad \square$

From PDL to SDL. The reverse translation from PDL to SDL can be performed in one step. The following translation τ simultaneously transforms both formulas and programs of PDL to SDL formulas.

$$\tau(\kappa) = \kappa, \text{ for any atom } \kappa \in \Pi \cup P$$

$$\tau(\phi \,\&\, \psi) = \tau(\phi) \wedge \tau(\psi)$$

$$\tau(\neg\phi) = \sim\tau(\phi)$$

$$\tau([\alpha]\phi) = \tau(\alpha) \to \tau(\phi)$$

$$\tau(\alpha; \beta) = \tau(\alpha) \wedge \tau(\beta)$$

$$\tau(\alpha \cup \beta) = \tau(\alpha) \vee \tau(\beta)$$

$$\tau(?\phi) = \tau(\phi).$$

First, we have the following technical result (see the appendix for the proof).

Lemma 10.26 *For any PDL formula ϕ, any program α, and any model M,*

1. $\tau(\phi)$ *is a static proposition, that is, $\mathcal{F}(\tau(\phi)) \subseteq Id$.*
2. $s \in V(\phi)$ *iff $(s, s) \in \mathcal{F}(\tau(\phi))$.*
3. $R(\alpha) = \mathcal{F}(\tau(\alpha))$.

As a conclusion, the next result shows that the translation τ preserves the inference relation of PDL.

Theorem 10.27 $\phi_1, \ldots, \phi_n \vDash_M^{pdl} \psi \quad$ iff $\quad \tau(\phi_1) \ldots \tau(\phi_n) \Rightarrow_M \tau(\psi)$.

The above two translations show that SDL and PDL are equivalent formal systems already at the level of semantics.

It turns out that, under this general correspondence, arbitrary PDL formulas will correspond precisely to static formulas of SDL. The latter correspondence can be described more precisely as follows. Recall that PDL formulas are constructed from static atoms using the classical connectives and the modal construct $[\pi]F$, which corresponds to the implication $\pi \to F$ in the SDL notation. Accordingly, let us say that a static SDL formula is of *PDL type* if it is constructed recursively from static atoms using the propositional connectives and implications of the form $\pi \to \phi$, where π is an action atom and ϕ a PDL-type formula. Then (using the above mutual translations between SDL-formulas and PDL-formulas and programs) it can be shown first that any static formula of SDL can be transformed into an equivalent PDL-formula, and then the latter can be transformed back to a static SDL-formula of PDL type. We will use this correspondence in what follows.

Remark. The above correspondence between the languages of SDL and PDL could be viewed as a dynamic generalization of the correspondence between plain causal rules $A \Rightarrow B$ and their associated modal formulas $A \to \Box B$ (this is a consequence of lemma 4.29 in chapter 4). Indeed, by this correspondence, dynamic causal rules of the form $F\pi \Rightarrow F_1$ correspond to basic modal formulas of dynamic logic, namely *Hoare* formulas of the form $F \to [\pi]F_1$ (see Harel, Kozen, and Tiuryn 2000).

10.2.3 Sequential Dynamic Calculus

Small letters $a, b, c, \ldots$ will denote finite sequences of formulas. Such sequences will be called *processes*. As usual, ab will denote the concatenation of sequences a and b (and similarly for aA, aAb, etc.).

As is customarily in sequent calculi, we will use both ordinary rules of the form $a \Rightarrow A$ and special rules $a \Rightarrow$ with an empty succedent (meaning that a is *inconsistent*). Note, however, that just as in the classical case, such rules are not strictly needed since they are equivalent to $a \Rightarrow \bot$, where $\bot$ is an arbitrary contradiction. As we mentioned earlier, a formula $\sim A \wedge A$ can play this role in our case. We will use a common description $a \Rightarrow X$ for rules of both kinds. In other words, X below is either empty or a single formula.

In the rules below, we will use $\{\phi, \psi, \ldots\}$ to denote static propositions.

A *dynamic consequence relation* is a set of causal rules $a \Rightarrow X$ that is closed with respect to the following inference rules:

- General structural rules:

$$\frac{a \Rightarrow X}{Aa \Rightarrow X} \quad \text{Left Monotonicity} \qquad \frac{a \Rightarrow}{aA \Rightarrow} \quad \text{Right Monotonicity}$$

- Structural rules for static propositions:

$$\frac{}{\phi \Rightarrow \phi} \quad \text{S-Reflexivity}$$

$$\frac{ab \Rightarrow X}{a\phi b \Rightarrow X} \quad \text{S-Monotonicity} \qquad \frac{a \Rightarrow \phi \quad a\phi b \Rightarrow X}{ab \Rightarrow X} \quad \text{S-Cut}$$

- Rules for negation:

$$\frac{a \sim A \Rightarrow}{a \Rightarrow A} \; (\sim L) \qquad \frac{a \Rightarrow \sim A}{aA \Rightarrow} \; (\sim R)$$

- Rules for conjunction:

$$\frac{aABb \Rightarrow X}{a\,A{\wedge}B\,b \Rightarrow X} \quad (\wedge)$$

- Rules for disjunction:

$$\frac{aAb \Rightarrow X \quad aBb \Rightarrow X}{a\,A{\vee}B\,b \Rightarrow X} \quad (\vee)$$

The double line in the above rules for connectives indicates that the rule is valid in both directions. The reader should note the similarity of these rules with the inference rules for connectives in classical Scott consequence relations that have been used in chapter 2 (see section 2.3.2).

Already the very form of our sequents involving *sequences* of propositions (instead of usual sets) indicates that the sequential dynamic logic is substructural, that is, it does not satisfy, in general, the usual structural rules for consequence relations such as contraction, permutation, and weakening. Nevertheless, all these structural rules can be shown to hold for static propositions.

Lemma 10.28 *If ϕ, ψ are static propositions, then the following rules hold for dynamic consequence relations:*

$$\frac{a\,\phi\,\phi\,b \Rightarrow X}{a\,\phi\,b \Rightarrow X} \quad \text{S-Contraction} \qquad\qquad \frac{a\,\phi\,\psi\,b \Rightarrow X}{a\,\psi\,\phi\,b \Rightarrow X} \quad \text{S-Permutation.}$$

Proof. (1) $a\,\phi \Rightarrow \phi$ by S-Reflexivity and Left Monotonicity, so if $a\,\phi\,\phi\,b \Rightarrow X$, then $a\,\phi\,b \Rightarrow X$ is derivable by S-Cut.

(2) $\psi \Rightarrow \psi$ by S-Reflexivity, so $\psi\,\phi \Rightarrow \psi$ by S-Monotonicity, and hence $a\,\psi\,\phi \Rightarrow \psi$ by Left Monotonicity. Now if $a\,\phi\,\psi\,b \Rightarrow X$ holds, then $a\,\psi\,\phi\,\psi\,b \Rightarrow X$ by S-Monotonicity. Together with $a\,\psi\,\phi \Rightarrow \psi$, this gives precisely $a\,\psi\,\phi\,b \Rightarrow X$ by S-Cut. $\square$

Remark. It has been shown in van Benthem (1996, proposition 7.4) that in a sequential setting, the rules S-Reflexivity, S-Monotonicity, S-Cut, and S-Contraction, viewed as rules that hold for all propositions, completely determine the structural properties of classical inference. However, in view of the above result, S-Contraction is derivable from the rest of the rules, so our static structural rules are sufficient for determining a "classical static subinference" inside a general causal dynamic inference.

The paper (van Eijck 1999) contains an axiomatization of the first-order DPL which is closely related to the above formalism. Using our notation, van Eijck's axiomatization includes Left Monotonicity, S-Reflexivity, and S-Permutation, two rules for negation introduction (on left and right), and two rules for double negation elimination (on both sides). It can be shown that these rules for negation are equivalent to our rules. In addition, van Eijck's axiomatization includes "first-order" rules that deal with variables and quantifiers. Two of these rules, however, also have a propositional import, namely Transitivity and

Right Conjunction:

$$\frac{A \Rightarrow B \quad B \Rightarrow C}{A \Rightarrow C} \quad \text{Transitivity} \quad \frac{A \Rightarrow B \quad A \Rightarrow C}{A \Rightarrow B \wedge C} \quad \text{R-Conjunction.}$$

These rules hold in van Eijck's system only under certain restrictions on the variables occurring in the relevant formulas. The restrictions are satisfied, however, when B in each rule above is a static proposition, in which case the above rules are derivable in our formalism.

The following admissible rules and properties of our calculus will be needed for the proof of the completeness theorem.

Lemma 10.29 1. $\sim\!A\,A \Rightarrow$.

2. $aA \Rightarrow$ *if and only if* $a \sim\!\sim\!A \Rightarrow$.

3. *For any static proposition* ϕ, *if* $a\,\phi\,b \Rightarrow$ *and* $a \sim\!\phi\,b \Rightarrow$, *then* $ab \Rightarrow$.

Strong completeness.   In this section we are going to show that dynamic consequence relations are strongly complete for the relational semantics.

To begin with, we have the following.

Lemma 10.30 *If* M *is a transition model, then* $\Rightarrow_M$ *is a dynamic consequence relation.*

The proof amounts to a straightforward check of all the rules of a dynamic consequence relation. In order to show that dynamic consequence relations are complete for this semantics, we are going to construct a canonical transition model of a consequence relation.

As a preparation, we will introduce the following key notion:

Definition 10.17

- A trace *is finite sequence* $T = X_1 X_2 \ldots X_n$, *where each* X_i *is either a formula, or a (possibly infinite) set of static propositions.*
- *An* instantiation *of a trace* $T = X_1 X_2 \ldots X_n$ *is a process* $x_1 x_2 \ldots x_n$ *such that, for each* $i \leq n$, *if* X_i *is a single formula, then* $x_i = X_i$ *or else* x_i *is a finite sequence of (static) propositions taken from the set* X_i.
- *A trace* T *will be said to be* inconsistent *with respect to a dynamic consequence relation* $\Rightarrow$ *(notation* $T \Rightarrow$*) if it has an inconsistent instantiation; otherwise it will be called* consistent *(notation* $T \not\Rightarrow$*).*

Now, given a dynamic consequence relation $\Rightarrow$, we will construct a *canonical transition model* $M_\Rightarrow = (S_\Rightarrow, \Pi, P, \mathcal{F})$ as follows:

- The set of states $S_\Rightarrow$ is a set of all maximal consistent sets of static propositions (viewed as one-element traces).
- For any $p \in \Pi \cup P$, $\mathcal{F}(p) = \{(X, Y) \mid X, Y \in S_\Rightarrow \ \& \ XpY \not\Rightarrow\}$.

We have to verify first that the above construction satisfies the "static" constraint of a transition model.

Lemma 10.31 $M_\Rightarrow$ *is a transition model.*

Proof. Let α be a static atom, and assume that $X\alpha Y \not\Rightarrow$. Then $X\alpha \not\Rightarrow$ (note that all instantiations of $X\alpha$ are instantiations of $X\alpha Y$), and therefore $\alpha \in X$, since X is a maximal consistent set of static propositions. Consequently all instantiations of $X\alpha Y$ are instantiations of XY, and hence $XY \not\Rightarrow$. The latter implies that $X \cup Y$ is a consistent trace (by S-Contraction and S-Permutation), and therefore $X = Y$ due to maximality of X and Y. Hence $\mathcal{F}(\alpha) \subseteq Id$, as required. □

As in classical constructions of canonical models for modal logics, the following lemma shows that the above canonical valuation function can be uniformly extended to all formulas of the language. The proof can be found in the appendix.

Lemma 10.32 (Basic Lemma) *In a canonical model $M_\Rightarrow$, for any formula A and any states $X, Y \in S_\Rightarrow$,*

$$X, Y \vDash A \text{ if and only if } XAY \not\Rightarrow .$$

The basic lemma gives us everything we need for proving the main theorem.

Theorem 10.33 *If $M_\Rightarrow$ is a canonical transition model of a dynamic consequence relation $\Rightarrow$, then $\Rightarrow \; = \; \Rightarrow_{M_\Rightarrow}$.*

Proof. To begin with, for a proposition A and a sequence of propositions a, let A_0 denote the conjunction of the sequence $a \sim A$. Then it is easy to verify that $a \Rightarrow_{M_\Rightarrow} A$ does not hold if and only if there exist states $X, Y \in S_\Rightarrow$ such that $X, Y \vDash A_0$. By the basic lemma, the latter is equivalent to $XA_0Y \not\Rightarrow$. By definition, this implies $A_0 \not\Rightarrow$. Moreover, if $A_0 \not\Rightarrow$ holds, then (using [3] of lemma 10.29) A_0 can be extended to a consistent trace XA_0Y, where X and Y are maximal consistent sets of static propositions. Therefore, $a \Rightarrow_{M_\Rightarrow} A$ holds if and only if $A_0 \Rightarrow$. But A_0 is a conjunction of $a \sim A$, so $A_0 \Rightarrow$ is equivalent to $a \sim A \Rightarrow$ by the rules for conjunction while the latter is equivalent to $a \Rightarrow A$ by ($\sim L$). This completes the proof. □

As a summary, we conclude with the following.

Corollary 10.34 $\Rightarrow$ *is a dynamic consequence relation if and only if $\Rightarrow \; = \; \Rightarrow_M$, for some transition model M.*

The above corollary provides a formal expression for a strong completeness of SDL with respect to transition models.

10.2.4 Iteration

So far, we have intentionally omitted iteration from our language for a number of reasons. First, the iteration operation on programs is of course essential for programming and computation, but it appears to be less needed in applications outside computation theory. Second, it is not expressible in our formalism already because it is not a bisimulation-safe operation (due to van Benthem's safety theorem). Moreover, the iteration operation describes transitive closures of relations, which are not first-order definable, and this makes it a difficult operation from a proof-theoretical point of view. Thus, one of the immediate effects of adding iteration to the language of PDL makes the latter a non-compact formalism. More precisely, the corresponding relational semantics makes valid some irreducibly infinite inference rules (Renardel de Lavalette, Kooi, and Verbrugge 2008). Due to the two-way correspondence between SDL and PDL established earlier, this will hold also for our formalism.

Despite all said above, we are going to show that SDL can also be extended with iteration in a relatively transparent way and that the corresponding formalization and completeness proof are no more complex (and often significantly less so) than in PDL.

Semantics and axiomatization. We extend the language of SDL with an *iteration* connective * that has the following semantic interpretation in transition models:

- $s, t \vDash A^*$ iff there exists a finite sequence of A-transitions that connects s with t.

We will use SDL^* to denote the extended formalism. The corresponding sequential dynamic calculus for SDL^* can be defined as follows.

A *dynamic consequence relation with iteration* is a dynamic consequence relation that satisfies the following rules for the iteration connective:

$$\frac{a\,A^*b \Rightarrow X}{ab \Rightarrow X} \qquad \frac{a\,A^*b \Rightarrow X}{aAA^*b \Rightarrow X} \quad (^*E)$$

$$\frac{\phi A \Rightarrow \phi}{\phi A^* \Rightarrow \phi} \quad (^*I)$$

Remark. The formalism of SDL^* turns out to be surprisingly similar to the sequential system S described in Kozen and Tiuryn (2003). Ignoring some inessential differences concerning the choice of connectives, the only important difference between the formalisms themselves amounts to the fact that the language of S imposes certain syntactic restrictions on admissible formulas in conclusions of sequents. Roughly, it requires that only static propositions can be conclusions. In this sense, SDL^* can be seen as a conservative extension of S that is free of such language restrictions. On the other hand, the restrictions of S are not as severe as they seem since any sequent of SDL can be transformed into an equivalent sequent that satisfies the restrictions of S. We should mention, however, that the completeness proof for S, given in Kozen and Tiuryn (2003), relied on highly nontrivial algebraic results obtained elsewhere. In this respect, the completeness proof for SDL^*, given below, seems to fulfill an explicit request for a direct completeness proof made at the end of their paper.

Filtration and weak completeness. The general strategy of proving completeness for SDL with iteration below will be similar to the completeness proof for the "full" PDL given, for example, in Kozen and Parikh (1981) and Harel, Kozen, and Tiuryn (2000). This strategy can be described as follows.

To begin with, the basic lemma 10.32 ceases to hold for the language with iteration. As a consequence, the canonical transition model $M_\Rightarrow$ constructed earlier cannot serve as a model of the source consequence relation $\Rightarrow$, which blocks, in effect, the possibility of proving the corresponding strong completeness theorem. Still, we will prove below that a rule $a \Rightarrow X$ is derivable in SDL from a *finite* set of rules Δ if and only if any model of Δ validates also $a \Rightarrow X$. Moreover, this *weak completeness* claim can be refined by restricting the set of transition models to finite models only. As for PDL, the way of proving this will consist in filtering the model $M_\Rightarrow$ with respect to a certain finite set of formulas determined by Δ and $a \Rightarrow X$. In our case, this set is defined as follows.

Definition 10.18 *A Fischer-Ladner closure of a set Q of SDL formulas is the least set $FL(Q)$ of formulas containing Q and such that:*

- *$FL(Q)$ is closed under subformulas;*
- *$FL(Q)$ is closed under single negations;*
- *If $(A \wedge B) \wedge C \in FL(Q)$ then $A \wedge (B \wedge C) \in FL(Q)$;*
- *If $A \in FL(Q)$ then $A \wedge \top \in FL(Q)$;*
- *If $A^* \wedge B \in FL(Q)$, then $A \wedge (A^* \wedge B) \in FL(Q)$; and*
- *If $(A \vee B) \wedge C \in FL(Q)$, then $A \wedge C \in FL(Q)$ and $B \wedge C \in FL(Q)$.*

Just as for the PDL language, both the number and the size of the formulas in $FL(Q)$ are linearly bounded by the size of Q. In particular, if Q is finite, $FL(Q)$ is also finite.

Let $M_\Rightarrow$ be a canonical model of a consequence relation $\Rightarrow$ and Q a finite set of formulas. We define a finite transition model M_Q, called a *filtration* of M by $FL(Q)$, as follows.

Let us say that states X and Y from $M_\Rightarrow$ are *equivalent*, if $X \cap FL(Q) = Y \cap FL(Q)$. Note that only static formulas in $FL(Q)$ are relevant for this relation. For any state X, $[X]$ will denote the equivalence class containing X. Then we define M_Q as a transition model $(S_Q, \Pi \cup P, \mathcal{F}_Q)$, where

- $S_Q = \{[X] \mid X \in S\}$;
- $\mathcal{F}_Q(\kappa) = \{([X], [Y]) \mid X\kappa Y \not\Rightarrow\}$, for any $\kappa \in \Pi \cup P$.

Then the following key result can be established.

Lemma 10.35 (Filtration Lemma) *For any states X, Y from $S_\Rightarrow$,*

(i)   *For any static $\phi \in FL(Q)$, $[X], [X] \vDash_{M_Q} \phi$ iff $\phi \in X$;*

(ii) (a)   *For any other formula $A \in FL(Q)$, if $XAY \not\Rightarrow$, then $[X], [Y] \vDash_{M_Q} A$;*

(b)   *If $[X], [Y] \vDash_{M_Q} A$ and $\sim(A \wedge B) \in X \cap FL(Q)$, then $\sim B \in Y$.*

The above filtration lemma provides the main step in proving the corresponding completeness result.

Due to the Horn form of the inference rules characterizing a dynamic consequence relation, intersections of dynamic consequence relations are again dynamic consequence relations. Consequently, for any set Δ of causal rules there exists a least dynamic consequence relation containing it. We will denote this consequence relation by $\Rightarrow_\Delta$. If Δ is finite, then it can be easily verified that $\Rightarrow_\Delta$ is the set of all sequents derivable from Δ. In other words, $a \Rightarrow_\Delta A$ holds if and only if $a \Rightarrow A$ is derivable from Δ using the rules for a dynamic consequence relation.

Theorem 10.36 (Weak Completeness Theorem) *If Δ is a finite set of sequents, then $a \Rightarrow_\Delta A$ iff $a \Rightarrow A$ is valid in every finite transition model that validates Δ.*

Proof. Let M_Δ be the canonical model of $\Rightarrow_\Delta$ (its description precedes lemma 10.31). Let Q denote the following (finite) set of formulas:

$$Q = \{ \wedge a_i \to A_i \mid a_i \Rightarrow A_i \in \Delta \} \cup \{ \wedge a \to A \}.$$

Finally, let M_Q be the finite transition model obtained by filtration of M_Δ by $L(Q)$.

Assume that $a \not\Rightarrow_\Delta A$. Since any dynamic sequent $b \Rightarrow B$ is equivalent to $\Rightarrow \wedge b \to B$, we have that if $a_i \Rightarrow A_i \in \Delta$, then $\wedge a_i \to A_i$ is included in every state from M_Δ, but $\wedge a \to A$ does not belong to at least one state of M_Δ. Now, due to (i) of the filtration lemma, the same will hold for the model M_Q, which means that M_Q validates Δ but does not validate $a \Rightarrow A$. This completes the proof. $\Box$

It has been shown above that the causal formulation of dynamic logic provides a transparent and convenient representation of dynamic inference. Hopefully, we have persuaded the reader that dynamic logic can be naturally viewed as a certain causal logic. Moreover, this causal logic is not an extension of classical logic with new operators but a direct *generalization* of the latter, suitable for dynamic discourses.

Of course, the difference between the causal SDL and traditional dynamic logic should not be overestimated; witness the mutual translations between the two established above. Still, the very possibility or, more precisely, the viability of the causal approach to describing dynamic domains obviously suggests new perspectives and new directions of research.

10.A   Appendix: Proofs of the Main Results

Proofs for section 10.1.

The following simple lemma provides a direct syntactic description of the least theory containing a single process. It will be used in what follows.

Lemma (A) $\mathrm{Th}(a) = \{ ab \mid a \Rightarrow b \}.$

Proof. Let T_a denote the set $\{ab \mid a \Rightarrow b\}$. It is easy to see that any theory containing a should include also T_a. Note also that $a \in T_a$ since $a \Rightarrow \epsilon$ by Identity and Left Monotonicity. Hence it is sufficient to show that T_a is a theory. Assume that $c \Rightarrow d$ and $sc \in T_a$ for some process s. Then $sc = ab$, for some b such that $a \Rightarrow b$. Now, $c \Rightarrow d$ implies $sc \Rightarrow d$ by Left Monotonicity, and hence $ab \Rightarrow d$. Therefore $a \Rightarrow bd$ by Cut, and consequently $abd = scd \in T_a$. Thus, T_a is a theory of $\Rightarrow$. This completes the proof. □

Lemma (10.1) $\Rightarrow_D$ *is a dynamic consequence relation.*

Proof. Identity is immediate.

Left Monotonicity. If $Aa \not\Rightarrow_D b$, then, for some dynamic model in D, there exist u and x such that $ux \in U, x \in V(Aa)$, but there is no y such that $uxy \in U$ and $y \in V(b)$. By definition, $x = zt$, where $z \in V(A)$ and $t \in V(a)$. Now if $u_1 = uz$, we have $u_1 t = ux \in U$ and $uxy = u_1 ty$ and consequently $a \not\Rightarrow_D b$.

Cut. If $a \not\Rightarrow_D bc$, then, for some dynamic model in D, there must exist u and x such that $ux \in U$, $x \in V(a)$, but there is no y such that $uxy \in U$ and $y \in V(bc)$. But if $a \Rightarrow_D b$, then $uxz \in U$ for some $z \in V(b)$, and hence $xz \in V(ab)$. On the other hand, if it were the case that $uxzy_0 \in U$ for some $y_0 \in V(c)$, then we would have $zy_0 \in V(bc)$, contrary to the supposition. Hence $ab \not\Rightarrow_D c$. □

Theorem (Representation Theorem 10.2) *If $\Rightarrow$ is a dynamic consequence relation, then* $\Rightarrow = \Rightarrow_{T_\Rightarrow}$.

Proof. We have to show that $a \Rightarrow b$ holds iff for any theory U of $\Rightarrow$ and any sequence s, if $sa \in U$, then $sab \in U$. The direction from left to right follows directly from the definition of a theory. In the other direction, we take s to be an empty sequence, and choose $U = \mathrm{Th}(a)$. Then $ab \in \mathrm{Th}(a)$ and therefore $a \Rightarrow b$ by lemma A. This completes the proof. □

Theorem (10.6) *If M_D is the relational model corresponding to the dynamic semantics D, then $\Rightarrow_D = \Rightarrow_{M_D}$.*

Proof. Note that, for any process a, $(s, t) \in F(a)$ iff $(s, t) = (x_U, y_U)$ for some x, y such that $y = xz$ for some $z \in F(a)$ and $y \preceq y_0$ for some $y_0 \in U \in D$.

Assume first that $a \Rightarrow_D b$ and $(s, t) \in F(a)$. Hence $(s, t) = (x_U, y_U)$ for some x, y such that $y = xz$ for some $z \in F(a)$ and $y \preceq y_0$ for some $y_0 \in U \in D$. Since U is an admissible set of a dynamic model in D, we have $xzz_1 = yz_1 \in U$ for some $z_1 \in F(b)$. Let $r = (yz_1)_U$. Then clearly $r \in \checkmark$ and $(t, r) \in F(b)$. This gives the direction from left to right.

Now assume that $a \Rightarrow_{M_D} b$ and $ux \in U \in D$ for some $x \in V(a)$. Then let us put $s = u_U$ and $t = (ux)_U$. Clearly $t \in \checkmark$ and $(s, t) \in F(a)$. Since $a \Rightarrow_{M_D} b$, there exists $r \in \checkmark$ such that $(t, r) \in F(b)$. By the definition of F, this can happen only if $r = (uxy)_U$ for some $y \in V(b)$, and therefore $uxy \in U$, which shows that $a \Rightarrow_D b$ holds. This completes the proof. □

Lemma (10.9) *If $\mathcal{D}$ is a sequential dynamic semantics, then $\Rightarrow_{\mathcal{D}}$ is a sequential consequence relation.*

Proof. We need only to check Right Anti-Monotonicity. If $a \not\Rightarrow_{\mathcal{D}} b$, then there exist a dynamic model and $s \in P$ such that $sx \in U$, for $x \in \mathcal{V}(a)$, and there is no y such that $sxy \in U$ and $y \in \mathcal{V}(b)$. Assume that $uxz \in U$, for some $z \in \mathcal{V}(bB)$. Then $z = z_1 z_2$, where $z_1 \in \mathcal{V}(b)$ and $z_2 \in \mathcal{V}(B)$, and hence $sxz_1 \in U$ by sequentiality, which contradicts the supposition. Therefore, $a \not\Rightarrow_{\mathcal{D}} bB$. $\square$

Theorem (10.13) *If $M_{\mathcal{D}}$ is a (plain) relational model corresponding to the homogeneous sequential semantics $\mathcal{D}$, then $\Rightarrow_{\mathcal{D}} = \Rightarrow_{M_{\mathcal{D}}}$.*

Proof. For any $s \in S$, $\hat{s}$ will denote the underlying path, that is, $\hat{u} = u$ and $\hat{u_U} = u$. Also, to simplify the notation, for a process a, we will use $[a]$ to denote $\mathcal{V}(a)$. Note, in particular, that if $(s, t) \in \mathcal{F}(a)$, then always $\hat{t} = \hat{s}[a]$.

Assume first that $a \Rightarrow_{\mathcal{D}} b$ and $(s, t) \in \mathcal{F}(a)$. We have to consider two cases. If $t \in P$, then let $r = t[b]$. Clearly, $r \in S$ and $(t, r) \in \mathcal{F}(b)$, as required. So assume now that $t = x_U$, for some $x \in U \in \mathcal{D}$. In this case we have $x = \hat{s}[a]$, and therefore $x[b] = \hat{s}[ab] \in U$ (since $a \Rightarrow_{\mathcal{D}} b$). Now we put $r = (x[b])_U$. Suppose that $b = B_1 \ldots B_m$. Then we define $t_i = (t[B_1 \ldots B_i])_U$, for every $1 \leq i < m$ (note that sequentiality secures that $\hat{t}_i \in U$). It is easy to see that $t\mathcal{F}(B_1)t_1 \ldots t_{m-1}\mathcal{F}(B_m)r$, which implies $(t, r) \in \mathcal{F}(b)$. This gives the direction from left to right.

Now assume that $a \Rightarrow_{M_{\mathcal{D}}} b$ and $s[a] \in U$. Let us put $t = (s[a])_U$. If $a = A_1 \ldots A_n$, we define $s_i = s[A_1 \ldots A_i]$ for any $1 \leq i < n$, and then we have $s\mathcal{F}(A_1)s_1 \ldots s_{n-1}\mathcal{F}(A_n)t$, which implies $(s, t) \in \mathcal{F}(a)$. Since $a \Rightarrow_{M_{\mathcal{D}}} b$, there exists $r \in S$ such that $(t, r) \in \mathcal{F}(b)$. By the definition of $\mathcal{F}$, this can happen only if $r = (s[a][b])_U$, and therefore $s[a][b] \in U$. This completes the proof. $\square$

Theorem (10.16) $\Rightarrow$ *is a deterministic consequence relation iff* $\Rightarrow = \Rightarrow_0^m$, *for some singular consequence relation* $\Rightarrow_0$.

Proof. For the direction from right to left, it is sufficient to show that if $\Rightarrow_0$ is a singular consequence relation, then $\Rightarrow_0^m$ is a deterministic consequence relation. Identity is immediate, while both Right Anti-Monotonicity and Cumulativity follow directly from the inductive construction. For the two remaining postulates of deterministic inference, we will show that $\Rightarrow_0^m$ is closed with respect to each by induction on the number of propositions in the consequents.

Left Monotonicity. Assume that $a \Rightarrow_0^m bB$. If the consequent is singular (that is, $b = \epsilon$), then $Aa \Rightarrow_0^m bB$ holds due to the fact that $\Rightarrow_0$ satisfies Left Monotonicity. Otherwise by construction of $\Rightarrow_0^m$, we have $a \Rightarrow_0^m b$ and $ab \Rightarrow_0^m B$. By the inductive assumption, we have $Aa \Rightarrow_0^m b$ while $Aab \Rightarrow_0^m B$ holds by Left Monotonicity for $\Rightarrow_0$. Therefore $Aa \Rightarrow_0^m bB$ by the construction.

Cut. Note first that $a \Rightarrow_0^m b$ and $ab \Rightarrow_0^m C$ imply $a \Rightarrow_0^m bC$ by the inductive construction itself. Assume now that $a \Rightarrow_0^m b$ and $ab \Rightarrow_0^m cC$. Then $ab \Rightarrow_0^m c$ and $abc \Rightarrow_0^m C$ by the inductive construction. The inductive assumption says that $a \Rightarrow_0^m b$ and $ab \Rightarrow_0^m c$ jointly imply $a \Rightarrow_0^m bc$ by Cut. Taken together with $abc \Rightarrow_0^m C$, this gives us $a \Rightarrow_0^m bcC$ by the inductive construction.

For the direction from left to right, assume that $\Rightarrow$ is deterministic, and let $\Rightarrow_0$ be the set of singular sequents from $\Rightarrow$. Clearly, $\Rightarrow_0^m$ is a singular consequence relation. Also, $\Rightarrow_0^m$ is included in $\Rightarrow$ (since $\Rightarrow$ is closed with respect to Cumulativity). Moreover, due to lemma 10.15, any sequent from $\Rightarrow$ can be obtained, ultimately, from the singular sequents of $\Rightarrow_0$ by applying the inductive construction. Thus, $\Rightarrow$ coincides with $\Rightarrow_0^m$. This completes the proof.   □

Lemma (10.17) *If $\mathcal{D}$ is a deterministic dynamic semantics, then $\Rightarrow_\mathcal{D}$ is a deterministic consequence relation.*

Proof. Note first that, due to singularity of $\mathcal{V}$, we have $\mathcal{V}(a) \in P$ for any $a \in L^*$. As before, to simplify the notation, we will write $[a]$ instead of $\mathcal{V}(a)$.

We need only to check Cumulativity. If $aB \not\Rightarrow_\mathcal{D} c$, then there exists a dynamic model with an admissible set U such that $s[aB] \in U$, but $s[aB][c] \notin U$. But then $s[a] \in U$, since $s[a]$ is a prefix of $s[aB]$. Moreover, $s[aB][c] = s[a][Bc]$ by associativity of concatenation, and therefore $s[a][Bc] \notin U$. Thus, $a \not\Rightarrow_\mathcal{D} Bc$. This completes the proof.   □

As a technical preparation for the next representation theorem, the lemma below gives a direct description of the least deterministic theory containing a given process.

Lemma (B) *$D(a)$ is a least deterministic theory of a deterministic consequence relation $\Rightarrow$ containing a process a iff*

$$D(a) = \{bc \mid b \preceq a \ \& \ b \Rightarrow c\}.$$

Proof. It is easy to check that any deterministic theory containing a should contain also $D(a)$. So we check only that $D(a)$ is a deterministic theory.

Assume that $d \Rightarrow e$ and $sd \in D(a)$, that is, $sd = bc$ for some b, c such that $b \preceq a$ and $b \Rightarrow c$. Then $d \Rightarrow e$ implies $sd \Rightarrow e$ by Left Monotonicity, and hence $bc \Rightarrow e$. Given $b \Rightarrow c$, this implies $b \Rightarrow ce$ by Cut, and therefore $bce \in D(a)$. But $bce = sde$, and hence $sde \in D(a)$. Thus, $D(a)$ is a theory of $\Rightarrow$.

Assume that $x \preceq y$ and $y \in D(a)$, Then $y = bc$ for some b, c such that $b \preceq a$ and $b \Rightarrow c$. Now if $x \preceq b$, then $x \preceq a$, and therefore $x \in D(a)$ (since $x \Rightarrow \epsilon$). Otherwise $b \prec x$, in which case $x = bd$ for some $d \preceq c$. In this case $b \Rightarrow c$ implies $b \Rightarrow d$ by Right Anti-Monotonicity, and hence again $x = bd \in D(a)$. Thus, $D(a)$ is a deterministic theory. This completes the proof.   □

Now we are ready to prove the following.

Theorem (Representation Theorem 10.18) *If $\mathcal{T}^d_{\Rightarrow}$ is a set of deterministic theories of a dynamic consequence relation $\Rightarrow$, then $\Rightarrow$ is deterministic iff $\Rightarrow$ $=$ $\Rightarrow_{\mathcal{T}^d_{\Rightarrow}}$.*

Proof. The direction from right to left follows from lemma 10.17. So let $\Rightarrow$ be a deterministic consequence relation. If $a \Rightarrow b$, then $a \Rightarrow_{\mathcal{T}^d_{\Rightarrow}} b$ directly from the definition of a theory. Assume then that $a \Rightarrow_{\mathcal{T}^d_{\Rightarrow}} b$, that is, for any s and any $U \in \mathcal{T}^d_{\Rightarrow}$, if $sa \in U$, then $sab \in U$. We take $s = \epsilon$ and U to be the least deterministic theory containing a. By lemma B, we obtain $ab \in D(a)$, and therefore $ab = cd$ for some c, d such that $c \preceq a$ and $c \Rightarrow d$. Let $a = ce$, for some e. Then $d = eb$, and hence $c \Rightarrow eb$. By Cumulativity we conclude $ce \Rightarrow b$, that is, $a \Rightarrow b$. This completes the proof. $\square$

Theorem (10.21) *If $M_{\mathcal{D}}$ is the relational model corresponding to the deterministic dynamic semantics $\mathcal{D}$, then $\Rightarrow_{\mathcal{D}}$ $=$ $\Rightarrow_{M_{\mathcal{D}}}$.*

Proof. In our present (simpler) case we have that $(s, t) \in \mathcal{F}(a)$ iff $(s, t) = (x_U, y_U)$ for some $x, y \in U \in \mathcal{D}$ such that $y = xa$.

Assume first that and $(s, t) \in \mathcal{F}(a)$. Hence $(s, t) = (x_U, y_U)$ for some $x, y \in U \in \mathcal{D}$ such that $y = xa$. Since U is an admissible set, we have $xab = yb \in U$. Let $r = (yb)_U$. Then clearly $(t, r) \in \mathcal{F}(b)$. This gives the direction from left to right.

Now assume that $a \Rightarrow_{M_{\mathcal{D}}} b$ and $ma \in U$ for some $m \in L^*$ and $U \in \mathcal{D}$. Since U is prefix-closed, we also have $m \in U$. So let us put $s = m_U$ and $t = (ma)_U$. Clearly $(s, t) \in \mathcal{F}(a)$. Since $a \Rightarrow_{M_{\mathcal{D}}} b$, there exists $r \in S$ such that $(t, r) \in \mathcal{F}(b)$. By the definition of $\mathcal{F}$, this can happen only if $r = (mab)_U$, and therefore $mab \in U$, which shows that $a \Rightarrow_{\mathcal{D}} b$ holds. $\square$

Proofs for Section 10.2.

Lemma (10.26) *For any PDL formula ϕ, any program α, and any model M,*

1. $\tau(\phi)$ is a static proposition, that is, $\mathcal{F}(\tau(\phi)) \subseteq Id$.
2. $s \in V(\phi)$ iff $(s, s) \in \mathcal{F}(\tau(\phi))$.
3. $R(\alpha) = \mathcal{F}(\tau(\alpha))$.

Proof. All claims are proved by simultaneous induction on the complexity of ϕ and α. The base cases $p \in P$ (for [1] and [2]) and $\pi \in \Pi$ (for [3]) follow directly from the definitions.

For the case $\neg\phi$, (1) follows from the fact that $\sim\tau(\phi)$ is a negated proposition. Moreover, we have $\mathcal{F}(\tau(\neg\phi)) = \mathcal{F}(\sim\tau(\phi)) = Id\backslash\mathcal{F}(\tau(\phi))$, so by the inductive assumption for ϕ, $V(\neg\phi) = S\backslash V(\phi) = \{s \mid (s, s) \in \mathcal{F}(\tau(\neg\phi))\}$.

For the case $\phi \,\&\, \psi$, (1) follows from the fact that conjunction ($\wedge$) of static propositions is a static proposition. Moreover, $\mathcal{F}(\tau(\phi \,\&\, \psi)) = \mathcal{F}(\tau(\phi) \wedge \tau(\psi)) = \mathcal{F}(\tau(\phi)) \cap \mathcal{F}(\tau(\psi))$ due to the fact that $\tau(\phi)$ and $\tau(\psi)$ are static propositions. This immediately gives (2).

Finally, for the case $[\alpha]\phi$, (1) follows again from the fact that $\tau(\alpha) \to \tau(\phi)$ ($= \sim(\tau(\alpha) \wedge \sim\tau(\phi))$) is a negated proposition. Then $(s, s) \in \mathcal{F}(\tau([\alpha]\phi))$ iff $(t, t) \in \mathcal{F}(\tau(\phi))$ for any t such that $(s, t) \in \mathcal{F}(\tau(\alpha))$ (since $\tau(\phi)$ is a static proposition). Now, by the inductive assumption (3), $R(\alpha) = \mathcal{F}(\tau(\alpha))$. Therefore, $(s, s) \in \mathcal{F}(\tau([\alpha]\phi))$ iff $s \in V([\alpha]\phi)$. This proves (2).

To complete the proof, we have to verify (3).

For the case $?\phi$, $\mathcal{F}(\tau(?\phi)) = \mathcal{F}(\tau(\phi))$ while $s \in V(\phi)$ iff $(s, s) \in \mathcal{F}(\tau(\phi))$ by (2). Hence (3) holds.

For the case $\alpha; \beta$, $\mathcal{F}(\tau(\alpha; \beta)) = \mathcal{F}(\tau(\alpha) \wedge \tau(\beta))) = \mathcal{F}(\tau(\alpha)) \circ \mathcal{F}(\tau(\beta))$. By the inductive assumption, the latter is equal to $R(\alpha) \circ R(\beta) = R(\alpha; \beta)$, as required.

Finally, we have

$$R(\alpha \cup \beta) = R(\alpha) \cup R(\beta) = \mathcal{F}(\tau(\alpha)) \cup \mathcal{F}(\tau(\beta)) = \mathcal{F}(\tau(\alpha) \wedge \tau(\beta)) = \mathcal{F}(\tau(\alpha \cup \beta)).$$

This completes the proof.   □

Theorem (10.27) $\phi_1, \ldots, \phi_n \vDash_M^{pdl} \psi$ *iff* $\tau(\phi_1) \ldots \tau(\phi_n) \Rightarrow_M \tau(\psi)$.

Proof. Let ϕ denote the formula $\phi_1 \& \ldots \& \phi_n \& \neg\psi$. Then $\phi_1, \ldots, \phi_n \vDash_M^{pdl} \psi$ does not hold iff ϕ is consistent iff there exists $s \in V(\phi)$. By (2) of the preceding lemma, $(s, s) \in \mathcal{F}(\tau(\phi))$. Moreover,

$$\mathcal{F}(\tau(\phi)) = \mathcal{F}(\tau(\phi_1) \wedge \cdots \wedge \tau(\phi_n) \wedge \sim\tau(\psi)),$$

where the conjuncts permute because each $\tau(\phi_i)$ is static. By the definition of $\mathcal{F}$, (s, s) belongs to this set, for some s, iff $\tau(\phi_1) \ldots \tau(\phi_n) \nRightarrow_M \tau(\psi)$.   □

Lemma (10.29)   1. $\sim\!A\,A \Rightarrow$.

2. $aA \Rightarrow$ *iff* $a \sim\!\sim\!A \Rightarrow$.

3.   *For any static proposition ϕ,* $\dfrac{a\,\phi\,b \Rightarrow \quad a\sim\!\phi\,b \Rightarrow}{ab \Rightarrow}$.

Proof.

1. $\sim\!A \Rightarrow \sim\!A$ by S-Reflexivity, so $\sim\!A\,A \Rightarrow$ by $(\sim R)$.

2.   Follows from the combination of $(\sim R)$ and $(\sim L)$ (in both directions).

3.   It is easy to see that conjunction is an associative connective. So assume that we have $a\,\phi\,b \Rightarrow$ and $a \sim\!\phi\,b \Rightarrow$, and let B be the conjunction of b. Using $(\wedge)$, it is easy to verify that $a\,\phi\,b \Rightarrow$ implies $a\,\phi\,B \Rightarrow$, while $a \sim\!\phi\,b \Rightarrow$ implies $a \sim\!\phi\,B \Rightarrow$. Then the following derivation proves $aB \Rightarrow$:

$$\cfrac{\cfrac{\cfrac{\cfrac{a \sim\!\phi\,B \Rightarrow}{a \sim\!\phi, \sim\!\sim\!B \Rightarrow}\,{}_{(2)}}{a \sim\!\sim\!B, \sim\!\phi \Rightarrow}\,\text{S-Permutation}}{a \sim\!\sim\!B \Rightarrow \phi}\,(\sim L) \qquad \cfrac{\cfrac{\cfrac{a\,\phi\,B \Rightarrow}{a\,\phi \sim\!\sim\!B \Rightarrow}\,{}_{(2)}}{a \sim\!\sim\!B\,\phi \Rightarrow}\,\text{S-Permutation}}{}\,\text{S-Cut}}{\cfrac{a \sim\!\sim\!B \Rightarrow}{aB \Rightarrow}\,{}_{(2)}}$$

Finally, $aB \Rightarrow$ implies $ab \Rightarrow$ by a number of (reverse) applications of $(\wedge)$.

□

Lemma (Basic Lemma 10.32) *In a canonical model* $M_\Rightarrow$, *for any formula A and any states* $X, Y \in S_\Rightarrow$,

$$X, Y \vDash A \; iff \; XAY \not\Rightarrow .$$

Proof. The proof is by induction on the complexity of the formula A. The inductive basis holds by the definition of V.

Assume that $A = {\sim}B$. Then $X, Y \vDash A$ holds iff $X = Y$ and there is no state Z such that $X, Z \vDash B$. By the inductive assumption, the latter holds iff $XBZ \Rightarrow$ for any Z. Now, the latter claim is reducible to $XB \Rightarrow$. Indeed, the direction from right to left is immediate while if $XB \not\Rightarrow$, we can apply the usual extension technics (using [3] of lemma 10.29) in order to extend the trace XB to a consistent trace XBZ for some maximal consistent set Z of static propositions. This will give us the direction from left to right. Hence, we have to show only that $X {\sim}BY \not\Rightarrow$ holds iff $X = Y$ and $XB \Rightarrow$.

Suppose first that $X = Y$ and $XB \Rightarrow$. Then $X {\sim}{\sim}B \Rightarrow$ by (2) of lemma 10.29, and therefore ${\sim}B \in X$ (since X is a maximal consistent set, either ${\sim}B \in X$, or ${\sim}{\sim}B \in X$). Consequently, we can safely extend X to a consistent trace $X {\sim}BX$ (due to S-Contraction and S-Permutation, any instantiation of $X {\sim}BX$ is an instantiation of X). Thus, $X {\sim}BY \not\Rightarrow$ holds. In the other direction, if $X {\sim}BY \not\Rightarrow$, then $X {\sim}B \not\Rightarrow$, and therefore ${\sim}B \in X$ (since X is a maximal consistent set). Consequently, $XB \Rightarrow$ by (1) of lemma 10.29. Moreover, in this case $XY \not\Rightarrow$, which is possible only if $X = Y$ (see the proof of lemma 10.31). This completes the proof for negation.

Assume now that $A = B \wedge C$. By definition, $X, Y \vDash B \wedge C$ iff there is a state Z such that $X, Z \vDash B$ and $Z, Y \vDash C$. By the inductive assumption, this is equivalent, respectively, to $XBZ \not\Rightarrow$ and $ZCY \not\Rightarrow$. On the other hand, by the rule $(\wedge)$ for conjunction, $X(B \wedge C)Y \not\Rightarrow$ is equivalent to $XBCY \not\Rightarrow$. Hence, it is sufficient to show that $XBCY \not\Rightarrow$ holds iff $XBZ \not\Rightarrow$ and $ZCY \not\Rightarrow$ for some state Z.

If $XBCY \not\Rightarrow$ holds, we can apply (3) of lemma 10.29 to extend the trace $XBCY$ to a consistent trace $XBZCY$, where Z is some maximal consistent set of static propositions. Then the two Monotonicity rules imply both $XBZ \not\Rightarrow$ and $ZCY \not\Rightarrow$, as required. In the other direction, assume that $XBZ \not\Rightarrow$ and $ZCY \not\Rightarrow$ for some state Z, but $XBCY \Rightarrow$. By definition, the latter implies that there exists an inconsistent sequence of formulas of the form $xBCy$, where x contains only formulas from X and y contains only formulas from Y. Let C_0 denote the conjunction of the subsequence Cy. Then $xBC_0 \Rightarrow$ by the rule $(\wedge)$ for conjunction, and hence $xB {\sim}{\sim}C_0 \Rightarrow$ by (2) of lemma 10.29. The latter implies that ${\sim}{\sim}C_0$ cannot belong to Z (since $XBZ \not\Rightarrow$), and therefore ${\sim}C_0 \in Z$. But now $ZCY \not\Rightarrow$ implies ${\sim}C_0Cy \not\Rightarrow$, and therefore we can apply $(\wedge)$ to derive ${\sim}C_0C_0 \not\Rightarrow$ — a contradiction with (1) of lemma 10.29. This completes the case of conjunction.

Finally, assume that $A = B \vee C$. By the inductive assumption, $X, Y \vDash B \vee C$ iff either $XBY \not\Rightarrow$, or $XCY \not\Rightarrow$. But the rules for disjunction immediately imply that this holds if and only if $X A \vee B Y \not\Rightarrow$. This completes the proof. $\square$

Lemma (Filtration Lemma 10.35) *For any states X, Y from $S_\Rightarrow$,*

(i)   *For any static $\phi \in FL(Q)$, $[X], [X] \vDash_{M_Q} \phi$ iff $\phi \in X$;*

(ii) (a)   *For any other formula $A \in FL(Q)$, if $XAY \not\Rightarrow$, then $[X], [Y] \vDash_{M_Q} A$;*

     (b)   *If $[X], [Y] \vDash_{M_Q} A$ and $\sim(A \wedge B) \in X \cap FL(Q)$, then $\sim B \in Y$.*

Proof. The proof is by induction on the complexity of formulas from $L(Q)$.

Case (i) is immediate if ϕ is an atom or a logical combination of static formulas. So we need only to verify the case $\phi = \sim A$. If $[X], [X] \vDash_{M_Q} \sim A$, then, for any state Y, $[X], [Y] \nvDash_{M_Q} A$, and therefore $XAY \Rightarrow$ by (a). The latter implies $XA \Rightarrow$ since if $XA \not\Rightarrow$, then the trace XA can be extended to a consistent trace XAY for some maximal consistent Y. Now, $XA \Rightarrow$ implies $X \sim \sim A \Rightarrow$, and therefore $\sim A \in X$, as required. In the other direction, assume that $\sim A \in X$, but $[X], [Y] \vDash_{M_Q} A$, for some Y. Then $\sim(A \wedge \top) \in X$ (since $\sim A \Rightarrow \sim(A \wedge \top)$). Moreover, since $\sim A \in FL(Q)$, we have $\sim(A \wedge \top) \in FL(Q)$ by the properties of Fischer-Ladner closure. Therefore, $\sim \top \in Y$ by (b), a contradiction. This completes the case (i).

Case (a). If A is an atom, then (a) holds by definition.

Assume that $B \wedge C \in L(Q)$ and $X(B \wedge C)Y \not\Rightarrow$. Then $XBCY \not\Rightarrow$ by the rule for conjunction, so we can apply (3) of lemma 10.29 to extend the trace $XBCY$ to a consistent trace $XBZCY$, where Z is some maximal consistent set of static propositions. Then the two Monotonicity rules imply both $XBZ \not\Rightarrow$ and $ZCY \not\Rightarrow$. By the inductive assumption, $[X], [Z] \vDash_{M_Q} B$ and $[Z], [Y] \vDash_{M_Q} C$, so $[X], [Y] \vDash_{M_Q} B \wedge C$, as required.

Assume that $B \vee C \in L(Q)$, and $X(B \vee C)Y \not\Rightarrow$. Then by the rules for disjunction either $XBY \not\Rightarrow$ or $XCY \not\Rightarrow$, so by the inductive assumption either $[X], [Y] \vDash_{M_Q} B$ or $[X], [Y] \vDash_{M_Q} C$, and consequently $[X], [Y] \vDash_{M_Q} B \vee C$.

Finally, assume that $A = B^*$. For a state Z, let us consider the set of (static) formulas $(Z \cap L(Q)) \cup \{\sim D \mid D \in L(Q) \backslash Z\}$. This set is finite, and it uniquely determines the equivalence class $[Z]$, so if ψ_Z is a conjunction of this set, then $W \in [Z]$ iff $\psi_Z \in W$. Now assume that $[X], [Y] \nvDash_{M_Q} B^*$, and let ϕ be the formula $\vee \{\psi_Z \mid [Z], [Y] \nvDash_{M_Q} B^*\}$. Clearly, a state W includes ϕ iff $[W], [Y] \nvDash_{M_Q} B^*$. Now we will show that $\phi B \Rightarrow \phi$. Indeed, if $\phi B \not\Rightarrow \phi$, then $\phi B \sim \phi \not\Rightarrow$, and hence $ZBW \not\Rightarrow$ for some states Z, W such that $\phi \in Z$ and $\sim \phi \in W$. Then $[W], [Y] \vDash_{M_Q} B^*$ and also $[Z], [W] \vDash_{M_Q} B$ by the inductive assumption (a), so $[Z], [Y] \vDash_{M_Q} B^*$ by the semantics of iteration, which contradicts the assumption $\phi \in Z$. Thus, $\phi B \Rightarrow \phi$ holds, and hence $\phi B^* \Rightarrow \phi$ by the rule $(^*I)$ of iteration, and consequently $\phi B^* \sim \phi \not\Rightarrow$. Since $[X], [Y] \nvDash_{M_Q} B^*$, we have $\phi \in X$ and $\sim \phi \in Y$ (because $[Y], [Y] \vDash_{M_Q} B^*$), and therefore $XB^*Y \Rightarrow$, as required.

Case (b). If π is an atom, then $[X], [Y] \vDash_{M_Q} \pi$ amounts to $X'\pi Y' \nRightarrow$ for some $X' \in [X]$ and $Y' \in [Y]$. Suppose that $\sim(\pi \wedge B) \in X \cap FL(Q)$, but $\sim B \notin Y$. Then $\sim(\pi \wedge B) \in X'$ and $\sim B \notin Y'$ (since $\sim B \in FL(Q)$). Consequently $\sim\sim B \in Y'$, and therefore $X'\pi Y' \nRightarrow$ implies $\sim(\pi \wedge B)\pi\sim\sim B \nRightarrow$. But $\sim(\pi \wedge B)(\pi \wedge B) \Rightarrow$, so $\sim(\pi \wedge B)\pi B \Rightarrow$—a contradiction. This confirms the inductive basis for (b).

Assume that $[X], [Y] \vDash_{M_Q} B \wedge C$ and $\sim((B \wedge C) \wedge D) \in X \cap FL(Q)$. Then there exists a state Z such that $[X], [Z] \vDash_{M_Q} B$ and $[Z], [Y] \vDash_{M_Q} C$. But $\sim(B \wedge (C \wedge D)) \in X \cap FL(Q)$ (by associativity of conjunction and the properties of $FL(Q)$), so $[X], [Z] \vDash_{M_Q} B$ implies $\sim(C \wedge D) \in Z$ by the inductive assumption, and therefore $[Z], [Y] \vDash_{M_Q} C$ gives $\sim D \in Y$.

Assume that $[X], [Y] \vDash_{M_Q} B \vee C$ and $\sim((B \vee C) \wedge D) \in X \cap FL(Q)$. Note that $\sim((B \vee C) \wedge D) \Rightarrow \sim(B \wedge D)$ and $\sim((B \vee C) \wedge D) \Rightarrow \sim(C \wedge D)$. Consequently, $\sim(B \wedge D) \in X \cap FL(Q)$ and $\sim(C \wedge D) \in X \cap FL(Q)$ by the properties of $FL(Q)$. But we also have that either $[X], [Y] \vDash_{M_Q} B$ or $[X], [Y] \vDash_{M_Q} C$, so $\sim D \in X$ by the inductive assumption.

Finally, assume that $[X], [Y] \vDash_{M_Q} B^*$ and $\sim(B^* \wedge C) \in X \cap FL(Q)$. By the semantic definition, $[X], [Y] \vDash_{M_Q} B^*$ implies that either $[X] = [Y]$ or there is a sequence of states $X_1, \ldots, X_n = Y$ such that $[X], [X_1] \vDash_{M_Q} B$ and $[X_i], [X_{i+1}] \vDash_{M_Q} B$ for any $i < n$. For the first case, we have $\sim(B^* \wedge C)(B^* \wedge C) \Rightarrow$, which implies $\sim(B^* \wedge C) B^* C \Rightarrow$, and hence $\sim(B^* \wedge C) B^* \Rightarrow \sim C$. Applying $(*E)$, we obtain $\sim(B^* \wedge C) \Rightarrow \sim C$, and therefore $\sim C \in X$. For the second case, $\sim(B^* \wedge C) B^* C \Rightarrow$ implies $\sim(B^* \wedge C) B B^* C \Rightarrow$ by $(*E)$, and hence $\sim(B^* \wedge C) \Rightarrow \sim(B \wedge (B^* \wedge C))$. Consequently, $\sim(B^* \wedge C) \in X \cap FL(Q)$ implies $\sim(B \wedge (B^* \wedge C)) \in X \cap FL(Q)$ (by the properties of $FL(Q)$), and therefore $\sim(B^* \wedge C) \in X_1$ by the inductive assumption. Applying this step n times, we conclude that $\sim(B^* \wedge C) \in X_n$, that is, $\sim(B^* \wedge C) \in Y$, and hence $\sim C \in Y$ due to the fact that $\sim(B^* \wedge C) \Rightarrow \sim C$. This completes the proof. $\square$

11 Dynamic Markov Assumption

Using the causal framework of dynamic inference described in chapter 10, we provide in this chapter a semantic interpretation and logical (inferential) characterization of the dynamic Markov principle that underlies the main action theories in artificial intelligence. As is suggested already by the name, the dynamic Markov principle plays much the same role as the *causal* Markov principle in structural approaches to causality (Pearl 2000; Spirtes, Glymour, and Scheines 2000). This principle will be shown to constitute a non-monotonic, abductive assumption that justifies the actual restrictions on action descriptions in these AI theories as well as constraints on allowable queries. It will be shown, in particular, that the well-known regression principle is a consequence of the dynamic Markov principle and that it is valid also for nondeterministic domains.

Theories of action and change form one of the central subjects of AI. A wide range of such theories have been suggested in the literature, starting from low-level representations of actions in first-order languages (such as the situation calculus) and ending with higher-order action description languages, as well as representations of actions in general formalisms of propositional dynamic logic (PDL) and linear temporal logic (LTL). Moreover, one of the useful ways of providing transferability results among different theories involves designing still more general theories that subsume existing ones as special cases (see, e.g., Thielscher 2011).

In this last part of our study we are pursuing, however, a somewhat different, top-down approach, which amounts to finding common knowledge representation principles that underlie these action theories. Our hope is that it is possible to single out a relatively small number of AI-specific principles and assumptions in such a way that existing action theories could be viewed as implementations of the same principles in particular representation frameworks.

The starting point of our approach in this chapter is the fact that actual descriptions used for representing actions in all the above-mentioned formalisms employ only a modest part of expressive capabilities allowed by their host logical frameworks. This is clearly true for the action descriptions formulated in the classical first-order framework, but it pertains also to formalizations of actions in logic programming, PDL and LTL. As we are going to show, the informal Markov principle provides an ultimate basis for most of these restrictions, and it is responsible for many important special properties of these action theories, such as the executability assumption and the regression principle.

In accordance with the general terminology adopted in AI, static atoms will also be called *fluents* in what follows. By a *fluent proposition* we will mean a classical formula constructed from fluent atoms only. Clearly, any fluent proposition will be static (in the sense of SDL), though not vice versa.

The causal language of sequential dynamic logic (SDL), described in the preceding chapter, provides a very simple and transparent representation for common action descriptions. As a typical example of representation in this language, if *Shoot* is an action while *Loaded* and *Dead* are fluents, then an action effect rule[1] like

$$Shoot \text{ \textbf{causes} } Dead \text{ \textbf{if} } Loaded$$

is representable in SDL simply as a dynamic causal rule

$$Loaded \, Shoot \Rightarrow Dead,$$

which is equivalent to each of the following rules (note that the conjunction $\wedge$ is ordered here):

$$Loaded \wedge Shoot \Rightarrow Dead,$$

$$\Rightarrow (Loaded \wedge Shoot) \rightarrow Dead.$$

11.1 Dynamic Markov Principle

It is widely held (especially in the AI literature) that a proper description of dynamic domains should conform to the following *Markov principle*: in every state of a dynamic system, both an executability and particular effects of every action should depend only on fluent propositions that hold in this state. Accordingly, a complete fluent information about a particular state should exhaustively determine all possible future developments from this state. Note that this principle does not presuppose determinism, so it is compatible with existence of nondeterministic actions. Still, even in such cases, it states that the range of possible outcomes of such an action in every state is fully determined by the fluents that hold in this state.

It is important to note that the Markov principle is not an intrinsic, "ontological" property of action domains. Rather, it should be viewed as a general *knowledge representation principle*, or assumption, according to which a dynamic system ought to be described in a way that conforms to this principle. Thus, if an actual description of a dynamic system does not satisfy the Markov principle, this could be viewed as an evidence that this

---

1. See, for example, Gelfond and Lifschitz (1998).

description simply lacks expressivity since the fluent information it can articulate is insufficient for distinguishing states with different possibilities of actions. Still, from a broader, more philosophical point of view, the dynamic Markov principle plays the role of a general Principle of Sufficient Reason for dynamic domains.

11.1.1 Semantics: Transparent Models

Arbitrary transition models (that constitute the semantics of our language) do not satisfy, in general, the Markov principle. Indeed, in such models two states satisfying the same fluent facts may occupy different places in a relational structure and hence may have different dynamic properties such as executability of actions as well as their effects. This immediately suggests, however, that a natural way of imposing the Markov principle consists in restricting transition models to models in which different states are distinguishable by their "fluent content":

Definition 11.1 *A transition model will be called* transparent *if for any two distinct states* $s, t \in S$ *there is a fluent formula F that holds in s but does not hold in t.*

A dynamic consequence relation will be called transparent *if it is determined by some transparent transition model.*

It can be shown that the distinguishing formula F in the above description can always be chosen to be a fluent literal. In addition, it can be verified that a dynamic consequence relation $\Rightarrow$ is transparent if and only if its canonical model $M_\Rightarrow$ (as defined in the chapter 10) is transparent.

Any state of a transparent model is uniquely determined by the set of fluent formulas that hold in it. Accordingly, transparent models can be alternatively described as quadruples $(I, P, \Pi, \mathcal{F})$, where I is a set of *propositional valuations* assigning a truth value to each fluent atom from P while $\mathcal{F}$ assigns each action a binary relation on such valuations.

As a matter of fact, transparent models have been used explicitly or implicitly in overwhelming majority of action theories developed in AI.[2] Thus, it is a standard model for action description languages (Gelfond and Lifschitz 1998), logic programming (see, e.g., Baral and Gelfond 2005), causal theories of action (Giunchiglia et al. 2004), and even some representations of actions in the PDL framework (see, e.g., Herzig and Varzinczak 2007). And though this fact is less obvious for first-order representations of actions like the situation calculus, we will see in what follows that such action theories directly embody the Markov principle.

It should be noted, however, that transparency is not a simple logical constraint on dynamic consequence relations. In particular, it cannot be expressed in terms of additional axioms or inference rules of a usual kind. This follows already from the fact that

---

2. Transparent Kripke models are often called *standard* models in the literature on nonmonotonic reasoning. Unfortunately, the latter term has another meaning in the literature on PDL.

any transition model can be conservatively transformed into a transparent model simply by adding new propositional fluents that distinguish the states of the model (in the extended language). Nevertheless, the restriction to transparent models may well sanction additional conclusions that are not derivable in the underlying logical formalism. The following example illustrates this.

Example 11.1 *Let us consider a language that contains a single fluent p and single action π. Any transparent model for this language has at most two states (= propositional valuations), and it can be verified that in any such model, if there is a trace πpπ from some state of a model, then performing π in this state leads to a state in which π will always be executable whenever p holds in it. Formally, this description corresponds to the following rule of SDL:*

$$\sim\sim(\pi \wedge p \wedge \pi)\pi p \Rightarrow \pi.$$

However, this rule is not valid in SDL, and it is refuted in a nontransparent model with three states s, t, t_1 such that p holds in t and t_1 only, and $\mathcal{F}(\pi) = \{(s, t), (s, t_1), (t, t_1)\}$:

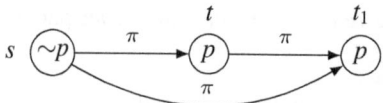

11.2 Basic Action Theories and Markov Completion

A key to a better understanding of the Markov principle is provided by the following restriction on dynamic rules:

Definition 11.2 *A dynamic causal rule will be called* basic *if it has the form $F\pi \Rightarrow F_1$ or the form $\Rightarrow F$, where π is an action atom and F, F_1 are fluent formulas. A set of basic rules will be called a* basic *action theory.*

Rules of the form $F\pi \Rightarrow F_1$ (or formulas $F \rightarrow [\pi]F_1$ in the PDL notation) correspond to action descriptions used in practically all action theories in AI. Such rules correspond also to the so-called *Hoare rules* $\{F\}\pi\{F_1\}$ in the theory of computation (see Harel, Kozen, and Tiuryn 2000). Note that these rules include *inexecutability conditions* $F\pi \Rightarrow \perp$ as a special case.

Rules of the form $\Rightarrow F$ correspond to the usual *static laws* or constraints. Note that in SDL a rule $F \Rightarrow F_1$ is equivalent to $\Rightarrow F \rightarrow F_1$, so such rules are reducible to basic static laws.

It should be noted that static laws in a basic action theory are completely independent of the action rules: any static law that is derivable from a basic action theory is derivable already from its static laws alone.[3] The situation is not symmetric, however, for action laws since static laws can be used to derive new action laws from given ones.

Our next result shows that transparent dynamic consequence relations are determined, in effect, by their basic rules.

Theorem 11.1 *Two transparent dynamic consequence relations coincide iff they have the same basic rules.*

Proof. Assume that M and M' are, respectively, transparent models of two dynamic consequence relations $\Rightarrow$ and $\Rightarrow'$ that have the same basic rules. Suppose first that they have different worlds, say, α is a world of M but not of M'. Recall that α in our case is just a (maximal consistent) set of propositional formulas. But then we have $\alpha \Rightarrow_{M'}$ and $\alpha \not\Rightarrow_M$, and therefore by compactness, there exists a fluent formula F such that $F \Rightarrow_{M'}$ and $F \not\Rightarrow_M$. But $F \Rightarrow$ is a basic rule, and this contradicts our assumption. Suppose now that these models have different transitions, for instance, $(\alpha, \beta) \in \mathcal{F}_M(\pi)$, for some worlds α, β and action π, but $(\alpha, \beta) \notin \mathcal{F}_{M'}(\pi)$. Then we will have $\alpha\pi\beta \not\Rightarrow_M$ and $\alpha\pi\beta \Rightarrow_{M'}$. Again, by compactness, $\alpha\pi\beta \Rightarrow_{M'}$ implies $F\pi F_1 \Rightarrow_{M'}$, for some $F \in \alpha$ and $F_1 \in \beta$, though $F\pi F_1 \not\Rightarrow_M$. Moreover, a rule of the form $F\pi F_1 \Rightarrow$ is equivalent to $F\pi \Rightarrow \sim F_1$, which is a basic rule, and consequently we obtain again a contradiction with our assumption that the two dynamic consequence relations have the same basic rules. Therefore, M and M' should coincide, and consequently they determine the same dynamic consequence relation. $\qquad\square$

The above theorem states that all the inferences sanctioned by a transparent consequence relation are determined ultimately by the basic rules that belong to it. In other words, if we accept the Markov principle, then a complete description of a dynamic model can always be achieved by using only basic rules. This formal result can be viewed as an ultimate logical justification for the widespread restriction of action descriptions in AI domains to basic action theories.

Actually, the correspondence between transparent dynamic consequence relations and basic action theories is bidirectional in the sense that any basic action theory determines a unique transparent consequence relation associated with it. This consequence relation can be constructed as follows.

Let Δ be a basic action theory. Then its *canonical transparent transition model* $M_\Delta^t = (S_\Delta, P, \Pi, \mathcal{F}_\Delta)$ is defined by taking S_Δ to be the set of all maximal sets of fluent formulas that are consistent with respect to (the static laws of) Δ and stipulating that, for any action atom π, $\mathcal{F}(\pi)$ is the set of all pairs (α, β) such that, for any basic rule $F\pi \Rightarrow F_1$ from Δ, if $F \in \alpha$, then $F_1 \in \beta$.

Then the following claim can be easily verified.

Lemma 11.2 M_Δ^t *is a transparent model of* Δ.

---

3. This will no longer hold if we would allow nonbasic rules; see, for example, Herzig and Varzinczak (2007).

It is interesting to note also that, for deterministic action theories, the above results can be strengthened even further.

Recall that a transition model $(S, \Pi, P, \mathcal{F})$ is *deterministic* if, for any action atom π, $\mathcal{F}(\pi)$ is a partial function: if $(s, t) \in \mathcal{F}(\pi)$ and $(s, r) \in \mathcal{F}(\pi)$, then $t = r$. We will say that a dynamic consequence relation is *deterministic* if it is generated by a deterministic transition model. Then the following result can be obtained.

Theorem 11.3 *Any transparent deterministic consequence relation is determined by the set of its static laws and the set of determinate action rules of the form $L\,\pi \Rightarrow l$, where l is a fluent literal and L a conjunction of fluent literals.*

Proof. For a deterministic dynamic consequence relation, $F\pi \Rightarrow F_1 \vee F_2$ holds only if either $F\pi \Rightarrow F_1$ or $F\pi \Rightarrow F_2$. Consequently, any basic rule of such a consequence relation is reducible to rules of the form $L\,\pi \Rightarrow l$.   □

The above result shows that, for describing deterministic action domains, we can safely use only determinate action rules of the form $L\,\pi \Rightarrow l$.

Let $\Rightarrow'_\Delta$ denote the dynamic consequence relation that is determined by the above canonical model M'_Δ while $\Rightarrow_\Delta$ will denote the least dynamic consequence relation including Δ. Clearly, $\Rightarrow_\Delta$ is the set of all dynamic causal rules that are logically derivable from Δ in SDL, and consequently $\Rightarrow_\Delta$ is included in $\Rightarrow'_\Delta$. However, in general these two consequence relations *do not coincide*. Nevertheless, we have the following theorem.

Theorem 11.4 $\Rightarrow_\Delta$ *has the same basic rules as* $\Rightarrow'_\Delta$.

Proof. Assume that $F\pi \not\Rightarrow_\Delta F_1$ for some basic rule. Then there exists a dynamic transition model M of Δ where $F\pi \not\Rightarrow_M F_1$. Now let us "collapse" all states in M that satisfy the same fluent formulas while preserving all transitions. Then we obtain a transparent model M', but it is easy to verify that M' has the same valid basic rules as M, and therefore M' is also a model of Δ, and $F\pi \not\Rightarrow_{M'} F_1$. Consequently, $F\pi \not\Rightarrow'_\Delta F_1$.   □

Moreover, since any transparent consequence relation is uniquely determined by its basic rules, we obtain the following corollary.

Corollary 11.5 *If Δ is a basic action theory, then $\Rightarrow'_\Delta$ is a unique transparent consequence relation that has the same basic rules as $\Rightarrow_\Delta$.*

The above results and correspondences allow us to be more precise about the exact impact of restricting the semantics of dynamic consequence relations to transparent models.

In general, $\Rightarrow'_\Delta$ contains more information than what is logically derivable from Δ. On the other hand, both consequence relations contain the same basic rules, so the additional information in $\Rightarrow^t_\Delta$ (as compared with $\Rightarrow_\Delta$) involves only nonbasic rules. Moreover, all these additional rules are determined ultimately only by the basic rules of Δ.

The above properties allow us to see $\Rightarrow^t_\Delta$ as a kind of *completion* of Δ that is based on the Markov assumption. Moreover, just as other known kinds of completion, this Markov completion is *nonmonotonic* in the sense that if Δ is extended with rules to another (even basic) action theory Δ', it does not necessarily hold that $\Rightarrow^t_\Delta$ is included in $\Rightarrow^t_{\Delta'}$. The following example illustrates this.

Example 11.2 *Consider a basic action theory* $\Delta = \{\pi \Rightarrow p\}$ *in a language containing only a fluent atom* p *and an action* π. *The corresponding canonical transparent model* M^t_Δ *involves only two worlds (valuations)* $s = \{\sim p\}$ *and* $t = \{p\}$, *whereas* $\mathcal{F}(\pi) = \{(s, t), (t, t)\}$.

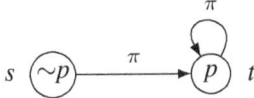

In this model, action π *can always be executed with an effect* p *(that is,* $\Rightarrow^t_\Delta \pi \wedge p$ *holds). This assertion does not follow, however, logically from* Δ *in SDL. Moreover, if we extend* Δ *to a basic theory* $\Delta' = \{\pi \Rightarrow p, \ p\pi \Rightarrow \sim p\}$, *we obtain a different canonical model in which* $\mathcal{F}(\pi) = \{(s, t)\}$.

$$s \ (\sim p) \xrightarrow{\ \ \pi\ \ } (p) \ t$$

In this model, the action π *is already not executable in* t, *and consequently* $\Rightarrow \pi \wedge p$ *will no longer hold.*

The executability assumption. One of the most significant descriptions that *cannot* be formulated in the restricted language of basic action rules are executability claims stating that, given some conditions, a certain action is executable. Such descriptions correspond to rules of the form $F_1 \Rightarrow \pi \wedge F$ in SDL (and to formulas $F_1 \rightarrow \langle \pi \rangle F$ in PDL). Moreover, the above example illustrates that a canonical transparent model actually maximizes such executability claims. In this sense it embodies a well-known *executability assumption* in the action literature according to which an action is always considered executable whenever this is consistent with other facts and action rules. Accordingly, the executability assumption can be viewed as a by-product of the Markov principle.

11.3 Admissible Queries

The above results create a peculiar ambiguity about the semantic interpretation and consequences of basic action theories. On the one hand, a basic theory Δ is interpreted "logically" by the consequence relation $\Rightarrow_\Delta$, the logical closure of Δ in SDL. On the other hand, the Markov principle suggests that Δ should be interpreted by its canonical transparent model, which corresponds to a larger consequence relation $\Rightarrow^t_\Delta$. Still, theorem 11.4 shows that the difference is inessential for basic rules. This naturally raises a question: What kinds of queries are invariant with respect to these two interpretations? In order to answer this question, we have to consider a broader class of dynamic inference rules.

By *a generalized basic rule* we will mean a dynamic rule of the form

$$F_0\,\pi_1\,F_1\ldots\pi_n\,F_n \Rightarrow F,$$

where F and each F_i is a fluent formula while π_i are action atoms. Then the following result can be proved (by induction) as a generalization of theorem 11.4.

Theorem 11.6 $\Rightarrow_\Delta$ *has the same generalized basic rules as* $\Rightarrow^t_\Delta$.

By the above result, derivability of generalized basic rules does not depend on the choice between the two interpretations, so each of them can be used for their verification.

As a matter of fact, generalized basic rules subsume practically all kinds of queries that are used in action theories. Actually, most of these theories use only a special kind of such rules, namely rules of the form $F_0\pi_1\ldots\pi_n \Rightarrow F_1$. However, though this restriction is reasonable for deterministic action domains, it is obviously insufficient for capturing properties of nondeterministic actions. This is because in the nondeterministic case there may be multiple ways of executing a sequence of actions $\pi_1\ldots\pi_n$, and they can be distinguished by putting fluent "tests" between these actions.

11.4   Inferential Markov Property

Now we are going to describe a syntactic, inferential counterpart of the Markov principle.

Definition 11.3 *A dynamic consequence relation will be said to satisfy the* Markov *property if, for any propositions A, B, if $A \Rightarrow B$ holds, then there is a fluent proposition F such that $A \Rightarrow F$ and $F \Rightarrow B$.*

According to the above description, the Markov property is a particular kind of interpolation property requiring that all inferences of a dynamic consequence relation are mediated by fluent propositions.

Remark. There is an obvious similarity between this property and the corresponding property of abductive production relations in which any causal inference is mediated by abducibles (see chapter 7). Note, in particular, that if F is a fluent formula, then the causal rule $F \Rightarrow F$ holds for dynamic inference! The analogy means, in effect, that fluent formulas play the same role in dynamic Markov consequence relations as abducibles in abductive causal inference.

We will denote by $\mathrm{Cf}(A)$ the set of fluent consequences of a proposition A:

$$\mathrm{Cf}(A) = \{F \mid F \text{ is a fluent proposition \& } A \Rightarrow F\}.$$

In this notation, the Markov property can be formulated (due to compactness) as follows:

(Markov)   If $A \Rightarrow B$, then $Cf(A) \Rightarrow B$.

The next result states an important consequence of the Markov property.

Lemma 11.7 *If a dynamic consequence relation $\Rightarrow$ satisfies the Markov property, then $Cf(\phi) \Rightarrow \phi$, for any static proposition ϕ.*

Proof. Follows from the fact that $\phi \Rightarrow \phi$ holds for any static proposition ϕ.   □

Remark. In the language of PDL, the above condition amounts to the requirement that for any PDL formula there exists an equivalent propositional formula.

The above result implies the following corollary.

Corollary 11.8 *If a dynamic consequence relation satisfies the Markov property, then for any fluent world α and any static proposition ϕ, either $\alpha \Rightarrow \phi$, or $\alpha \Rightarrow \sim\phi$.*

Proof. The preceding lemma said, in effect, that for any static proposition ϕ there is an equivalent fluent proposition F, that is, $\Rightarrow \phi \leftrightarrow F$. It follows that any fluent world either implies ϕ (when it includes F) or is incompatible with ϕ.   □

Recall that the states of the canonical model $M_\Rightarrow$ of a dynamic consequence relation (see the description before lemma 10.31 in chapter 10) are maximal consistent sets of static propositions. Now, the above corollary immediately implies that this model cannot have different states that coincide on their fluents. In other words, the canonical model will be transparent, and therefore we obtain this theorem:

Theorem 11.9 *If a dynamic consequence relation satisfies the Markov property, then it is transparent.*

In the other direction, we have the following result:

Lemma 11.10 *Any transparent dynamic consequence relation satisfies the Markov property.*

Proof. Let $\Rightarrow$ be a transparent consequence relation, and assume that $A \Rightarrow B$ holds, though $Cf(A) \not\Rightarrow B$. Then the canonical model of $\Rightarrow$ should contain a canonical state S that includes both $Cf(A)$ and $\sim B$. Since $A \Rightarrow B$, this state cannot be an "output" of A, and therefore $AS \Rightarrow$. By compactness, there exists a static proposition $\phi \in S$ such that $A\phi \Rightarrow$, and therefore $A \Rightarrow \sim\phi$, where $\sim\phi \notin S$. However, by lemma 11.7, there exists a fluent equivalent F of $\sim\phi$, and hence $A \Rightarrow F$. But then $F \in Cf(A)$ and therefore $F \in S$, contrary to the fact that $\sim\phi \notin S$. The obtained contradiction shows that the Markov property should hold.   □

Combining the above results, we conclude with the following.

Corollary 11.11 *A dynamic consequence relation is transparent iff it has the Markov property.*

11.5 Markov Property and Regression

Now we are going to show that Markov property implies the well-known regression principle for action domains. The basis of this principle can be found in the fact that the Markov property turns out to be equivalent to the following very special case of lemma 11.7:

Theorem 11.12 *A dynamic consequence relation has the Markov property iff it satisfies the following condition:*

$$\mathrm{Cf}(\pi{\to}F) \Rightarrow \pi{\to}F, \text{ for any atomic action } \pi \text{ and any fluent formula } F.$$

Proof. The above condition asserts an existence of a fluent equivalent for any implication $\pi{\to}F$, so it is a direct consequence of lemma 11.7. So, we need only to prove the opposite direction: namely that existence of fluent equivalents for implications $\pi{\to}F$ implies existence of such fluent equivalents for any static formula.

Recall that any static formula of SDL can be transformed into an equivalent formula of PDL type. Now, a fluent equivalent of any PDL-type formula ϕ can be obtained using the following simple *regression method*:

- *Recursively replace any subformula of ϕ of the form $\pi{\to}F$ with a corresponding fluent equivalent.*

The above algorithm always terminates and produces a fluent formula that is equivalent to ϕ. As has been shown earlier, this will imply that the corresponding dynamic consequence relation is transparent and hence will satisfy the Markov property. $\qquad\square$

The regression algorithm sketched in the above proof can be seen as an abstract generalization of the regression method in the situation calculus (see Reiter 1991, 2001). Moreover, it extends the latter to general nondeterministic domains that satisfy the Markov property.

The above regression algorithm can be used for verifying arbitrary queries concerning a Markov dynamic consequence relation. To see this, note that a question whether a rule $A \Rightarrow B$ holds for a Markov consequence relation is reducible to the question whether the corresponding static proposition $A{\to}B$ is derivable, which, by the above algorithm, is reducible to a validity of a certain (classical) fluent formula.

The existence of a fluent equivalent for any implication $\pi{\to}F$ implies, in particular, that for any action π there is a fluent formula F_π that is equivalent to $\pi \to \bot$. The formula F_π determines precise conditions for inexecutability of action π, so $\sim F_\pi$ provides us with exact fluent preconditions for *executability* of π,[4] while the conjunction $F_0 \wedge \sim F_\pi$ determines the conditions under which an action π is executable and produces effect F.

If we want, however, to transform the above regression algorithm into a working procedure for verifying inferences from action theories, we need a method of producing fluent equivalents for the "effect" implications $\pi \rightarrow F$ from the actual domain descriptions given in the form of basic action theories. In the finite case, this can be done as follows.

For a finite language, the canonical transparent model M^t_Δ of a basic action theory Δ is also finite, and each state s of the model is uniquely determined by the conjunction F_s of the fluent literals that hold in this state. Then it can be verified that the following formula gives a fluent equivalent of $\pi \rightarrow F$:

$$G_F^\pi = \bigvee \{F_s \mid F \in t, \text{ for any } t \text{ such that } s \pi t\}.$$

A proof-theoretic counterpart of the above semantic procedure can be found in Bochman (2013).

Actually, the task of constructing fluent equivalents for implications $\pi \rightarrow F$ is partly simplified by the fact that $\pi \rightarrow (F_1 \wedge F_2)$ is equivalent to $(\pi \rightarrow F_1) \wedge (\pi \rightarrow F_2)$, so if G_1 and G_2 are fluent equivalents of $\pi \rightarrow F_1$ and $\pi \rightarrow F_2$, respectively, then $G_1 \wedge G_2$ is a fluent equivalent of $\pi \rightarrow F_1 \wedge F_2$. Consequently, it is sufficient to find fluent equivalents for all implications of the form $\pi \rightarrow C$, where C is a fluent clause.

A further simplification can be obtained for *deterministic* consequence relations. As has been established earlier, action rules for such consequence relations can be restricted to fluent literals. Moreover, for such consequence relations we have that $\pi \rightarrow (F_1 \vee F_2)$ is equivalent to $(\pi \rightarrow F_1) \vee (\pi \rightarrow F_2)$, so the task reduces to finding fluent equivalents for implications $\pi \rightarrow l$, where l is a fluent literal. In addition, for such consequence relations it holds also that

$$\pi \rightarrow \sim F \text{ is equivalent to } (\pi \rightarrow \bot) \vee \sim (\pi \rightarrow F).$$

Consequently the task of determining fluent equivalents for any static proposition is reduced in this case to finding fluent equivalents for implications $\pi \rightarrow p$, where p is either a fluent atom, or a falsity constant.

It turns out that the above constructions are directly related to Reiter's *basic action theories* (BATs) in the situation calculus (see Reiter 2001). Indeed, BATs are defined in terms of static laws, action precondition axioms, and successor state axioms. Now, action precondition axioms are precisely assertions that determine fluent equivalents for implications $\pi \rightarrow \bot$ while the successor state axioms are nothing other than descriptions of fluent equivalents for implications $\pi \rightarrow p$ for each action π and every fluent atom p. Thus, as was rightly noted in Lin (2008), Reiter's basic action theories provide in this respect a precise realization of the Markov assumption in the setting of deterministic action theories.

---

4. Thus, it corresponds to a fluent equivalent of the predicate *Poss*(π) in Reiter's situation calculus.

11.6 Summary

It has been shown in this chapter that many of the actual restrictions and specific properties of action theories in AI can be obtained as consequences of the general Markov principle for action domains. Moreover, we have shown that the Markov principle constitutes a nonmonotonic assumption that allows us to "complete" information determined by the basic action rules and static laws. This assumption embodies, in particular, the executability assumption and provides an ultimate logical basis for the regression method by which dynamic information is abduced, or explained by, the static (fluent) information.

All the above constructions and results did not take into account, however, the famous *frame problem* in reasoning about actions. The omission of this problem from the above exposition was intentional since the main aim of this chapter has been to show that many important features and methods of current action theories arise ultimately from basic knowledge representation principles that are independent of, or complementary to, the frame problem. Still, an account of the frame problem is essential for an adequate description of action theories in AI.

Generally speaking, the frame problem is a by-product of an additional, and also highly reasonable *inertia assumption*, according to which the world changes only as a result of actions. This assumption is nothing other than a dynamic version of the law of causality; in our context, it says that any change in truth-values of fluents should be caused by some action.

Each particular action directly affects only a relatively small number of fluents. Therefore, each time a large set of fluent literals should retain their value in such a change. Now, in a descriptive setting of action theories, this means that any action description should involve (explicitly or implicitly) a large number of inertia, or frame, rules of the form $l\pi \Rightarrow l$, where l is a fluent literal. Accordingly, the frame problem amounts to the problem how such inertia rules (which are a special kind of basic rules) can be efficiently represented and handled.

Markov and inertia assumptions are completely independent knowledge representation principles, but they jointly determine the ultimate form of action theories in AI. In some simple cases (specifically, in the deterministic setting without static laws), the impact of the inertia assumption can be directly "compiled" into a fluent equivalent of an implication $\pi \rightarrow p$. This procedure is nothing other than a dynamic counterpart of propositional completion that we have described in section 4.5.2 of chapter 4. In fact, this is the essence of Reiter's "simple" solution to the frame problem.[5] Unfortunately, the task becomes much more complex in the presence of static laws, mainly because it immediately invokes both the qualification and ramification problems. These problems have been discussed in

---

5. The corresponding regression principle in PDL has been described in Demolombe, Herzig, and Varzinczak (2003).

Ginsberg and Smith (1988), Lin and Reiter (1994), and in many subsequent studies. A predominant way of resolving these problems that emerged from these studies amounted to augmenting domain descriptions with directional, static *causal* laws for fluents, combined with an appropriate nonmonotonic reasoning mechanism for dealing with such laws. Such a combined description has been implemented, for instance, in causal theories of actions—see Giunchiglia et al. (2004), as well as chapter 12.

The formalism of SDL, used as a logical basis of this chapter, naturally suggests a uniform approach to the problem of incorporating the inertia assumption. Indeed, SDL can be viewed as dynamic generalization of causal inference relations from chapter 4 while the inertia assumption is just a dynamic version of the law of causality. This immediately suggests that a prospective dynamic causal calculus could be constructed by assigning SDL an appropriate *dynamic nonmonotonic semantics* that would incorporate the inertia principle as an additional constraint on its causal models. This is what we are going to do in the last chapter of this book.

12 Dynamic Causal Calculus

Unlike general descriptions of temporal dynamics in temporal logics and related logical formalisms, action theories in artificial intelligence (AI) have to deal primarily with two quite specific reasoning tasks, namely the *prediction* task (what are the results of a given sequence of actions from an initial state) and the *planning* task (what sequence of actions could lead from an initial state to a target goal state).

The majority of theories for such a reasoning in AI have followed the lead of the situation calculus (McCarthy and Hayes 1969) in adopting the inertia assumption as a basis for an alternative representation of temporal dynamics. A salient advantage of the use of inertia in action descriptions is that it provides both a more succinct and more natural representation of such a dynamics from the commonsense point of view. However, the inertia assumption almost inevitably leads to a triple of its own notorious difficulties known as the frame, ramification, and qualification problems (see, e.g., Shanahan 1997). It has been realized quite early that classical logic and its temporal/dynamic extensions, taken by themselves, encounter difficulties in resolving these problems. More precisely, it has become clear that these problems have an essentially nonmonotonic character, so their proper solution requires augmenting purely logical, monotonic deductive reasoning with an appropriate mechanism for making nonmonotonic conclusions.[1]

In recent years a dominant approach to solving these problems in AI has been based, in one form or another, on causal reasoning. The corresponding causal closure assumption (see, e.g., Reiter 2001) is a particular form of the law of causality, which amounts to the requirement that all facts that hold in a situation should either be caused by actions or else preserve their truth-values in time (due to the accompanying *inertia* assumption). A direct incorporation of such causal assertions into the language of the situation calculus has been proposed in Lin (1995, 1996) and has been shown to provide a natural account of both the frame and ramification problems. And finally, the causal calculus itself has been

---

1. Even the well-known "monotonic" solution to the frame problem, described in Reiter (1991), can be seen as a result of compiling this nonmonotonic information into the successor state equivalences; see section 4.5.2 in chapter 4.

suggested in McCain and Turner (1997) as a general formal framework for this kind of causal reasoning.

In this last chapter of the book we will introduce a dynamic causal calculus, a general nonmonotonic theory that can be viewed as a primary formalism for dynamic causal reasoning. On a logical side, this formalism will incorporate and combine in an abstract form the basic insights of the preceding two chapters. Thus, we will adopt from chapter 10 the sequential form of causal rules that contain processes as their antecedents. However, we will also take into account the observation, made in chapter 11, that *basic* causal rules of the form $F\pi \Rightarrow F_1$ and corresponding basic action theories are sufficient for an adequate representation of dynamic domains that satisfy the Markov property. In accordance with this, we will restrict the antecedents of causal rules to *ordered pairs* of premises. We will "suppress," however, the distinction between actions and fluents in this setting and generalize such antecedents to ordered pairs of classical propositional formulas. As a result, the dynamic causal calculus will become a theory of causal rules of the form

$$B . C \Rightarrow E,$$

where B, C and E are classical propositions. It will be shown that the resulting formalism is sufficiently expressive to capture the main high-level action description languages in AI, described in Gelfond and Lifschitz (1998).

12.1 Dynamic Causal Calculus

Causal rules of the form $B . C \Rightarrow E$ will be called *dynamic causal rules*. An informal meaning of such rules will be *"After B, C causes E."* By the intended interpretation, such a rule describes a dynamic transition from a state that satisfies proposition B to a subsequent state in which E is caused by C. Such rules naturally correspond, for instance, to action rules, or dynamic action laws, of the form

caused E if C after B

that constitute a common syntactic core of many action description languages today (see, e.g., Lang, Lin, and Marquis 2003).

Remark. As a matter of fact, the above dynamic causal rules are somewhat ambiguous from a syntactic point of view. On a more abstract level, such a rule can be viewed simply as an instantiation of a primitive ternary propositional operator. There are, however, at least two other more articulated possibilities.[2] Thus, a dynamic causal rule $B . C \Rightarrow E$ could be viewed as a plain causal rule $C \Rightarrow E$ that is conditioned by a preceding context B. In this case, proposition B can be viewed as a *background condition*, or *precondition*, of the causal rule (see chapter 9). In fact, this "parsing" agrees with the informal reading of

---

2. Such possibilities have been discussed, for example, in Beall et al. (2012) in their compositional analysis of general and relevant conditionals.

such rules, given above. Moreover, we will see that this understanding of dynamic causal rules as conditional "static" rules provides a natural justification for the postulates of the associated dynamic causal inference that will be given below. Still, a different parsing possibility consists in viewing such rules as binary causal rules, though with complex premises consisting of pairs of propositions (B, C). This reading makes such rules similar to sequential dynamic causal rules that have been used in chapter 10, though strictly speaking, they cannot be identified with the latter since they have somewhat different logical properties. However, we will see later that this reading can also be given a formal support due to a possible translation of such rules as propositions of the form $(B \circ C) \to E$ in arrow logic.

In addition to the action rules, however, an important feature of the majority of action description languages as well as corresponding theories of action and change in AI consists in the explicit use of *static* causal rules or state constraints. In particular, it is these rules that allow us to provide a succinct and efficient description of both ramifications and qualifications in action descriptions.

Given the above notion of a dynamic causal rule, we will identify ordinary *static* causal rules with a special kind of dynamic rules of the form $\mathbf{t} . A \Rightarrow B$, where $\mathbf{t}$ is the truth constant. In other words, we will adopt the following definition:

$$A \Rightarrow B \equiv_{df} \mathbf{t} . A \Rightarrow B.$$

According to this definition, static causal rules are precisely rules that hold after any legitimate transition. The consequences and variations created by this identification will be discussed below.

The above reduction of static causal rules to a special kind of dynamic rules makes dynamic causal rules the only kind of rules of the dynamic causal calculus. In accordance with this, a *dynamic causal theory* will be defined below simply as a set of dynamic causal rules.

12.1.1 Nonmonotonic Transition Semantics

It is convenient this time to begin with describing the intended nonmonotonic semantics of the dynamic causal calculus since it will determine, ultimately, the corresponding logical formalism of causal inference. Actually, we will describe below two dynamic nonmonotonic semantics, one corresponding to a direct generalization of the causal nonmonotonic semantics for plain causal rules, described in chapter 4, the other a particular instantiation of the relative causal semantics, introduced in chapter 9. As we will see, these two semantics will provide, respectively, a formal interpretation for the action description languages $\mathcal{C}$ and $\mathcal{B}$ in the classification of Gelfond and Lifschitz (1998).

The guiding principle behind both nonmonotonic semantics will be a thorough enforcement of the dynamic law of causality, according to which every state of a dynamic model should be explained (i.e., caused) by a preceding state and the active causal rules. The difference will amount, however, to whether the inertia principle is captured by syntactic means or as an integral part of the nonmonotonic semantics.

Recall that by a world we mean a maximal consistent set of classical propositions, or fluents. Such worlds will also be called *states* in this chapter.

Dynamic causal semantics.   Given a dynamic causal theory Δ and states α, β, we will denote by $\Delta(\alpha . \beta)$ the set of propositions that are caused due to a transition from α to β.

$$\{C \mid A.B \Rightarrow C \in \Delta \text{ for some } A \in \alpha, B \in \beta \}.$$

Definition 12.1 *A pair* (α, β) *of worlds will be called a(n exact)* C-transition *with respect to a dynamic causal theory* Δ *if* β *is the unique model of* $\Delta(\alpha.\beta)$, *that is,*

$$\beta = \mathrm{Th}(\Delta(\alpha . \beta)).$$

A C-transition model *of a dynamic causal theory* Δ *is a set of states* $\mathfrak{I}$ *such that, for any* $\beta \in \mathfrak{I}$ *there is* $\alpha \in \mathfrak{I}$ *such that* (α, β) *is a* C-transition wrt Δ.

A C-transition is a transition between two states in which the resulting state is fully explained (caused), given the preceding state and the causal laws of the domain. In this respect, the above definition of a transition *model* implements the law of causality with respect to the states themselves by requiring, in effect, that every state should have sufficient reasons for its occurrence.

If (α, β) is a C-transition, then $\Delta(\alpha.\beta)$ is included in β. As a consequence, the output world β is always closed with respect to the static causal rules (for our definition of the latter): namely, if $A \Rightarrow B$ (that is, $\mathbf{t}.A \Rightarrow B$) belongs to Δ and $A \in \beta$, then $B \in \beta$. Moreover, note that any world of a C-transition model is also an output of some C-transition. Accordingly, we immediately obtain the following lemma.

Lemma 12.1 *Any world of a* C-transition model of a dynamic causal theory Δ is closed *with respect to the static causal rules of* Δ.

Remark. As a matter of fact, our definition of a C-transition almost coincides with the corresponding definition of a *causally explained transition* that has been given in Giunchiglia and Lifschitz (1998) for an action description language C, a predecessor of the language $C+$ described in Giunchiglia et al. (2004). In fact, the only difference between the two definitions is that Giunchiglia and Lifschitz (1998) required further that both the initial and resulting states of such a transition should be closed with respect to the static causal laws. On our construction, this additional requirement is accounted for, respectively, as a byproduct of our definition of static causal rules on the one hand (for the resulting states) and a C-transition model on the other hand (for the initial states).

It can be easily verified that the union of two C-transition models of a causal theory is also a C-transition model. Consequently, if a dynamic causal theory has at least one C-transition model, it has a unique maximal such model. We will call the latter the *canonical*

C-transition model of a dynamic causal theory. As can be verified, this model includes all legitimate dynamic chains of states that can be causally justified.

In the general case, it is difficult to guarantee that there exists a C-transition from a given state to some other state. This will change, however, if we will take into account the *inertia principle*. In the framework of C-transition models, this principle can be represented syntactically using the following dynamic explanatory rules for literal fluents:

(Inertia) $\quad l\,.\,l \Rightarrow l.$

The above rule implies that if a literal l holds in some state, it becomes a default for every state that can be reached from this state by a direct C-transition. In other words, l will continue to hold in such a subsequent state, unless an opposite literal will be caused to hold in it. Consequently, unless there is some inherent contradiction in the description of causal rules, any state of a canonical C-transition model will have at least one possible subsequent state (including possible persistence in the same state). As we will see below, this syntactic encoding of the inertia principle has been adopted in the action description languages C and $C+$ as an important part of their representation framework.

Remark. The restriction of the inertia principle to literals only is based on the same idea that has guided us in restricting actual causes to literals (in chapter 8), namely that compound logical formulas are completely determined *logically* by their constituting literals, so their temporal behavior is also fully dependent on the temporal behavior of the latter; they cannot have life (i.e., temporal evolution) of their own. It is important to note, however, that our general definitions of transitions are formulated for the full classical language, so they impose the law of causality on all propositional formulas. Consequently, since the reasons why a compound logical formula holds in a given state cannot be based directly on the inertia principle, they should be obtained either in terms of active static or dynamic causal rules or else as a "logical" consequence of the corresponding reasons for their constituents.

In contrast to the above syntactic encoding of the inertia principle, the notion of a B-transition, described below, will provide a direct semantic representation of inertia.

Inertial causal semantics. We are going to describe now a dynamic version of the relative nonmonotonic semantics that was presented in chapter 9.

In the following definition, $Lit(\alpha)$ denotes the set of literals that belong to a state α.

Definition 12.2 *A pair (α, β) of worlds will be called a(n exact) B-transition with respect to a dynamic causal theory Δ if*

$$\beta = \mathrm{Th}((Lit(\alpha) \cap \beta) \cup \Delta(\alpha\,.\,\beta)).$$

A B-transition model of a dynamic causal theory Δ is a set of worlds $\mathfrak{I}$ such that, for any $\beta \in \mathfrak{I}$, there is $\alpha \in \mathfrak{I}$ such that (α, β) is a B-transition wrt Δ.

In contrast to $\mathcal{C}$-transitions, $\mathcal{B}$-transitions are determined not only by active causal rules but also by literals that persist in the transition; such literals remain to hold due to inertia, and therefore they do not require causal explanation.

For determinate dynamic causal theories, the above description of $\mathcal{B}$-transitions can be simplified as follows.

Lemma 12.2 *A pair of states* (α, β) *is a $\mathcal{B}$-transition with respect to a determinate dynamic causal theory* Δ *iff*

$$Lit(\beta)\setminus\alpha \subseteq \Delta(\alpha\,.\,\beta) \subseteq \beta.$$

Proof. Since β is deductively closed, the inclusion $Th((Lit(\alpha)\cap\beta)\cup\Delta(\alpha\,.\,\beta)) \subseteq \beta$ amounts to $\Delta(\alpha\,.\,\beta) \subseteq \beta$. Note now that, if Δ is a determinate causal theory, then

$$(Lit(\alpha)\cap\beta)\cup\Delta(\alpha\,.\,\beta)$$

is just a set of literals, and consequently the reverse inclusion holds if and only if

$$Lit(\beta) \subseteq (Lit(\alpha)\cap\beta)\cup\Delta(\alpha\,.\,\beta).$$

The latter inclusion amounts to $Lit(\beta) \subseteq \alpha\cup\Delta(\alpha\,.\,\beta)$, which is equivalent to

$$Lit(\beta)\setminus\alpha \subseteq \Delta(\alpha\,.\,\beta). \qquad \square$$

As before, we have that if (α, β) is a $\mathcal{B}$-transition, then $\Delta(\alpha.\beta)$ is included in β. Therefore, the output world β is also closed with respect to the static causal rules. Moreover, since any world of a $\mathcal{B}$-transition model is also an output of some $\mathcal{B}$-transition, we obtain again the following lemma.

Lemma 12.3 *Any world of a $\mathcal{B}$-transition model of a dynamic causal theory* Δ *is closed with respect to the static causal rules of* Δ.

Again, the union of two $\mathcal{B}$-transition models of a causal theory is also a $\mathcal{B}$-transition model. Consequently, if a dynamic causal theory has at least one $\mathcal{B}$-transition model, it has a unique maximal such model. We will call the latter the *canonical $\mathcal{B}$-transition model* of a dynamic causal theory.

Our final result in this section will show that, modulo the inertia principle, the above two nonmonotonic semantics are essentially equivalent.

For a dynamic causal theory Δ, we will denote by Δ_I a theory obtained from Δ by adding all inertia rules of the form $l\,.\,l \Rightarrow l$.

Theorem 12.4 *A pair of states* (α, β) *is a $\mathcal{B}$-transition of a dynamic causal theory* Δ *if and only if it is a $\mathcal{C}$-transition of* Δ_I.

The proof follows immediately from the fact that, for any states $\alpha, \beta, \Delta_I(\alpha\,.\,\beta)$ coincides with $(Lit(\alpha)\cap\beta)\cup\Delta(\alpha\,.\,\beta)$.

12.1.2   Literal Dynamic Causal Theories

A common simplifying assumption in theories of action and change in AI amounts to a syntactic restriction of the underlying logical language to classical literals. This simplification corresponds to restriction of dynamic causal rules to *literal* causal rules of the form

$$L . L' \Rightarrow l,$$

where l is a literal while L and L' are sets of literals.

Just as for plain causal rules (see section 4.3 in chapter 4), under this restriction, it turns out to be convenient to represent states (worlds) as maximal consistent sets of literals. In what follows, we will use $s, t, \ldots$ for denoting "literal" states in this sense. Thus, a C-transition can now be characterized as a pair (s, t) of such states that satisfies the following simple condition:

$$t = \Delta(s . t).$$

Similarly, the description of the B-transition can be reduced to the following equality:

$$t = (s \cap t) \cup \Delta(s . t).$$

It can be easily verified that these simplified descriptions provide, respectively, equivalent characterizations of C- and B-transitions for the restricted case of literal dynamic causal theories:

Lemma 12.5 *If Δ is literal dynamic causal theory, then a pair of states (α, β) is a*

- *C-transition iff $Lit(\beta) = \Delta(Lit(\alpha) . Lit(\beta))$;*
- *B-transition iff $Lit(\beta) = \Delta(Lit(\alpha) . Lit(\beta)) \cup (Lit(\alpha) \cap Lit(\beta))$.*

12.2   Transition Inference Relation

As with other formalisms for nonmonotonic reasoning, the dynamic causal rules presuppose a certain underlying logic that agrees with the above nonmonotonic semantics. Such a logic will provide us with a formal description of the associated dynamic causal inference. Unlike the general dynamic formalisms we have been working with in chapter 10, however, this logic will describe only one-step transitions between states.

By a *transition inference relation* we will mean a set of dynamic causal rules $A . B \Rightarrow C$ that satisfies the postulates described below.

The first group of postulates states that a set of dynamic causal rules with a fixed first premise (B) should satisfy the postulates of an "ordinary" causal inference (see chapter 4):

(T-Strengthening) If $A \vDash C$ and $B . C \Rightarrow E$, then $B . A \Rightarrow E$;

(T-Weakening) If $E \vDash D$ and $B . C \Rightarrow E$, then $B . C \Rightarrow D$;

(T-And)   If $B . C \Rightarrow E$ and $B . C \Rightarrow D$, then $B . C \Rightarrow E \wedge D$;

(T-Or)   If $B . C \Rightarrow E$ and $B . D \Rightarrow E$, then $B . C \vee D \Rightarrow E$;

(T-Cut)   If $B . A \Rightarrow C$ and $B . A \wedge C \Rightarrow D$, then $B . A \Rightarrow D$;

(T-Truth) $\mathbf{t} \cdot \mathbf{t} \Rightarrow \mathbf{t}$;
(T-Falsity) $\mathbf{t} \cdot \mathbf{f} \Rightarrow \mathbf{f}$.

In view of the above postulates, the dynamic causal rules $B \cdot C \Rightarrow E$ can be seen as ordinary causal rules $C \Rightarrow E$ that are conditioned by the preceding (background) context B.

In addition, the next two postulates describe the logical properties of this preceding context in dynamic causal rules:

(B-Strengthening) If $A \vDash B$ and $B \cdot C \Rightarrow E$, then $A \cdot C \Rightarrow E$;
(B-Or) If $A \cdot C \Rightarrow E$ and $B \cdot C \Rightarrow E$, then $A \vee B \cdot C \Rightarrow E$.

The combined effect of the above postulates is that the associated semantic interpretation of dynamic causal inference (described in the next section) will be a possible world semantics in which both the two premises and conclusion of a dynamic causal rule are evaluated with respect to worlds (complete states).

As with plain causal rules, we will extend dynamic causal rules to rules having arbitrary sets of propositions as premises using compactness; for any sets u, v of propositions, we define $u \cdot v \Rightarrow A$ as follows:

$$u \cdot v \Rightarrow A \equiv \bigwedge a \cdot \bigwedge b \Rightarrow A, \text{ for some finite } a \subseteq u, b \subseteq v.$$

For a pair (u, v) of sets of propositions, $\mathcal{C}(u \cdot v)$ will denote the set of propositions caused by the pair, that is,

$$\mathcal{C}(u \cdot v) = \{A \mid u \cdot v \Rightarrow A\}.$$

The causal operator $\mathcal{C}$ can again be viewed as a derivability operator corresponding to a transition inference relation. Note that it is monotonic:

Monotonicity If $u \subseteq u'$ and $v \subseteq v'$, then $\mathcal{C}(u \cdot v) \subseteq \mathcal{C}(u' \cdot v')$.

Also, $\mathcal{C}(u \cdot v)$ will always be a deductively closed set:

$$\mathcal{C}(u \cdot v) = \text{Th}(\mathcal{C}(u \cdot v)).$$

For any dynamic causal theory Δ there exists a least transition inference relation that includes Δ. We will denote it by $\Rightarrow_\Delta$ while $\mathcal{C}_\Delta$ will denote the corresponding derivability operator. Clearly, $\Rightarrow_\Delta$ is the set of all dynamic causal rules that can be derived from Δ using the postulates for transition inference relations.

In the next section we will describe a possible-worlds semantics for the above logical formalism of dynamic inference.

12.2.1   A Possible Worlds Semantics

A possible worlds semantics of transition inference can be obtained by generalizing an accessibility relation on possible worlds to ternary relations.

A *causal possible world model* of a transition inference relation is a triple (W, R, V), where W is a set of possible worlds, R a ternary accessibility relation on W, and V a function assigning each world a propositional interpretation. The accessibility relation will be required to satisfy the following condition:

(Quasi-reflexivity) If $R\alpha\beta\gamma$, then $R\alpha\beta\beta$.

Definition 12.3 *A rule $A.B \Rightarrow C$ is* valid *in a model (W, R, V) if, for any worlds α, β, γ such that $R\alpha\beta\gamma$, if A holds in α and B holds in β, then C holds in γ.*

Given the above definition of validity, it is easy to verify the following:

Lemma 12.6 *The set of dynamic causal rules valid in a causal possible world model forms a transition inference relation.*

Moreover, using a suitable construction of a canonical semantics for a transition inference relation, the following completeness result can be established:

Theorem 12.7 *A set of dynamic causal rules forms a transition inference relation if and only if it is determined by a causal possible world model.*

Proof. Due to the connection between dynamic causal rules and the original, plain causal rules, the proof is a relatively straightforward generalization of the corresponding completeness proof for causal inference relations, given in Bochman (2004, theorem 7.4). More precisely, given a transition inference relation $\Rightarrow$, we can construct the corresponding canonical model (W, R_c) by taking W to be the set of maximal consistent sets of propositions, and defining R_c as follows:

$$R_c\alpha\beta\gamma \equiv C(\alpha . \beta) \subseteq \beta \cap \gamma.$$

Notice that this definition directly implies quasi-reflexivity of R_c. Thus, we only have to show that $A . B \Rightarrow C$ holds for the source dynamic causal relation if and only if it is valid in (W, R_c), namely, for any sets of propositions u, v and any A,

$$u . v \Rightarrow A \text{ iff } A \in \gamma, \text{ for any } \alpha, \beta, \gamma \in W \text{ such that } R_c\alpha\beta\gamma \text{ and } u \subseteq \alpha, v \subseteq \beta.$$

If $u . v \Rightarrow A$, $u \subseteq \alpha$ and $v \subseteq \beta$, for some worlds α, β, then clearly $A \in C(\alpha . \beta)$, and therefore $A \in \gamma$, for any world γ that includes $C(\alpha . \beta)$. Thus, $u . v \Rightarrow A$ is valid in the canonical semantics. In the other direction, if $u . v \nRightarrow A$, then u and v are included, respectively, in maximal sets α and β such that $\alpha . \beta \nRightarrow A$. By the two Or rules, T-Or and B-Or, both α and β should be worlds. Now, since $A \notin C(\alpha . \beta)$, there must exist a world γ containing $C(\alpha . \beta)$ and such that $A \notin \gamma$. Moreover, we will show that $C(\alpha . \beta) \subseteq \beta$.

Suppose that $C(\alpha . \beta) \nsubseteq \beta$. This means that $\alpha . \beta \Rightarrow D$, for some $D \notin \beta$. Since (α, β) is a maximal pair of sets that does not imply A, we have $\alpha . \beta, D \Rightarrow A$. But then we can apply T-Cut (using compactness) and obtain $\alpha . \beta \Rightarrow A$, contrary to our assumptions. Hence, $C(\alpha . \beta) \subseteq \beta$, and therefore $R_c\alpha\beta\gamma$, which means that $u . v \Rightarrow A$ is not valid in the canonical semantics. This completes the proof. $\square$

One of the interesting consequences of the above semantic characterization of transition inference is that, similarly to a straightforward modal translation of ordinary causal rules as formulas of the form $A \to \Box B$, dynamic causal rules can be represented as formulas of *arrow logic* (see, e.g., Venema 1997). As a matter of fact, one of the principal motivations behind arrow logic has also consisted in providing an abstract description of dynamic (transition) models (cf. van Benthem 1994). Moreover, semantic interpretation of arrow logic is also based on a possible world semantics with a ternary accessibility relation, and it can be easily verified that, by the above semantic description, a dynamic causal rule $A \cdot B \Rightarrow C$ turns out to be equivalent to a formula

$$A \circ B \to C$$

of arrow logic, where $\circ$ is a binary "arrow conjunction" operator having the following semantic interpretation:

$A \circ B$ holds in a world α if and only if there are worlds β, γ such that $R\beta\gamma\alpha$, A holds in β and B holds in γ.

Finally, both our definitions of a transition semantics involve also an additional, global requirement that any state should be an output of some transition. This requirement is captured by the following postulate:

(Transition) If $\mathbf{t} \cdot A \Rightarrow \mathbf{f}$, then $A \cdot \mathbf{t} \Rightarrow \mathbf{f}$.

Recall that we have decided to identify static causal rules $A \Rightarrow B$ with dynamic causal rules of the form $\mathbf{t} \cdot A \Rightarrow B$. Then the above postulate can be rewritten as:

(Transition 1) If $A \Rightarrow \mathbf{f}$, then $A \cdot \mathbf{t} \Rightarrow \mathbf{f}$.

On this reformulation, the above postulate stipulates, in effect, that any input state of a consistent transition should be (statically) causally consistent. Combined with the other postulates, this will immediately imply that both the input and output state of a transition should be closed with respect to the valid static laws.

A semantic counterpart of this postulate is given by the following constraint on the ternary accessibility relation:

(Permutation) If $R\alpha\beta\gamma$, then $R\delta\alpha\epsilon$, for some $\delta, \epsilon \in W$.

It can be verified that this constraint validates Transition. Moreover, given Transition, the canonical accessibility relation R_c will satisfy Permutation.

12.2.2 Correspondences

Since a transition inference relation can also be viewed as a large dynamic causal theory, the two definitions of transitions for the latter can be immediately extended to transition inference relations. Moreover, due to the logical properties of a transition inference, the definition of a C-transition can now be simplified; namely a pair of worlds (α, β) will be a C-transition with respect to a transition inference relation if and only if

$$\beta = C(\alpha . \beta).$$

Still, the definition of a B-transition for transition inference will remain much the same:

$$\beta = \mathrm{Th}((Lit(\alpha) \cap \beta) \cup C(\alpha . \beta)).$$

The following result will show, in effect, that the logic of transition inference is adequate for reasoning with respect to the nonmonotonic transition semantics of dynamic causal theories since it preserves the latter.

As before, $\Rightarrow_\Delta$ will denote a least transition inference relation containing a dynamic causal theory Δ.

Theorem 12.8 *If Δ is a dynamic causal theory, then*

- *C-transitions of Δ coincide with C-transitions of $\Rightarrow_\Delta$.*
- *B-transitions of Δ coincide with B-transitionss of $\Rightarrow_\Delta$.*

Proof sketch. If C_Δ is the provability operator corresponding to $\Rightarrow_\Delta$, then similarly to plain causal rules (see proposition 4.33) it can be shown that, for any "causally consistent" pair of worlds α, β, $C_\Delta(\alpha.\beta)$ coincides with $\mathrm{Th}(\Delta(\alpha.\beta))$. Consequently, $\alpha = C_\Delta(\alpha.\beta)$ iff $\alpha = \mathrm{Th}(\Delta(\alpha.\beta))$, and therefore C-transitions of Δ will coincide with C-transitions of $\Rightarrow_\Delta$. Moreover, since $\mathrm{Th}((Lit(\alpha) \cap \beta) \cup \Delta(\alpha . \beta))$ is classically equivalent to

$$\mathrm{Th}((Lit(\alpha) \cap \beta) \cup \mathrm{Th}(\Delta(\alpha . \beta))),$$

it is easy to see that B-transitions of Δ will also coincide with B-transitions of $\Rightarrow_\Delta$. $\square$

12.3 Comparisons with Action Languages A and B

In this section we will describe the relations between inertial causal semantics and the action languages $\mathcal{A}$ and $\mathcal{B}$.

A closest counterpart of our dynamic causal rules has been introduced in Pednault (1989) as part of his action description language (ADL) in the form of rules

$$A \textbf{ causes } l \textbf{ if } L,$$

where A is an action name, l is a literal, and L is a set (or conjunction) of literals. This language has been called language $\mathcal{A}$ in Gelfond and Lifschitz (1998).

The semantics of the language $\mathcal{A}$, given in Gelfond and Lifschitz (1998), can be described as a set of transitions (s, A, t), where A is an action name and s, t are states (maximal consistent sets of literals) that satisfy the following condition:

$$E(A, s) \subseteq t \subseteq E(A, s) \cup s, \tag{*}$$

where $E(A, s)$ is the set of heads of all action rules of the form A **causes** l **if** L in the action description such that $L \subseteq s$.

Now we will identify the action rules of $\mathcal{A}$ with literal dynamic causal rules of the form

$$L \cdot A \Rightarrow l.$$

In addition, we should restrict possible transitions to transitions produced by single actions; in other words, we should prevent concurrent actions. This can be achieved by accepting the following *static* causal rules:

$$A, B \Rightarrow \mathbf{f}$$

for any two different actions A and B.

Let Δ_D denote the dynamic causal theory that corresponds in this sense to an action description D in the language $\mathcal{A}$. Then we have the following theorem.

Theorem 12.9 *Transitions of an action description D in the language $\mathcal{A}$ coincide with $\mathcal{B}$-transitions of Δ_D.*

Proof sketch. The inertial $\mathcal{B}$-semantics for Δ_D can be described as a set of transitions (s, t) that satisfy the equality

$$t = (s \cap t) \cup \Delta(s \cdot A),$$

for some action name A (see lemma 12.5). Note, however, that $\Delta(s \cdot A)$ for this causal theory is just the set of direct effects of action A in a state s; in other words, it coincides with $E(A, s)$. Moreover, it is easy to show that the above equality is equivalent to the condition (*) above (cf. Gelfond and Lifschitz 1998, n13). Accordingly, the semantics of $\mathcal{B}$-transitions corresponds precisely to the semantic interpretation of $\mathcal{A}$.   □

The action description language $\mathcal{B}$ (see Gelfond and Lifschitz 1998) is obtained from the language $\mathcal{A}$ by adding *static laws* of the form

$$l \text{ if } L,$$

where l is a literal and L a set of literals. These static laws are viewed as plain inference rules that allow, in particular, to derive indirect effects of actions.

For a set Z of static laws, let $\mathbb{C}\mathrm{n}_Z(s)$ denote the least set of literals that contains s and is closed with respect to the rules from Z. Then the semantics of $\mathcal{B}$ is defined as a set of transitions (s, A, t), where s and t are literal states that are closed with respect to the static laws and satisfy the equation

$$t = \mathbb{C}\mathrm{n}_Z((s \cap t) \cup E(A, s)). \tag{**}$$

As before, $E(A, s)$ above is the set of heads of all action rules of the form A **causes** l **if** L from the dynamic description such that $L \subseteq s$.

The language $\mathcal{B}$ has been generalized to language $\mathcal{AL}$ (see, e.g., Baral and Gelfond 2005) which has lifted the restriction to single actions and thereby has allowed concurrency and has added explicit impossibility conditions for executability of actions; such executability conditions are covered in the dynamic causal calculus by constraints of the form

$$A . B \Rightarrow \mathbf{f},$$

where A is a fluent proposition and B is an action formula.

Now let us represent static laws as static causal rules $L \Rightarrow l$, as defined earlier in this chapter. Then it is easy to verify that any transition (s, t) that satisfy the above equation (**) will be a $\mathcal{B}$-transition in our sense. Still, the two semantics do not coincide because there are $\mathcal{B}$-transitions that are not transitions by the original definition for the language $\mathcal{B}$. The following example, adapted from Zhang and Lin (2017), illustrates this.

Example 12.1 *Let us consider a dynamic causal theory that consists of a single dynamic rule* $\mathbf{t} . A \Rightarrow f_2$ *and the following set of static causal rules:*

$$f_2, f_3 \Rightarrow f_1 \quad f_2, \neg f_3 \Rightarrow f_1 \quad f_1, f_2 \Rightarrow f_3.$$

It can be verified that the pair of states $(\{\neg f_1, \neg f_2, \neg f_3\}, \{f_1, f_2, f_3\})$ *will be a $\mathcal{B}$-transition with respect to this causal theory, though it is not an admissible transition for the original semantics of $\mathcal{B}$.*

In assessing this discrepancy, we should take into account that the underlying logic of static causal rules in the dynamic causal calculus is different from, and even incomparable with, the implicit logic of static laws in the language $\mathcal{B}$.

To begin with, static causal rules of the causal calculus admit the classical logical rule of disjunction in the antecedent:

(Or)   If $C \Rightarrow E$ and $D \Rightarrow E$, then $C \vee D \Rightarrow E$.

In the dynamic causal calculus, the above rule follows from the postulate *T-Or* of transition inference (see section 12.2). In contrast, static laws of the language $\mathcal{B}$ do not admit this rule. For instance, the first two static rules in the above example imply $f_2 \Rightarrow f_1$ by the rule *Or*, but if we would add this static rule to the theory, the above transition would become admissible for the language $\mathcal{B}$. Actually, the *Or* rule has played a key role in the characterization of the difference between the languages $\mathcal{B}$ and $\mathcal{C}$ that has been made in Zhang and Lin (2017) (see their postulate 5).

The discrepancy between our inertial semantics of $\mathcal{B}$-transitions and the "official" semantics of the language $\mathcal{B}$ does not arise in cases when the set of static rules is acyclic. Indeed, Gelfond and Lifschitz (2012) have shown that the semantics of $\mathcal{B}$ and $\mathcal{C}$ coincide, in effect, on action theories in which the dependence graph of their static rules is acyclic. This result can be immediately adapted for showing that, for such causal theories, the standard semantics of $\mathcal{B}$ will coincide with our semantics of $\mathcal{B}$-transitions.

This naturally brings us to the second crucial aspect of the difference between the languages $\mathcal{B}$ and $\mathcal{C}$. In the language $\mathcal{B}$, static laws are viewed as plain inference rules, and therefore they freely admit the logical postulate of reflexivity, namely $A \Rightarrow A$. In contrast, in the causal calculus and language $\mathcal{C}$ such rules have nontrivial content, and they are actually used in a formal representation of defaults, exogenous propositions, and actions (see below). On the other hand, the noncausal, inferential understanding of static laws in the framework of $\mathcal{B}$ has allowed to employ them, for instance, for describing defined propositions and predicates, including recursive definitions that play an important role in the general representation methodology behind the use of the languages $\mathcal{B}$ and especially $\mathcal{AL}$. Thus, the ability to use such recursive constructs has been crucial for modeling systems and for the development of industrial size planning and diagnostic applications (see, e.g., Balduccini, Gelfond, and Nogueira 2006; Son et al. 2006; Tu et al. 2011).

A comprehensive treatment of the latter discrepancy falls beyond the scope of this study already because it would require a generalization of our formalism to a first-order logical language. Nevertheless, a proper resolution of this problem could be provided only if we abandon an unjustified "purist" presumption that a theory of reasoning about actions and change should be based exclusively on causal reasoning. As we have argued from the very beginning, causal reasoning is not a replacement of logic but its extension, or complement, for situations where we do not have logically sufficient knowledge. It is the logical background of a causal formalism and its underlying logic that should be a proper place for defining new predicates and connectives, including recursive definitions and meaning postulates. All this could be done while still retaining a separate category of *causal* static rules when they are appropriate. Of course, this could create obvious problems for implementing such combined descriptions, for instance, in logic programming. Still, such implementation problems should not detract us from a no less important task of providing an *adequate representation* of reasoning in dynamic domains.

12.4   Representing Action Language C

An elaborate implementation of the causal approach to reasoning about actions has been given in Giunchiglia et al. (2004). The formalism of Giunchiglia et al. (2004) is a multi-sorted and multilayered representation framework. As its top layer, it employs a causal action description language $\mathcal{C}+$ that provides high-level descriptions of action domains in terms of three kinds of propositional atoms (actions, simple fluents, and statically determined fluents) and three different kinds of causal laws (static laws, action dynamic laws, and fluent dynamic laws). Domain descriptions in this language are then instantiated by assigning temporal stamps to propositions and incorporating the resulting descriptions into the causal calculus of McCain and Turner (1997). The models of the resulting causal theories are viewed then as intended models of the source, higher-level action descriptions.

In this section we are going to single out and "streamline" the logical framework behind the language C+. More specifically, we will show that the dynamic causal calculus, coupled with the semantics of C-transitions, provides a direct and uniform description for this action language. It will be shown, in particular, that it will allow us to alleviate most of the syntactic distinctions between propositional atoms, maintained by C+, as well as type restrictions imposed on its causal laws.

12.4.1 An Overview of C+

The underlying propositional language of the action description language C+, described in Giunchiglia et al. (2004), is somewhat more general than the standard classical propositional language in that it is based on a multivalued propositional signature[3] that consists of a set of constants, along with a function *Dom* assigning every constant c a nonempty finite set $Dom(c)$ of values. Propositional atoms in this signature are expressions of the form $c = v$, where c is a constant while v is one of its possible values. Still, propositional formulas in this language are defined as usual combinations of atoms with the help of the ordinary classical connectives. Moreover, ordinary propositional (Boolean) constants are defined in this setting as a special kind of constants whose domain is the set $\{\mathbf{f}, \mathbf{t}\}$ of truth values.

We have provided a general description of such multivalued propositions in causal contexts in chapter 8 (see section 8.8), and it will be implicitly presupposed in our general representation of the language C+.

An *action description* in C+ is defined as a set of causal laws. There are, however, three kinds of causal laws in C+, and the differences between them are based, ultimately, on a distinction between three kinds of constants, and thereby three kinds of atoms, stipulated by the theory. First, propositional atoms are partitioned into *action* atoms and *fluent* atoms while the latter are further partitioned into *simple* and *statically determined* fluents. A *fluent formula* is defined as a formula such that all constants occurring in it are fluent constants, whereas an *action formula* is a formula that contains at least one action constant and no fluent constants.

Granted the above syntactic distinctions among propositional atoms, the following three kinds of causal laws are defined in C+:

- *Static laws* are expressions of the form

$$\textbf{caused } F \textbf{ if } G,$$

where F and G are fluent formulas;

- *Action dynamic laws* are expressions of the form

$$\textbf{caused } F \textbf{ if } G,$$

---

3. Similar to the language of structural equation models—see chapter 5.

where F is an action formula and G is a formula;

* *Fluent dynamic laws* are expressions of the form

caused F if G after H,

where F is a fluent formula that does not contain statically determined constants, G is a fluent formula, and H is an arbitrary formula.

Static laws are used in $C+$ to talk about causal dependencies between fluents in the same state while action dynamic laws are purported to express causal dependencies between concurrently executed actions. In accordance with their very name, statically determined fluent constants are allowed in the heads of static laws but not in the heads of dynamic laws. As we will see in what follows, the necessity of introducing a separate syntactic sort of statically determined fluents stems from a particular definition of a state of a transition system, which has been used in interpreting action descriptions in $C+$.

The language $C+$ employs also a number of abbreviations and names for special kinds of causal laws. Two such abbreviations play an especially important role in the descriptions of action domains that serve to illustrate the general theory. If c is a simple fluent constant, then

inertial c

stands for the fluent dynamic laws

caused $c = v$ if $c = v$ after $c = v$

for all $v \in Dom(c)$. These laws provide an encoding of the *inertia assumption* for (simple) fluent atoms. Similarly, if c is an action constant, the expression

exogenous c

stands for the action dynamic laws

caused $c = v$ if $c = v$

for all $v \in Dom(c)$. These laws make action atoms exogenous in an action domain, which exempts them, in effect, from explanation.

Interpretations and models of action descriptions in $C+$ are defined indirectly by translating them into plain causal theories. To begin with, for every natural number m, an action description D is transformed into an atemporal causal theory D_m as follows. First, time stamps i: for $i \in \{0, \ldots, m\}$ are inserted in front of every occurrence of every atom in propositional formulas. Then each static law is translated into the following set of causal rules for every $i \leq m$:

$$i : G \Rightarrow i : F,$$

where $i : A$ is the result of inserting i: in front of every occurrence of every atom in a formula A. Similarly, any action dynamic law is translated into a set of causal rules of the same form but only for $i < m$.

Finally, any fluent dynamic law is translated into the following set of causal rules:

$$(i:H) \wedge (i+1:G) \Rightarrow i+1:F$$

for every $i < m$.

As a concluding step, to deal with the initial states, the following causal rules are added to the resulting causal theory:

$$0:l \Rightarrow 0:l$$

for every *simple* fluent atom l. These rules make simple fluent atoms exogenous (self-explained) in the initial state. As a result, we obtain an ordinary causal theory, and the exact models of this theory are considered to be the models of the original action description in $\mathcal{C}+$. Such models can be visualized as histories of length m of the source dynamic domain. More precisely, these are histories in which the initial state is "self-explainable" (except for statically determined fluents), but every subsequent state is already causally explained by the preceding state and actions taken in it.

Giunchiglia et al. (2004) contains also a more general semantic construction, according to which an action description D in $\mathcal{C}+$ describes, in effect, a *transition model* (i.e., a set of states with a set of transitions among them) in which states are the models of the "smallest" (static) causal theory D_0 (which is a theory D_m for $m=0$)[4] while transitions correspond precisely to the models of the minimal dynamic causal theory D_1 (that is, a theory D_m for $m=1$). It has been shown in Giunchiglia et al. (2004, proposition 8) that, for any $m>0$, models of a causal theory D_m are exactly histories (paths) of length m in this transition model.

On statically determined fluents. The above construction of a canonical transition model for an action description implicitly relied on the property (stated as proposition 7 in Giunchiglia et al. 2004) that for any transition $\langle s, e, t \rangle$ in the above sense (that is, for any model of D_1), both s and t are states (i.e., models of D_0). As has been observed in the paper, the validity of this property depends essentially on the fact that the heads of fluent dynamic laws were not allowed to contain statically determined fluent constants, which explained the very need for a *syntactic* (type) separation between simple and statically determined fluent constants in the framework of the general theory described in Giunchiglia et al. (2004).

The distinction between simple and statically determined fluents is only a very special case of a broader distinction between inertial and non-inertial propositions. This distinction plays an important role in reasoning about actions and change in AI, so it should not be obliterated or neglected. Still, it need not be defined as a syntactic distinction among types of fluents. Instead, we suggest to view statically determined fluents as propositional atoms

---

4. Note that, by the definition of the translation, D_0 includes no dynamic laws and only "zero-stamped" static causal laws.

that are not required to comply with the inertia principle (on a pair with compound logical formulas).[5]

Unfortunately, such a solution is unfeasible in the framework of Giunchiglia et al. (2004) because statically determined fluents play an essential role in the definition of the very notion of a state of a transition model that determines, in turn, the semantics of C+. Accordingly, in order to make the language of C+ syntactically uniform, we should make changes also to its semantic description.

12.4.2   The Representation and Comparisons

To begin with, we will rewrite a fluent dynamic law **caused** F **if** G **after** H as a dynamic causal rule

$$H \cdot G \Rightarrow F.$$

As a second step, we will represent both static and action dynamic laws of the form **caused** F **if** G as static causal rules

$$G \Rightarrow F.$$

As a result, any action description in the action description language C+ will be immediately transformed into a dynamic causal theory in our sense. So, as a final step, we will assign this dynamic causal theory a C-transition semantics as its intended (nonmonotonic) interpretation.

Despite obvious similarities, there are some perceptible differences between the original action formalism of C+ and our translation. To begin with, the language of the dynamic causal calculus is thoroughly uniform in that it does not presuppose any a priori, syntactic distinctions among its propositional atoms, and it employs only a single form of causal rules instead of three kinds of causal laws used in C+. On the other hand, the notion of a C-transition model is somewhat different and apparently more restricted than the corresponding notion of a model for C+. In what follows we will discuss these differences in more detail.

Fluents versus actions.   Many action theories in AI, as well as some general dynamic formalisms described in chapter 10, maintain rigid semantic and syntactic distinctions between *fluent* propositions that describe particular states or situations and *actions* that describe transitions between states. Thus, the situation calculus (Reiter 2001) treats actions essentially as modifiers, or functions on, situations. Similarly, propositional dynamic logic interprets actions as binary relations on states, though the corresponding formalism of sequential dynamic logic, introduced in chapter 10, has "relegated" the difference between actions and fluents to a purely semantic distinction between kinds of propositions.

---

5. This suggestion both corrects and simplifies a somewhat more convoluted solution concerning statically determined fluents that has been suggested in Bochman (2014).

In contrast to the above formalisms, it can be immediately observed that there are no inherent syntactic differences between fluent and action atoms in $\mathcal{C}+$; both kinds are translated as (temporally indexed) propositional atoms in the underlying causal calculus. Of course, there are still important differences between these two kinds of propositions: (simple) fluents are governed by the inertia principle while action atoms are normally treated in $\mathcal{C}+$ as exogenous (see above). However, these distinctions are semantical, and they can be secured by incorporating causal rules that make action propositions exogenous:

(Exogeneity) $l \Rightarrow l$,

where l is an action literal. It can be easily shown, in particular, that if Exogeneity holds for all action literals, it will hold for all compound action propositions that are composed of action atoms.

Once we add such rules to action descriptions, there is no need to maintain a syntactic distinction between fluents and actions in the dynamic causal calculus, and there is no need to maintain a separate category of action dynamic laws in addition to static laws.

The treatment of actions in $\mathcal{C}+$ makes the latter much similar to temporal formalisms for analyzing general computation processes, such as linear temporal logic (LTL). In the latter, computational processes are represented as plain temporal sequences of states (which are characterized by fluent propositions), and they do not use explicit action descriptions. Still, actions are encoded in these formalisms by using associated "action fluents" that appropriately constrain such temporal sequences (see, e.g., Calvanese, De Giacomo, and Vardi 2002; De Giacomo and Vardi 2013). However, an important difference of this latter encoding, compared with the representation in $\mathcal{C}+$, is that these action fluents are included as conditions of effects in the "next" state, namely in LTL formulas of the form $\circ A \to E$ (where A is an action name and E an effect). In contrast, in $\mathcal{C}+$ action formulas are included in the preceding state. Note in this respect that the expression A **causes** E **if** C, where A is an action formula, is used in Giunchiglia et al. (2004) only as an abbreviation of **caused** E **after** $A \wedge C$, whereas the latter is in turn an abbreviation of **caused** E **if** T **after** $A \wedge C$. Accordingly, in our translation it corresponds to the dynamic causal rule

$$A \wedge C.\mathbf{t} \Rightarrow E.$$

Speaking more generally, nonvacuous **if** conditions in fluent dynamic rules are used in $\mathcal{C}+$ almost exclusively for describing only special conditions of inertia and constraints.

We suggest, however, that a more natural representation of the expression A **causes** E **if** C could be provided by a full-fledged dynamic causal rule

$$C.A \Rightarrow E.$$

This representation would make it more similar to the above representation in LTL. A more important advantage of this representation, however, is that it appears to be more in accord with the concept of relative causality that has been introduced in chapter 9. On this understanding, the above rule can be viewed as providing a syntactic expression for the distinction between background conditions (C) and proximate causes (A) of effect E (cf. also Denecker, Bogaerts, and Vennekens 2019, for a similar formal representation). We will leave this topic, however, to future studies.

State constraints. A C-transition model of a dynamic causal theory has been defined earlier as a set of states in which every state is caused as a result of some exact transition. As a result, any state of such a model will be closed with respect to the static laws on our reformulation of the latter. Moreover, it can be verified that any such state will be a state in the sense of Giunchiglia et al. (2004). Consequently, we will obtain that any C-transition model in our sense will correspond to a model (transition system) in the sense of Giunchiglia et al. (2004):

Theorem 12.10 *If Δ is an action description in $C+$ and Δ_C its corresponding dynamic causal theory, then any C-transition model of Δ_C is a transition model of Δ.*

Still, our semantics for dynamic causal theories is more restrictive since it requires that any state of the model, *including the initial state*, should be an output of some transition, whereas Giunchiglia et al. (2004) have exempted, in effect, initial states from the need of causal explanation by making them exogenous.

As we have seen earlier, the construction of a causal theory for a given action description in $C+$ necessarily involves an addition of causal rules that make simple fluents exogenous (self-explainable) in the initial state. As a result, the framework Giunchiglia et al. (2004) exempts to some extent initial states from the need of explanation, though it still requires that such states should be closed with respect to the static laws (which are completely separated from dynamic ones) and, moreover, that any statically determined fluent literal that holds in an initial state should still be explained (caused) by the static laws.

Remark. In some sense, the difference between these two interpretations amounts to choosing different "horns" of the basic Agrippan trilemma in causal reasoning (see chapter 1): whereas our definition implements this time infinite regress of causation with respect to states, the approach of Giunchiglia et al. (2004) is based on allowing self-explained states.

To make a proper assessment of the above discrepancy, we should distinguish between two aspects of the difference: the conceptual one and the practical one. On the conceptual side, we believe that an ultimate reason for imposing even the above minimal restrictions on initial states in $C+$ stems from a broader requirement that such states should be somehow accessible in accordance with the laws of the domain. In other words, any state of a dynamic system should be consistently viewed as a result of some legitimate transition (including possible "loops" in this state). Speaking more generally, we contend that static laws and constraints should be viewed as constraints that are effective after every legitimate transition, and vice versa: any constraint that happens to hold after any possible transition should be considered as a static law of the domain.

The suggested semantics of C-transition models allows us to treat static causal rules as a special case of general dynamic causal rules. In addition, it allows us to remove the syntactic distinction between simple and statically determined fluents. Moreover, the law of causality, used as a conceptual basis of this semantics, implies also some important consequences concerning general principles of reasoning in dynamic domains.

Generally speaking, the law of causality (alias the principle of universal causation) provides logical foundations for *abductive reasoning*, namely for backward reasoning from effects to their causes (see chapter 7). This kind of reasoning constitutes an essential part of our commonsense reasoning, especially in reasoning about actions and change, and it occupies also an important part of reasoning in current action theories in AI. For example, the well-known regression method (Reiter 2001) can be viewed as a systematic implementation of abductive reasoning in the situation calculus (see chapter 11).

Due to its causal foundations, abductive inference is also an essential, though implicit, part of the representation framework of $C+$; this is because every state of a transition model for an action description, *except the initial one*, is explained as an output of some causal transition. The associated abductive explanation may sanction, in particular, some further static constraints for such states, constraints that arise as a by-product of the fact that the state in question is a result of a particular action with some further effects. By the same token, however, the initial states in $C+$ are exempt from the abductive explanation of this kind, which, at least in some regular domains, may lead to a loss of important information about these states.

12.4.3 Persistent Action Domains

In case we don't accept the above understanding of static laws, we should maintain two separate kinds of state constraints, dynamic and purely static ones, as is actually done in Giunchiglia et al. (2004) as well as in many other action formalisms. This separation would allow us to include a broader class of transition models as admissible models for action descriptions, namely models that involve (initial) states that need not be a result of some transition, though they still satisfy the static laws of the domain. Moreover, we can easily construct some artificial though logically consistent action domains for which such a distinction would be necessary, namely action domains in which there are distinguished initial states that lack some property that holds for any state that results from a transition. Still, to find "real" examples of such domains is really difficult (even the famous "big bang" exception in physics has become debatable). This naturally brings us to the practical side of the difference between the two formalisms, namely whether, and if so, how much, we miss in restricting the semantics of action descriptions to C-transition models.

It turns out that for a fairly broad class of action descriptions in the language $C+$, including all the examples given in Giunchiglia et al. (2004), we can *guarantee* in advance that any transition model of $C+$ that satisfies a given action description can be extended to a C-transition model in our sense. For such action descriptions, our dynamic causal calculus provides the same answers to the queries as the original theory of Giunchiglia et al. (2004). In what follows, we will illustrate this correspondence for a broad class of what we will call persistent action domains.

Informally speaking, a persistent action domain is a dynamic domain that does not involve involuntary, "natural" actions that lead to unavoidable, necessary changes of some state. For such action domains, any legitimate state either remains persistent in the absence of any further actions upon it (alias after an action *Wait*), or else it can be forced to persist by using suitable (voluntary) actions.

Formally, by a *persistent action domain* we will mean any action description D in the language $C+$ such that for any state s of D (that is, for any model of D_0), there is a consistent transition (i.e., a model of D_1) from s to s.

It turns out to be surprisingly difficult to express the above semantic property of transition models in terms of some syntactic restrictions on action descriptions in $C+$. Still, it is easy to verify the validity of this persistence property for many descriptions used in action theories. Thus, in the central Monkey and Bananas example from Giunchiglia et al. (2004), any state of the transition model can be shown to persist in the absence of actions. On the other hand, for the case of non-inertial fluents that tend to change by themselves, the corresponding action descriptions usually contain actions that make them persist. Thus, the pendulum domain from Giunchiglia et al. (2004) involves an action *Hold* that keeps the position of the pendulum (which tends to sway otherwise).

Now, in persistent action domains, any state of a transition model can be consistently viewed as a result of some transition, which immediately leads to the following theorem.

Theorem 12.11 *If Δ is a persistent action description in $C+$ and Δ_C its corresponding dynamic causal theory, then the canonical transition model of Δ coincides with the exact transition model of Δ_C.*

As a matter of fact, the above correspondence between $C+$ and the dynamic causal calculus can be extended, though in a somewhat weaker form, even beyond persistent action domains. To this end, we should note that practically all queries that are usually formulated in action formalisms are "future-oriented," namely they ask whether there is a path in a transition model from a given initial state that satisfies certain further requirements (such as whether it ends in a target goal state). Now, in many cases we can extend the source action description in $C+$ with some auxiliary actions that will allow us again to reconstruct a given initial state as an output of some exact transition (though in the extended action description). Moreover, this can be done without changing the future of the original states of the transition model, but only by augmenting their "past." As an immediate consequence, we will obtain that, for such action domains, the dynamic causal calculus will provide the same answer to such queries as the original action descriptions in $C+$.

12.5 Summary

The primary objective of this chapter consisted in showing that causal reasoning in dynamic domains of AI can be given a direct and concise formal representation. Being combined with the wealth of representation capabilities of such a reasoning, demonstrated in Giunchiglia et al. (2004) and corresponding studies based on the languages $\mathcal{B}$ and $\mathcal{AL}$ (see Baral and Gelfond 2005), the results of this chapter strongly indicate that a theory of dynamic causal inference can be viewed as a self-subsistent logical theory that provides an adequate representation framework for reasoning in dynamic domains. Of course, much work still has to be done in order to extend this causal framework to description of more complex processes that involve, for instance, temporally extended actions and events, concurrency, and triggered (natural) events. Such a formalism should also be capable of providing more concise description of causal interactions ("collisions") between different causal processes that have occupied so much place in our examples discussed in chapters 8 and 9. We see these issues as important topics for subsequent stages of the logical theory of causality that has been presented in this book.

Bibliography

Alchourrón, C., P. Gärdenfors, and D. Makinson. 1985. "On the logic of theory change: Partial meet contraction and revision functions." *Journal of Symbolic Logic* 50:510–530.

Amgoud, L., and P. Besnard. 2013. "Logical limits of abstract argumentation frameworks." *Journal of Applied Non-Classical Logics* 23 (3): 229–267.

Angioni, L. 2018. "Causality and Coextensiveness in Aristotle's Posterior Analytics 1. 13." In *Oxford Studies in Ancient Philosophy,* edited by V. Caston, 54:159–185. Oxford University Press.

Anscombe, G. E. M. 1981. "Causality and Determination." In *The Collected Philosophical Papers of G. E. M. Anscombe,* 133–147. Basil Blackwell.

Armstrong, D. M. 1997. *A World of States of Affairs.* Cambridge University Press.

Balduccini, M., M. Gelfond, and M. Nogueira. 2006. "Answer set based design of knowledge systems." *Annals of Mathematics and Artificial Intelligence* 47:183–219.

Baldwin, R. A., and E. Neufeld. 2004. "The structural model interpretation of the NESS test." In *Advances in Artificial Intelligence,* 3060:297–307. Lecture Notes in Computer Science. Springer.

Baral, C. 2003. *Knowledge Representation, Reasoning and Declarative Problem Solving.* Cambridge University Press.

Baral, C., and M. Gelfond. 2005. "Logic programming and reasoning about actions." In *Handbook of Temporal Reasoning in Artificial Intelligence,* edited by M. Fisher, D. Gabbay, and L. Vila, 389–426. Elsevier.

Barnes, J. 1984. *The Complete Works of Aristotle: The Revised Oxford Translation.* Princeton University Press.

Barnes, J. 1993. *Aristotle: Posterior Analytics.* Clarendon Press.

Baroni, P., and M. Giacomin. 2007. "On principle-based evaluation of extension-based argumentation semantics." *Artificial Intelligence* 171 (10–15): 675–700.

Barwise, J., D. Gabbay, and C. Hartonas. 1995. "On the logic of information flow." *Bulletin of the IGPL* 3:7–50.

Barwise, J., and J. Perry. 1983. *Situations and Attitudes.* MIT Press.

Baumgartner, M. 2013. "A Regularity Theoretic Approach to Actual Causation." *Erkenntnis* 78:85–109.

Beall, J., R. Brady, J. M. Dunn, A.P. Hazen, E. Mares, R. K. Meyer, G. Priest, et al. 2012. "On the Ternary Relation and Conditionality." *Journal of Philosophical Logic* 41 (3): 595–612.

Beckers, S., and J. Vennekens. 2018. "A principled approach to defining actual causation." *Synthese* 195:835–862.

Beebee, H. 2006. *Hume on Causation.* Routledge.

Beebee, H. 2014. "Causation." In *The Bloomsbury Companion to Analytic Philosophy,* edited by B. Dainton and H. Robinson. Bloomsbury Press.

Beebee, H. 2016. "Hume and the problem of causation." In *The Oxford Handbook of Hume,* edited by P. Russell. Oxford University Press.

Belnap, N. D., Jr. 1977. "A useful four-valued logic." In *Modern Uses of Multiple-Valued Logic,* edited by M. Dunn and G. Epstein, 8–41. D. Reidel.

Benferhat, S., C. Cayrol, D. Dubois, J. Lang, and H. Prade. 1993. "Inconsistency Management and Prioritized Syntax-Based Entailment." In *Proceedings of the Thirteenth International Joint Conference on Artificial Intelligence, IJCAI'93,* 640–645. Morgan Kaufmann.

Bennett, J. 1988. *Events and Their Names.* Clarendon Press.

Van Benthem, J. 1986. "Partiality and Monotonicity in Classical Logic." *Logique et analyse* 19 (114): 225–247.

Van Benthem, J. 1994. "A Note on Dynamic Arrow Logic." In *Logic and Information Flow,* edited by J. van Eijck and A. Visser, 15–29. MIT Press.

Van Benthem, J. 1996. *Exploring Logical Dynamics.* CSLI Publications.

Van Benthem, J. 2009. "The information in intuitionistic logic." *Synthese* 167 (2): 251–270.

Van Benthem, J. 2019. "Implicit and Explicit Stances in Logic." *Journal of Philosophical Logic* 48 (3): 571–601.

Beth, E. W. 1956. "Semantic construction of intuitionistic logic." *Mededelingen Koninklijke Nederlandse Akademie van Wetenschappen* 19 (11): 357–388.

Blackburn, P., and Y. Venema. 1995. "Dynamic squares." *Journal of Philosophical Logic* 24:469–523.

Blanchard, T., and J. Schaffer. 2017. "Cause without default." In *Making a Difference,* edited by H. Beebee, C. Hitchcock, and H. Price, 175–214. Oxford University Press.

Bochman, A. 1980. "Mereology as a Theory of Part-Whole." *Logique et Analyse* 129/130:75–101.

Bochman, A. 1992. "Mereological Semantics." PhD diss., Tel-Aviv University.

Bochman, A. 2001. *A Logical Theory of Nonmonotonic Inference and Belief Change.* Springer.

Bochman, A. 2003a. "A Logic for Causal Reasoning." In *IJCAI-03: Proceedings of the 18th International Joint Conference on Artificial Intelligence,* 141–146. Morgan Kaufmann.

Bochman, A. 2003b. "Collective Argumentation and Disjunctive Logic Programming." *Journal of Logic and Computation* 9:55–56.

Bochman, A. 2003c. "On disjunctive causal inference and indeterminism." In *Proceedings of the IJCAI-03 Workshop on Nonmonotonic Reasoning, Action and Change, NRAC'03,* edited by G. Brewka and P. Peppas, 45–50.

Bochman, A. 2004. "A causal approach to nonmonotonic reasoning." *Artificial Intelligence* 160:105–143.

Bochman, A. 2005. *Explanatory Nonmonotonic Reasoning.* World Scientific.

Bochman, A. 2007a. "A Causal Theory of Abduction." *Journal of Logic and Computation* 17:851–869.

Bochman, A. 2007b. "Non-monotonic reasoning." In *The Handbook of the History of Logic,* vol. 8: The Many Valued and Non-monotonic Turn in Logic. Elsevier.

Bochman, A. 2008. "Default logic generalized and simplified." *Annals of Mathematics and Artificial Intelligence* 53:21–49.

Bochman, A. 2011. "Logic in Nonmonotonic Reasoning." In *Nonmonotonic Reasoning. Essays Celebrating Its 30th Anniversary,* edited by G. Brewka, V. W. Marek, and M. Truszczynski, 25–61. College Publications.

Bochman, A. 2013. "The Markov Assumption: Formalization and Impact." In *Proceedings of the Twenty-Third International Joint Conference on Artificial Intelligence.* AAAI Press.

Bochman, A. 2014. "Dynamic Causal Calculus." In *Principles of Knowledge Representation and Reasoning: Proceedings of the Fourteenth International Conference, KR 2014,* edited by C. Baral, G. De Giacomo, and T. Eiter. AAAI Press.

Bochman, A. 2016a. "Abstract Dialectical Argumentation Among Close Relatives." In *Computational Models of Argument: Proceedings of COMMA 2016, Potsdam, Germany,* edited by P. Baroni, T. F. Gordon, T. Scheffler, and M. Stede, 127–138. IOS Press.

Bochman, A. 2016b. "On Logics and Semantics of Indeterminate Causation." In *Principles of Knowledge Representation and Reasoning: Proceedings of the Fifteenth International Conference, KR 2016,* 401–410. AAAI Press.

Bochman, A. 2018a. "Actual causality in a logical setting." In *Proceedings of the 27th International Joint Conference on Artificial Intelligence and the 23rd European Conference on Artificial Intelligence, IJCAI-ECAI-18,* 1730–1736.

Bochman, A. 2018b. "On Laws and Counterfactuals in Causal Reasoning." In *Principles of Knowledge Representation and Reasoning: Proceedings of the Sixteenth International Conference, KR 2018,* 494–503. AAAI Press.

Bochman, A., and D. M. Gabbay. 2012a. "Causal dynamic inference." *Annals of Mathematics and Artificial Intelligence* 66 (1-4): 231–256.

Bochman, A., and D. M. Gabbay. 2012b. "Sequential dynamic logic." *Journal of Logic, Language and Information* 21 (3): 279–298.

Bochman, A., and V. Lifschitz. 2015. "Pearl's Causality in a Logical Setting." In *Proceedings of the 29th AAAI Conference on Artificial Intelligence,* 1446–1452. AAAI Press.

Bogaerts, B., J. Vennekens, M. Denecker, and J. Van den Bussche. 2014. "FO(C) and Related Modelling Paradigms." *CoRR* abs/1404.6394.

Bondarenko, A., P. M. Dung, R. A. Kowalski, and F. Toni. 1997. "An Abstract, Argumentation-Theoretic Framework for Default Reasoning." *Artificial Intelligence* 93:63–101.

Brewka, G., J. Dix, and K. Konolige. 1997. *Nonmonotonic Reasoning: An Overview.* CSLI Publications.

Brewka, G., H. Strass, S. Ellmauthaler, J. P. Wallner, and S. Woltran. 2013. "Abstract Dialectical Frameworks Revisited." In *IJCAI 2013: Proceedings of the 23rd International Joint Conference on Artificial Intelligence.* AAAI Press.

Brewka, G., and S. Woltran. 2010. "Abstract Dialectical Frameworks." In *Principles of Knowledge Representation and Reasoning: Proceedings of the Twelfth International Conference, KR 2010.*

Bromberger, S. 1966. "Why-Questions." In *Mind and Cosmos,* edited by R. Colodny, 86–111. University of Pittsburg Press.

Burgess, J. P. 1977. "Forcing." In *Handbook of Mathematical Logic,* edited by J. Barwise. North-Holland.

Burgess, J. P. 1981. "Quick Completeness Proofs for some Logics of Conditionals." *Notre Dame Journal of Formal Logic* 22:76–84.

Burnyeat, M. 1981. "Aristotle on Understanding Knowledge." In *Aristotle on Science: The Posterior Analytics,* edited by E. Berti, 97–140. Antenore.

Byrne, R. 2005. *The Rational Imagination: How People Create Alternatives to Reality.* MIT Press.

Calvanese, D., G. De Giacomo, and M. Y. Vardi. 2002. "Reasoning about Actions and Planning in LTL Action Theories." In *Proceedings of the 8th International Conference on Principles of Knowledge Representation and Reasoning, (KR-02),* 593–602. Morgan Kaufmann.

Caminada, M., and L. Amgoud. 2007. "On the evaluation of argumentation formalisms." *Artificial Intelligence* 171 (5–6): 286–310.

Caminada, M. W. A., and D. M. Gabbay. 2009. "A Logical Account of Formal Argumentation." *Studia Logica* 93 (2–3): 109–145.

Carnap, R. 1943. *Formalization of Logic.* Harvard University Press.

Carroll, J. W. 2016. "Laws of Nature." In *The Stanford Encyclopedia of Philosophy,* edited by E. N. Zalta. Metaphysics Research Lab, Stanford University.

Cartwright, N. 1983. *How the Laws of Physics Lie.* Clarendon Press.

Castilho, M., A. Herzig, and I. Varzinczak. 2002. "It depends on the context! A decidable logic of actions and plans based on a ternary dependence relation." In *Proceedings of the 9th International Workshop on Non-Monotonic Reasoning, NMR'2002.*

Ciardelli, I., J. Groenendijk, and F. Roelofsen. 2018. *Inquisitive Semantics.* Oxford University Press.

Ciardelli, I., and F. Roelofsen. 2011. "Inquisitive logic." *Journal of Philosophical Logic* 40 (1): 55–94.

Clatterbaugh, K. 1999. *The Causation Debate in Modern Philosophy, 1637–1739.* Routledge.

Collins, J. 2000. "Preemptive Prevention." *The Journal of Philosophy* 97 (4): 223–234.

Console, L., D. Theseider Dupre, and P. Torasso. 1991. "On the Relationship Between Abduction and Deduction." *Journal of Logic and Computation* 1:661–690.

Corcoran, J. 2009. "Aristotle's Demonstrative Logic." *History and Philosophy of Logic* 30 (1): 1–20.

Cox, P. T., and T. Pietrzykowski. 1987. "General diagnosis by abductive inference." In *Proceedings of the 1987 Symposium on Logic Programming,* 183–189. IEEE.

Cresswell, M. J. 2004. "Possibility Semantics for Intuitionistic Logic." *Australasian Journal of Logic* 2:11–29.

Darwiche, A. 1995. "Model-based diagnosis using causal networks." In *Proceedings of the International Joint Conference on Artificial Intelligence, IJCAI-95,* 211–217. Morgan Kaufmann.

Darwiche, A., and P. Marquis. 2002. "A Knowledge Compilation Map." *Journal of Artificial Intelligence Research* 17 (1): 229–264.

Darwiche, A., and J. Pearl. 1994. "Symbolic causal networks." In *Proceedings of the Twelfth National Conference on Artificial Intelligence (AAAI-94),* 238–244.

Dash, D., and M. Druzdzel. 2001. "Caveats for causal reasoning with equilibrium models." In *European Conference on Symbolic and Quantitative Approaches to Reasoning and Uncertainty (ECSQARU),* edited by S. Benferhat and P. Besnard, 192–203. Springer.

Davidson, D. 1980. *Essays on Actions and Events.* Clarendon Press.

De Giacomo, G., and M. Y. Vardi. 2013. "Linear Temporal Logic and Linear Dynamic Logic on Finite Traces." In *IJCAI 2013: Proceedings of the 23rd International Joint Conference on Artificial Intelligence.*

De Pierris, G., and M. Friedman. 2018. "Kant and Hume on Causality." In *The Stanford Encyclopedia of Philosophy,* edited by E. N. Zalta. Metaphysics Research Lab, Stanford University.

Dekker, P. J. E. 2008. *A Guide to Dynamic Semantics.* Technical report. ILLC Prepublications.

Demolombe, R., A. Herzig, and I. J. Varzinczak. 2003. "Regression in Modal Logic." *Journal of Applied Non-Classical Logics* 13 (2): 165–185.

Denecker, M., B. Bogaerts, and J. Vennekens. 2019. "Explaining Actual Causation in Terms of Possible Causal Processes." In *Proceedings of the 16th European Conference on Logics in Artificial Intelligence, JELIA,* 214–230. Springer.

Detel, W. 2012. "Aristotle's Logic and Theory of Science." In *A Companion to Ancient Philosophy,* 245–269. Blackwell.

Dietz Saldanha, E. A. 2017. "From Logic Programming to Human Reasoning: How to be Artificially Human." PhD diss., Dresden University of Technology, Germany.

Dix, J., G. Gottlob, and V. Marek. 1994. "Causal Models for Disjunctive Logic Programs." In *Proceedings of the Eleventh International Conference on Logic Programming,* edited by P. Van Hentenryck, 290–302. MIT Press.

Došen, K. 1989. "Logical constants as punctuation marks." *Notre Dame Journal of Formal Logic* 30 (3): 362–381.

Dowe, P. 2000. *Physical Causation.* Cambridge University Press.

Doyle, J. 1979. "A truth maintenance system." *Artificial Intelligence* 12:231–272.

Doyle, J. 1994. "Reasoned Assumptions and Rational Psychology." *Fundamenta Informaticae* 20:35–73.

Dummett, M. 2000. *Elements of Intuitionism.* Clarendon Press.

Duncombe, M. 2014. "Irreflexivity and Aristotle's syllogismos." *The Philosophical Quarterly* 64 (256): 434–452.

Dung, P. M. 1995a. "An argumentation-theoretic foundation for logic programming." *Journal of Logic Programming* 22:151–177.

Dung, P. M. 1995b. "On the Acceptability of Arguments and its Fundamental Role in Non-Monotonic Reasoning, Logic Programming and N-Persons Games." *Artificial Intelligence* 76:321–358.

Dung, P. M., and P. M. Thang. 2014. "Closure and Consistency In Logic-Associated Argumentation." *Journal of Artificial Intelligence Research (JAIR)* 49:79–109.

Ehring, D. 1987. "Causal Relata." *Synthese* 73:319–328.

Van Eijck, J. 1999. "Axiomatising dynamic logics for anaphora." *Journal of Language and Computation* 1:103–126.

Van Eijck, J., and M. Stokhof. 2006. "The gamut of dynamic logics." In *Handbook of the History of Logic,* edited by D. M. Gabbay and J. Woods, 7:499–600. North-Holland.

Eiter, T., G. Gottlob, and Y. Gurevich. 1993. "Curb Your Theory! A Circumscriptive Approach for Inclusive Interpretation of Disjunctive Information." In *Proceedings of the International Joint Conference on Artificial Intelligence, IJCAI-93,* 634–639. Morgan Kaufman.

Ellmauthaler, S. 2012. "Abstract Dialectical Frameworks: Properties, Complexity, and Implementation." Master's thesis, Technische Universität Wien, Institut für Informationssysteme.

Etherington, D., and R. Reiter. 1983. "On Inheritance Hierarchies with Exceptions." In *Proceedings of the Third National Conference on Artificial Intelligence (AAAI-83),* 104–108.

Fine, K. 1975. "Vagueness, truth and logic." *Synthese* 30 (3-4): 265–300.

Fischer, D. A. 1992. "Causation in fact in omission cases." *Utah Law Review,* 1335–84.

Fischer, D. A. 2006. "Insufficient Causes." *University of Kentucky Law Review* 94:277–317.

Forbus, K. D. 1984. "Qualitative process theory." *Artificial Intelligence* 24:85–168.

Van Fraassen, B. C. 1971. *Formal Semantics and Logic.* Macmillan.

Gabbay, D. M. 1981. *Semantical Investigations in Heyting's Intuitionistic Logic.* D. Reidel.

Galles, D., and J. Pearl. 1998. "An Axiomatic Characterization of Causal Counterfactuals." *Foundations of Science* 3 (1): 151–182.

Gardner, J., and A. Honoré. 2017. "Causation in the Law." In *The Stanford Encyclopedia of Philosophy,* edited by E. N. Zalta. Unpublished ed. Metaphysics Research Lab, Stanford University.

Gardner, M. 1970. "Mathematical Games—The Fantastic Combinations of John Conway's New Solitaire Game 'Life'." *Scientific American* 223 (October): 120–123.

Garson, J. W. 2001. "Natural Semantics: Why Natural Deduction is Intuitionistic." *Theoria* 67:114–139.

Garson, J. W. 2013. *What Logics Mean: From Proof Theory to Model-Theoretic Semantics.* Cambridge University Press.

Geffner, H. 1992. *Default Reasoning. Causal and Conditional Theories.* MIT Press.

Geffner, H. 1997. "Causality, constraints and the indirect effects of actions." In *Proceedings of the International Joint Conference on Artificial Intelligence, IJCAI'97,* 555–561.

Geffner, H., and J. Pearl. 1992. "Conditional entailment: bridging two approaches to default reasoning." *Artificial Intelligence* 53 (2–3): 209–244.

Gelfond, M., and V. Lifschitz. 1998. "Action languages." *Electronic Transactions on Artificial Intelligence* 3:195–210.

Gelfond, M., and V. Lifschitz. 2012. "The common core of action languages B and C." In *Working Notes of the International Workshop on Nonmonotonic Reasoning (NMR)*.

Gelfond, M., V. Lifschitz, H. Przymusińska, and M. Truszczyński. 1991. "Disjunctive defaults." In *Proceedings of the 2nd International Conference on Principles of Knowledge Representation and Reasoning, KR'91*, 230–237.

Gerstenberg, T., N. D. Goodman, D. Lagnado, and J. B. Tenenbaum. 2014. "From couterfactual simulation to causal judgment." In *Proceedings of the 36th Annual Conference of the Cognitive Science Society (CogSci 2014)*, 523–528.

Ginsberg, M. L., and D. E. Smith. 1988. "Reasoning About Action II: The Qualification Problem." *Artificial Intelligence* 35 (3): 311–342.

Giunchiglia, E., J. Lee, V. Lifschitz, N. McCain, and H. Turner. 2004. "Nonmonotonic Causal Theories." *Artificial Intelligence* 153:49–104.

Giunchiglia, E., and V. Lifschitz. 1998. "An Action Language Based on Causal Explanation: Preliminary report." In *Proceedings of the Fifteenth National Conference on Artificial Intelligence (AAAI-98)*, 623–630. AAAI Press.

Van Glabbeek, R. J., and G. D. Plotkin. 2009. "Configuration structures, event structures and Petri nets." *Theoretical Computer Science* 410:4111–4159.

Glymour, C. 2016. "Responses." *Synthese* 193:1251–1285.

Glymour, C., D. Danks, B. Glymour, F. Eberhardt, J. Ramsey, R. Scheines, P. Spirtes, C. M. Teng, and J. Zhang. 2010. "Actual causation: A stone soup essay." *Synthese* 175:169–192.

Glymour, C., and F. Wimberly. 2007. "Actual causes and thought experiments." In *Causation and Explanation*, edited by J. K. Campbell, M. O'Rourke, and H. S. Silverstein, 43–67. MIT Press.

Goodman, N. 1947. "The Problem of Counterfactual Conditionals." *Journal of Philosophy* 44:113–128.

Groenendijk, J., and M. Stokhof. 1991. "Dynamic predicate logic." *Linguistics and Philosophy* 14:39–101.

Hall, N. 2000. "Causation and the Price of Transitivity." *Journal of Philosophy* 97:98–222.

Hall, N. 2003. "Causation." In *The Oxford Handbook of Contemporary Philosophy*, edited by F. Jackson and M. Smith. Oxford University Press.

Hall, N. 2004a. "The Intrinsic Character of Causation." In *Studies in Metaphysics*, edited by D. Zimmerman, 1:255–300. Oxford University Press.

Hall, N. 2004b. "Two Concepts of Causation." In *Causation and Counterfactuals*, edited by J. Collins, N. Hall, and L. Paul, 225–276. MIT Press.

Hall, N. 2007. "Structural equations and causation." *Philosophical Studies* 132:109–136.

Halpern, J. Y. 2000. "Axiomatizing causal reasoning." *Journal of Artificial Intelligence Research* 12:317–337.

Halpern, J. Y. 2008. "Defaults and normality in causal structures." In *Principles of Knowledge Representation and Reasoning: Proceedings of the 11th International Conference (KR'08)*, 198–208.

Halpern, J. Y. 2015. "A modification of the Halpern-Pearl definition of causality." In *Proceedings of the 24th International Joint Conference on Artificial Intelligence (IJCAI 2015)*, 3022–3033.

Halpern, J. Y. 2016a. *Actual Causality*. MIT Press.

Halpern, J. Y. 2016b. "Appropriate Causal Models and the Stability of Causation." *Review of Symbolic Logic* 9 (1): 76–102.

Halpern, J. Y., and C. Hitchcock. 2015. "Graded causation and defaults." *The British Journal for the Philosophy of Science* 66 (2): 413–457.

Halpern, J. Y., and J. Pearl. 2001. "Causes and Explanations: A Structural-Model Approach. Part I: Causes." In *Proceedings of the 7th Conference on Uncertainty in Artificial Intelligence (UAI'01)*, 194–202. Morgan Kaufmann.

Halpern, J. Y., and J. Pearl. 2005. "Causes and explanations: A structural-model approach. Part I: Causes." *British Journal for Philosophy of Science* 56 (4): 843–887.

Hamblin, C. L. 1970. *Fallacies*. Methuen.

Hamblin, C. L. 1971. "Mathematical Models of Dialogue." *Theoria* 37:130–155.

Harel, D., D. Kozen, and J. Tiuryn. 2000. *Dynamic Logic*. MIT Press.

Hart, H. L. A., and T. Honoré. 1985. *Causation in the Law*. 2nd. Oxford University Press.

Hausman, D. M., R. Stern, and N. Weinberger. 2014. "Systems without a graphical causal representation." *Synthese* 191:1925–1930.

Hempel, C. G. 1965. *Aspects of Scientific Explanation*. Free Press.

Hempel, C. G. 1988. "Provisos: A Problem Concerning the Inferential Function of Scientific Theories." *Erkenntnis* 28 (2): 147–164.

Henne, P., L. Niemi, Á. Pinillos, F. De Brigard, and J. Knobe. 2019. "A counterfactual explanation for the action effect in causal judgment." *Cognition* 190:157–164.

Herzig, A., and I. Varzinczak. 2007. "Metatheory of actions: Beyond consistency." *Artificial Intelligence* 171 (16–17): 951–984.

Hiddleston, E. 2005. "Causal powers." *British Journal for Philosophy of Science* 56:27–59.

Hitchcock, C. 2001. "The intransitivity of causation revealed in equations and graphs." *Journal of Philosophy* 98 (6): 273–299.

Hitchcock, C. 2007. "Prevention, preemption, and the principle of sufficient reason." *Philosophical Review* 116:495–532.

Hitchcock, C. 2011. "The Metaphysical Bases of Liability: Commentary on Michael Moore's Causation and Responsibility." *Rutgers Law Journal* 42 (2): 377–404.

Hitchcock, C. 2012. "Events and times: A case study in means-ends metaphysics." *Philosophical studies* 160 (1): 79–96.

Hoare, T., G. Struth, and J. Woodcock. 2019. "A Calculus of Space, Time, and Causality: Its Algebra, Geometry, Logic." In *Unifying Theories of Programming,* edited by P. Ribeiro and A. Sampaio, 3–21. Springer.

Hollenberg, M. 1997. "An Equational Axiomatization of Dynamic Negation and Relational Composition." *Journal of Logic, Language and Information* 6:381–401.

Hopkins, M., and J. Pearl. 2003. "Clarifying the usage of structural models for commonsense causal reasoning." In *Proceedings of the AAAI Spring Symposium on Logical Formalizations of Commonsense Reasoning.*

Horty, J. F. 1994. "Some Direct Theories of Nonmonotonic Inheritance." In *Handbook of Logic in Artificial Intelligence and Logic Programming, Vol. 3: Nonmonotonic Reasoning and Uncertain Reasoning,* edited by D. M. Gabbay, C. J. Hogger, and J. A. Robinson. Oxford University Press.

Humberstone, I. L. 1981. "From Worlds to Possibilities." *Journal of Philosophical Logic* 10:313–339.

Humberstone, I. L. 2011. *The Connectives.* MIT Press.

Hume, D. [1739–1740]1978. *A Treatise of Human Nature.* Edited by L. A. Selby-Bigge. 2nd edn, rev. and ed. P. H. Nidditch. Clarendon Press.

Hume, D. [1748]1975. *Enquiries Concerning Human Understanding and Concerning the Principles of Morals.* Edited by L. A. Selby-Bigge. 3rd edn, rev. and ed. P. H. Nidditch. Clarendon Press.

Jakobovits, H., and D. Vermeir. 1999. "Robust Semantics for Argumentation Frameworks." *Journal of Logic and Computation* 9:215–261.

Janhunen, T. 1999. "On the intertranslatability of non-monotonic logics." *Annals of Mathematics and Artificial Intelligence* 27:791–828.

Kahneman, D., and D. Miller. 1986. "Norm Theory: Comparing Reality to its Alternatives." *Psychological Review* 93:136–153.

Kanazawa, M. 1994. "Completeness and decidability of the mixed style of inference with composition." In *Proceedings of the Ninth Amsterdam Colloquium,* edited by P. Dekker and M. Stokhof, 377–390.

Kant, I. [1787]1998. *Critique of Pure Reason.* Edited by P. Guyer and A. Wood. Cambridge University Press.

Kim, J. 1971. "Causes and Events: Mackie on Causation." *Journal of Philosophy* 68:426–41.

De Kleer, J. 1986. "An assumption-based TMS." *Artificial Intelligence* 28:127–162.

Kneale, W. C. 1956. "The Province of Logic." In *Contemporary British Philosophy: Third Series,* edited by H. D. Lewis, 237–261. Allen / Unwin.

Konolige, K. 1992. "Abduction versus Closure in Causal Theories." *Artificial Intelligence* 53:255–272.

Konolige, K. 1994. "Using Default and Causal Reasoning in Diagnosis." *Annals of Mathematics and Artificial Intelligence* 11:97–135.

Konolige, K., and K. L. Myers. 1989. "Representing defaults with epistemic concepts." *Computational Intelligence* 5:32–44.

Kozen, D., and R. Parikh. 1981. "An elementary proof of the completeness of PDL." *Theoretical Computer Science* 14:113–118.

Kozen, D., and J. Tiuryn. 2003. "Substructural logic and partial correctness." *ACM Transactions on Computational Logic* 4 (3): 355–378.

Kratzer, A. 1981. "Partition and Revision: The Semantics of Counterfactuals." *Journal of Philosophical Logic* 10 (2): 201–216.

Kratzer, A. 1989. "An Investigation of the Lumps of Thought." *Linguistics and Philosophy* 12:607–653.

Kratzer, A. 2012. *Modals and Conditionals.* Oxford University Press.

Kraus, S., D. Lehmann, and M. Magidor. 1990. "Nonmonotonic reasoning, preferential models and cumulative logics." *Artificial Intelligence* 44:167–207.

Van Lambalgen, M., and F. Hamm. 2004. *The Proper Treatment of Events.* Wiley-Blackwell.

Lang, J., F. Lin, and P. Marquis. 2003. "Causal Theories of Action: A Computational Core." In *IJCAI-03: Proceedings of the 8th International Joint Conference on Artificial Intelligence,* 1073–1078.

Lee, J. 2004. "Nondefinite Vs. Definite Causal Theories." In *Proceedings of the 7th International Conference on Logic Programming and Nonmonotonic Reasoning, LPNMR 2004,* edited by V. Lifschitz and I. Niemelä, 141–153. Springer.

Lehmann, D. 1995. "Another Perspective on Default Reasoning." *Annals of Mathematics and Artificial Intelligence* 15:61–82.

Lehmann, D., and M. Magidor. 1992. "What does a conditional knowledge base entail?" *Artificial Intelligence* 55:1–60.

Lent, J., and R. H. Thomason. 2015. "Action Models for Conditionals." *Journal of Logic, Language and Information* 24 (2): 211–231.

Lewis, D. 1973a. "Causation." *Journal of Philosophy* 70:556–567.

Lewis, D. 1973b. *Counterfactuals.* Harvard University Press.

Lewis, D. 1979. "Counterfactual Dependence and Time's Arrow." *Noûs* 13:455–476.

Lewis, D. 1981. "Ordering Semantics and Premise Semantics for Counterfactuals." *Journal of Philosophical Logic* 10:217–234.

Lewis, D. 1986a. *On the Plurality of Worlds.* Oxford University Press.

Lewis, D. 1986b. *Philosophical Papers.* Vol. II. Oxford University Press.

Lewis, D. 2000. "Causation as Influence." *The Journal of Philosophy* 97 (4): 181–197.

Lifschitz, V. 1985. "Computing Circumscription." In *Proceedings of the 9th International Joint Conference on Artificial Intelligence, IJCAI-85,* 121–127. Morgan Kaufmann.

Lifschitz, V. 1994. "Minimal belief and negation as failure." *Artificial Intelligence* 70:53–72.

Lifschitz, V. 1997. "On the Logic of Causal Explanation." *Artificial Intelligence* 96:451–465.

Lifschitz, V. 2019. *Answer Set Programming.* Springer.

Lifschitz, V., D. Pearce, and A. Valverde. 2001. "Strongly Equivalent Logic Programs." *ACM Transactions on Computational Logic* 2:526–541.

Lifschitz, V., and F. Yang. 2013. "Functional completion." *Journal of Applied Non-Classical Logics* 23 (1–2): 121–130.

Lin, F. 1995. "Embracing causality in specifying the inderect effect of actions." In *Proceedings of the International Joint Conference on Artificial Intelligence, IJCAI-95,* 1985–1991. Morgan Kaufmann.

Lin, F. 1996. "Embracing Causality in Specifying the Indeterminate Effects of Actions." In *Proceedings of the Thirteenth National Conference on Artificial Intelligence: AAAI-96,* 670–676.

Lin, F. 2008. "Situation Calculus." In *Handbook of Knowledge Representation,* edited by F. van Harmelen, V. Lifschitz, and B. Porter, 649–669. Elsevier.

Lin, F., and R. Reiter. 1994. "State Constraints Revisited." *Journal of Logic and Computation* 4 (5): 655–678.

Lin, F., and Y. Shoham. 1989. "Argument systems: A uniform basis for nonmonotonic reasoning." In *Proceedings of the 1st International Conference on Principles of Knowledge Representation and Reasoning,* 245–255.

Lin, F., and Y. Shoham. 1992. "A logic of knowledge and justified assumptions." *Artificial Intelligence* 57:271–289.

Livengood, J. 2013. "Actual Causation and Simple Voting Scenarios." *Noûs* 47 (2): 316–345.

Lobo, J., J. Minker, and A. Rajasekar. 1992. *Foundations of Disjunctive Logic Programming.* MIT Press.

Lobo, J., and C. Uzcátegui. 1997. "Abductive Consequence Relations." *Artificial Intelligence* 89:149–171.

Lombard, L. B. 1990. "Causes, enablers, and the counterfactual analysis of events causation." *Philosophical Studies* 59:195–211.

Lombard, L. B., and T. Hudson. 2020. "Causation by Absence: Omission Impossible." *Philosophia* 48 (2): 625–641.

Lorenzen, P., and K. Lorenz. 1978. *Dialogische Logik.* Wissenschaftliche Buchgesellschaft.

Mackie, J. L. 1974. *The Cement of the Universe. A Study of Causation.* Clarendon Press.

Makinson, D., and L. van der Torre. 2000. "Input/output Logics." *Journal of Philosophical Logic* 29:383–408.

Mandel, D. R. 2003. "Judgment dissociation theory: An analysis of differences in causation, counterfactual, and covariational reasoning." *Journal of Experimental Psychology: General* 137:419–34.

Marek, V. W., G. F. Schwarz, and M. Truszchinski. 1993. "Modal nonmonotonic logics: Ranges, characterization, computation." *Journal of the ACM* 40:963–990.

Marek, W., A. Nerode, and J. Remmel. 1990. "A theory of nonmonotonic rule systems." *Annals of Mathematics and Artificial Intelligence* 1:241–273.

Marek, W., and M. Truszczyński. 1989. "Relating autoepistemic and default logics." In *Proceedings of the International Conference on Principles of Knowledge Representation and Reasoning, KR'89,* 276–288. Morgan Kaufmann.

Marek, W., and M. Truszczyński. 1993. *Nonmonotonic Logic, Context-Dependent Reasoning.* Springer.

Maslen, C. 2012. "Regularity Accounts of Causation and the Problem of Pre-emption: Dark Prospects Indeed." *Erkenntnis* 77 (3): 419–434.

Maudlin, T. 2004. "Causation, Counterfactuals, and the Third Factor." In *Counterfactuals and Causation,* edited by J. Collins, N. Hall, and L. A. Paul. MIT Press.

McCain, N., and H. Turner. 1997. "Causal Theories of Action and Change." In *Proceedings of the Fourteenth National Conference on Artificial Intelligence (AAAI-97),* 460–465.

McCarthy, J. 1980. "Circumscription—a form of non-monotonic reasoning." *Artificial Intelligence* 13:27–39.

McCarthy, J. 1986. "Applications of circumscription to formalizing common sense knowledge." *Artificial Intelligence* 13:27–39.

McCarthy, J. 1998. "Elaboration tolerance." In *The 1998 Symposium on Logical Formalizations of Commonsense Reasoning (Common Sense-98).*

McCarthy, J., and P. Hayes. 1969. "Some philosophical problems from the standpoint of artificial intelligence." In *Machine Intelligence,* edited by B. Meltzer and D. Michie, 463–502. Edinburg University Press.

McDermott, D. 1982. "Nonmonotonic logic II: Nonmonotonic modal theories." *Journal of the ACM* 29:33–57.

McDermott, D., and J. Doyle. 1980. "Nonmonotonic logic." *Artificial Intelligence* 13:41–72.

McDermott, M. 1995. "Redundant Causation." *British Journal for the Philosophy of Science* 46 (4): 523–544.

McDonnell, N. 2018. "Transitivity and proportionality in causation." *Synthese* 195:1211–1229.

McLaughlin, J. A. 1925. "Proximate cause." *Harvard Law Review* 39:149–99.

Melamed, Y. Y., and M. Lin. 2018. "Principle of Sufficient Reason." In *The Stanford Encyclopedia of Philosophy,* edited by E. N. Zalta. Metaphysics Research Lab, Stanford University.

Mellor, D. M. 1995. *The Facts of Causation.* Routledge.

Menzies, P. 2004. "Difference-making in context." In *Causation and Counterfactuals,* edited by J. Collins, N. Hall, and L. A. Paul, 139–180. MIT Press.

Menzies, P. 2006. "A Structural Equations Account of Negative Causation." In *Contributed Papers of the Philosophy of Science Association 20th Biennial Meeting.* http://philsci-archive.pitt.edu/2962/.

Menzies, P. 2007. "Causation in context." In *Causation, Physics, and the Constitution of Reality: Russell's Republic Revisited,* edited by H. Price and R. Corry, 191–223. Clarendon Press.

Menzies, P. 2011. "The role of counterfactual dependence in causal judgements." In *Understanding Counterfactuals, Understanding Causation,* edited by C. Hoerl, T. McCormack, and S. R. Beck. Oxford University Press.

Menzies, P. 2017. "The Problem of Counterfactual Isomorphs." In *Making a Difference: Essays on the Philosophy of Causation,* edited by H. Beebee, C. Hitchcock, and H. Price, 153–174. Oxford University Press.

Mill, J. S. 1872. *A System of Logic, Ratiocinative and Inductive.* Eighth ed. Harper & Bros.

Minker, J. 1993. "An Overview of Nonmonotonic Reasoning and Logic Programming." *Journal of Logic Programming* 17:95–126.

Minsky, M. 1974. *A Framework for Representing Knowledge.* Tech. Report 306. Artificial Intelligence Laboratory, MIT.

Moore, M. S. 2009. *Causation and Responsibility: An Essay in Law, Morals, and Metaphysics.* Oxford University Press.

Nayak, P. P. 1994. "Causal approximations." *Artificial Intelligence* 70 (1): 277–334.

Nute, D. 1994. "Defeasible logic." In *Handbook of Logic for Artificial Intelligence and Logic Programming,* edited by D. Gabbay and C. Hogger, III:353–395. Oxford University Press.

Paul, L. A., and N. Hall. 2013. *Causation: A User's Guide.* Oxford University Press.

Pearl, J. 1987. "Embracing Causality in Formal Reasoning." In *Proceedings of the Sixth National Conference on Artificial Intelligence (AAAI-87),* 369–373.

Pearl, J. 1990. "System Z: A Natural Ordering of Defaults with Tractable Applications to Default Reasoning." In *Proceedings of the Third Conference on Theoretical Aspects of Reasoning About Knowledge (TARK'90),* 121–135. Morgan Kaufmann.

Pearl, J. 2000. *Causality: Models, Reasoning and Inference.* 1st ed. (2nd ed. 2009). Cambridge University Press.

Pearl, J. 2012. "The Causal Foundations of Structural Equation Modeling." In *Handbook of Structural Equation Modeling,* edited by R. H. Hoyle, 68–91. Guilford Press.

Pearl, J. 2017. *The Eight Pillars of Causal Wisdom.* Technical report R-470. Department of Computer Science, University of California, Los Angeles.

Pearl, J. 2019. "Sufficient Causes: On Oxygen, Matches, and Fires." *Journal of Causal Inference* 7 (2).

Pednault, E. P. D. 1989. "ADL: Exploring the Middle Ground Between STRIPS and the Situation Calculus." In *Proceedings of the 1st International Conference on Knowledge Representation and Reasoning, KR-89,* 324–332.

Peregrin, J. 2006. "Meaning As An Inferential Role." *Erkenntnis* 64 (1): 1–36.

Peregrin, J. 2015. "Logic Reduced To Bare (Proof-Theoretical) Bones." *Journal of Logic, Language and Information* 24:193–209.

Perry, J. 1986. "From Worlds to Situations." *Journal of Philosophical Logic* 15:83–107.

Pollock, J. L. 1987. "Defeasible reasoning." *Cognitive Science* 11 (4): 481–518.

Pollock, J. L. 1995. *Cognitive Carpentry: A Blueprint for How to Build a Person.* MIT Press.

Poole, D. 1988. "A logical framework for default reasoning." *Artificial Intelligence* 36:27–47.

Poole, D. 1994. "Representing Diagnosis Knowledge." *Annals of Mathematics and Artificial Intelligence* 11:33–50.

Pratt, V. R. 2003. "Transition and Cancellation in Concurrency and Branching Time." *Mathematical Structures in Computer Science* 13:485–529.

Ramsey, F. P. 1925. "General Propositions and Causality." In *The Foundations of Mathematics and Other Logical Essays,* 233–255. Routledge & Kegan Paul.

Reiter, R. 1978. "On Closed World Data Bases." In *Logic and Data Bases,* edited by H. Gallaire and J. Minker, 119–140. Plenum Press.

Reiter, R. 1980. "A logic for default reasoning." *Artificial Intelligence* 13:81–132.

Reiter, R. 1987a. "A Theory of Diagnosis from First Principles." *Artificial Intelligence* 32:57–95.

Reiter, R. 1987b. "Nonmonotonic reasoning." *Annual Review of Computer Science* 2:147–186.

Reiter, R. 1991. "The Frame Problem in the Situation Calculus: A Simple Solution (Sometimes) and a Completeness Result for Goal Regression." In *Artificial Intelligence and Mathematical Theory of Computation: Papers in Honor of John McCarthy,* edited by V. Lifschitz, 318–420. Academic Press.

Reiter, R. 2001. *Knowledge in Action: Logical Foundations for Specifying and Implementing Dynamic Systems.* MIT Press.

Reiter, R., and G. Criscuolo. 1981. "On interacting defaults." In *Proceedings of the International Joint Conference on Artificial Intelligence, IJCAI-81,* 270–276.

Renardel de Lavalette, G., B. Kooi, and R. Verbrugge. 2008. "Strong completeness and limited canonicity for PDL." *Journal of Logic, Language and Information* 17:69–87.

Reutlinger, A., G. Schurz, A. Hüttemann, and S. Jaag. 2019. "Ceteris Paribus Laws." In *The Stanford Encyclopedia of Philosophy,* edited by E. N. Zalta. Metaphysics Research Lab, Stanford University.

Rosenberg, I., and C. Glymour. 1918. "Review of Joseph Halpern, Actual Causality." *The British Journal for the Philosophy of Science.*

Ross, W. D. 1949. *Aristotle's Prior and Posterior Analytics.* Oxford University Press.

Rumfitt, I. 2015. *The Boundary Stones of Thought.* Oxford University Press.

Russell, B. 1912. "On the Notion of Cause." *Proceedings of the Aristotelian Society* 7:1–26.

Sakama, C. 1989. "Possible Model Semantics for Disjunctive Databases." In *Proceedings of the 1st International Conference on Deductive and Oject-Oriented Databases,* edited by W. Kim, J.-M. Nicolas, and S. Nishio, 369–383. Elsevier.

Sambin, G., G. Battilotti, and C. Faggian. 2000. "Basic Logic: Reflection, Symmetry, Visibility." *Journal of Symbolic Logic* 65 (3): 979–1013.

Sanford, D. H. 1989. *If P then Q: Conditionals and the Foundations of Reasoning.* Routledge.

Sartorio, C. 2015. "Resultant Luck and the Thirsty Traveler." *Methode* 4 (6): 153–171.

Schaffer, J. 2000. "Trumping Preemption." *The Journal of Philosophy* 97 (4): 165–181.

Schaffer, J. 2004. "Causes need not be physically connected to their effects: The case for negative causation." In *Contemporary Debates in Philosophy of Science,* edited by C. Hitchcock. Blackwell.

Schaffer, J. 2010. "Contrastive Causation in the Law." *Legal Theory* 16:259–97.

Schaffer, J. 2012. "Causal contextualism." In *Contrastivism in Philosophy,* edited by M. Blaauw, 35–63. Routledge.

Schlipf, J. S. 1994. "A Comparison of Notions of Negation as Failure." In *Advances in Logic Programming Theory,* edited by G. Levi, 1–53. Clarendon Press.

Schulz, K. 2011. "'If you'd wiggled A, then B would've changed'—Causality and counterfactual conditionals." *Synthese* 179 (2): 239–251.

Scott, D. 1974. "Completeness and axiomatizability in many-valued logic." In *Proceedings of Symposia in Pure Mathematics,* 431–435. 25.

Segerberg, K. 1982. *Classical Propositional Operators.* Clarendon Press.

Shanahan, M. P. 1997. *Solving the Frame Problem.* The MIT Press.

Shoesmith, D. J., and T. J. Smiley. 1978. *Multiple-Conclusion Logic.* Cambridge University Press.

Simon, H. A. 1953. "Causal ordering and identifiability." In *Studies in Econometric Method,* edited by Wm. C. Hood and T. C. Koopmans, 49–74. Wiley.

Son, T. C., C. Baral, N. Tran, and S. Mcilraith. 2006. "Domain-Dependent Knowledge in Answer Set Planning." *ACM Transactions on Computational Logic* 7 (4): 613–657.

Spirtes, P., C. Glymour, and R. Scheines. 2000. *Causation, Prediction, and Search.* 2nd ed. MIT Press.

Steedman, M. 2005. "The productions of time: Temporality and causality in linguistic semantics." Unpublished manuscript, Draft 5.0.

Steinberger, F. 2011. "Why Conclusions Should Remain Single." *Journal of Philosophical Logic* 40:333–355.

Stenning, K., and M. van Lambalgen. 2008. *Human Reasoning and Cognitive Science.* MIT Press.

Strass, H. 2013. "Approximating operators and semantics for abstract dialectical frameworks." *Artificial Intelligence* 205:39–70.

Strevens, M. 2007. "Mackie Remixed." In *Causation and Explanation,* edited by J. K. Campbell, M. O'Rourke, and H. S. Silverstein. MIT Press.

Thielscher, M. 1997. "Ramification and Causality." *Artificial Intelligence* 89:317–364.

Thielscher, M. 2011. "A unifying action calculus." *Artificial Intelligence* 175 (1): 120–141.

Thomason, R. 2003. "Logic and artificial intelligence." In *The Stanford Encyclopedia of Philosophy,* edited by E. N. Zalta. Metaphysics Research Lab, Stanford University.

Tichy, P. 1976. "A counterexample to the Stalnaker-Lewis analysis of counterfactuals." *Philosophical Studies* 29:271–273.

Touretzky, D. S. 1986. *The Mathematics of of Inheritance Systems.* Morgan Kaufmann.

Tu, P. H., T. C. Son, M. Gelfond, and A. R. Morales. 2011. "Approximation of action theories and its application to conformant planning." *Artificial Intelligence* 175 (1): 79–119.

Turner, H. 1999. "A Logic of Universal Causation." *Artificial Intelligence* 113:87–123.

Turner, R. 1981. "Counterfactuals without Possible Worlds." *Journal of Philosophical Logic* 10 (4): 453–493.

Veltman, F. 1986. "Data Semantics and the Pragmatics of Indicative Conditionals." In *On Conditionals,* edited by E. Traugott, A. ter Meulen, J. Reilly, and Ch. Ferguson. Cambridge University Press.

Veltman, F. 2005. "Making Counterfactual Assumptions." *Journal of Semantics* 22 (2): 159–180.

Venema, Y. 1997. "A crash course in arrow logic." In *Arrow Logic and Multi-Modal Logic,* edited by M. Marx, L. Pólos, and M. Masuch, 3–34. Center for the Study of Language / Information, Stanford University.

Vennekens, J., M. Bruynooghe, and M. Denecker. 2010. "Embracing Events in Causal Modelling: Interventions and Counterfactuals in CP-Logic." In *Logics in Artificial Intelligence: Proceedings of the 12th European Conference, JELIA 2010,* edited by T. Janhunen and I. Niemelä, 313–325. Springer.

Vennekens, J., M. Denecker, and M. Bruynooghe. 2009. "CP-logic: A language of causal probabilistic events and its relation to logic programming." *Theory and Practice of Logic Programming* 9 (3): 245–308.

Walsh, C. R., and S. A. Sloman. 2005. "The meaning of cause and prevent: The role of causal mechanism." In *Proceedings of the 27th Annual Conference of the Cognitive Science Society,* edited by B. G. Bara, L. Barsalou, and M. Bucciarelli, 2331–2336. Lawrence Erlbaum Associates.

Walsh, C. R., and S. A. Sloman. 2011. "The Meaning of Cause and Prevent: The Role of Causal Mechanism." *Mind & Language* 26 (1): 21–52.

Weslake, B. 2015. "A partial theory of actual causation." Unpublished manuscript.

Wittgenstein, L. [1921]1961. *Tractatus Logico-Philosophicus.* Trans. D. F. Pears and B. F. McGuinness. Routledge & Kegan Paul.

Wojcicki, R. 1988. *Theory of Logical Calculi.* Kluwer Academic.

Woodward, J. 2003. *Making Things Happen: A Theory of Causal Explanation.* Oxford University Press.

Woodward, J. 2016. "Causation and Manipulability." In *The Stanford Encyclopedia of Philosophy,* edited by E. N. Zalta. Metaphysics Research Lab, Stanford University.

Wright, R. W. 1985. "Causation in Tort Law." *California Law Review* 1735:1788–1791.

Wright, R. W. 2001. "Once more into the bramble bush: Duty, causal contribution, and the extent of legal responsibility." *Vanderbilt Law Review* 54:1071–1132.

Wright, R. W. 2007. "Acts and Omissions as Positive and Negative Causes." In *Emerging Issues in Tort Law,* edited by J. W. Neyers, E. Chamberlain, and S. G. A. Pitel, 287–307. Hart Publishing.

Wright, R. W. 2008. "The Nightmare and the Noble Dream: Hart and Honoré on Causation and Responsibility." In *The Legacy of H. L. A. Hart: Legal, Political and Moral Philosophy,* edited by M. H. Kramer, C. Gran, B. Colburn, and A. Hatzistavrou. Oxford University Press.

Wright, R. W. 2013. "The NESS Account of Natural Causation: A Response to Criticisms." In *Causation and Responsibility: Critical Essays,* edited by B. Kahmen and M. Stepanians. De Gruyter.

Yablo, S. 2004. "Advertisement for a Sketch of an Outline of a Proto-Theory of Causation." In *Causation and Counterfactuals,* edited by J. Collins, N. Hall, and L. A. Paul, 119–138. MIT Press.

Zhang, H., and F. Lin. 2017. "Characterizing causal action theories and their implementations in answer set programming." *Artificial Intelligence* 248:1–8.

Zhang, Y., and N. Y. Foo. 1996. "Updating Knowledge Bases with Disjunctive Information." In *Proceedings of the Thirteenth National Conference on Artificial Intelligence (AAAI-96),* 562–568. AAAI/MIT Press.

Index